G. Prodi (Ed.)

Problems in Non-Linear Analysis

Lectures given at a Summer School of the
Centro Internazionale Matematico Estivo (C.I.M.E.),
held in Varenna (Como), Italy,
August 20-29, 1970

 Springer

FONDAZIONE
CIME
ROBERTO CONTI

C.I.M.E. Foundation
c/o Dipartimento di Matematica "U. Dini"
Viale margagni n. 67/a
50134 Firenze
Italy
cime@math.unifi.it

ISBN 978-3-642-10996-6   e-ISBN: 978-3-642-10998-0
DOI:10.1007/978-3-642-10998-0
Springer Heidelberg Dordrecht London New York

Printed on acid-free paper

Springer.com

CENTRO INTERNAZIONALE MATEMATICO ESTIVO

(C. I. M. E.)

4º Ciclo - Varenna - dal 20 al 29 Agosto 1970

PROBLEMS IN NON-LINEAR ANALYSIS

Coordinatore: Prof. G. PRODI

CENTRO INTERNAZIONALE MATEMATICO ESTIVO

(C. I. M. E.)

H. BREZIS

PROPRIÉTÉS RÉGULARISANTES DE CERTAINS SEMIGROUPES ET
APPLICATIONS

Corso tenuto a Varenna dal 20 al 29 Agosto 1970

PROPRIÉTÉS RÉGULARISANTES DE CERTAINS SEMIGROUPES ET APPLICATIONS

par

H. Brezis

(Université - Paris)

I. Propriétés régularisantes de certains semigroupes nonlinéaires.

Soit H un espace de Hilbert et soit $\varphi : H \to]-\infty, +\infty]$ une fonction convexe semi continue inférieurement, $\varphi \not\equiv +\infty$.

On pose $Au = \partial\varphi(u) = \left\{ f \in H; \ \varphi(v) - \varphi(u) \geqslant (f, v-u) \ \forall v \in H \right\}$.

Il est bien connu que A est maximal monotone et donc $-A$ engendre un semigroupe continu de contractions $S(t)$ sur $\overline{D(A)}$ (pour les notations et résultats standards relatifs aux semigroupes nonlinéaires voir par exemple les exposés de A. Pazy à ce séminaire

On se propose de montrer que $S(t)$ possède une propriété "régularisante" comparable en un certain sens à celle des semigroupes linéaires analytiques. Nous indiquons ici seulement le principe des démonstrations; pour les détails et diverses extensions, on pourra se reporter à [2] .

<u>Théorème</u> 1. Soit $u_o \in \overline{D(A)}$, alors $S(t)u_o \in D(A)$ pour tout $t > 0$ et on a

$$| A^o S(t)u_o | \leqslant (1 + \varepsilon) | A^o v | + \left(\frac{1 + \varepsilon^2}{\varepsilon}\right) \frac{|v - u_o|}{t}$$

$$\forall \varepsilon > 0 , \quad \forall v \in D(A) , \quad \forall t > 0 .$$

H. Brezis

Autrement dit, pour $u_o \in \overline{D(A)}$, il existe une fonction

$u \in C([0, +\infty[; H)$ unique vérifiant

(1) u est dérivable p.p. sur $]0, +\infty[$

(2) u est dérivable à droite en tout $t > 0$

(3) $u(t) \in D(A)$ pour tout $t > 0$

(4) $\dfrac{d^+u}{dt} + A^{\circ}u(t) = 0$ pour tout $t > 0$

(5) $u(0) = u_o$.

De plus on a

(6) $\left| \dfrac{d^+u}{dt}(t) \right| \leq (1 + \varepsilon)|A^{\circ}v| + \left(\dfrac{1+\varepsilon^2}{\varepsilon} \right) \dfrac{|v-u_o|}{t}$

$\qquad\qquad \forall \varepsilon > 0, \ \forall v \in D(A), \ \forall t > 0$.

<u>Principe de la démonstration</u>. Soit A_λ la régularisée Yosida de

A, i.e. $A_\lambda = \dfrac{I-(I+\lambda A)^{-1}}{\lambda}$;   on montre que $A_\lambda = \partial \varphi_\lambda$ avec $\varphi_\lambda(u) =$

$= \underset{v \in H}{\text{Inf}} \left\{ \dfrac{1}{2\lambda} |u-v|^2 + \varphi(v) \right\}$ et que φ_λ est Fréchet – différentia

ble.

Soit u_λ la solution de l'équation $\dfrac{du_\lambda}{dt} + A_\lambda u_\lambda = 0$,

$u_\lambda(0) = u_o$.

Comme $\underset{\lambda \to 0}{\lim} u_\lambda(t) = S(t)u_o$, on se ramène donc à établir

H. Brezis

(6). Pour simplifier les notations on supprime dorénavant λ

et on cherche à prouver (6) dans le cas où φ est Fréchet

différentiable avec $A = \partial \varphi$ lipschitzien.

Soit $v \in H$ __fixé__ et soit $f = -Av$; on a

$$\varphi(u) - \varphi(v) \geq -(f,\ u-v)\ .$$

Posant $\widetilde{\varphi}(u) = \varphi(u) - \varphi(v) + (f,u-v)$, il vient $\widetilde{\varphi}(u) \geq 0$,

$\forall u \in H$ et $\widetilde{\varphi}(v) = 0$.

De plus l'équation (4) s'écrit

(7) $$\frac{du}{dt} + \partial \widetilde{\varphi}(u) = f$$

__Première estimation.__ On a

(8) $\quad \widetilde{\varphi}(v) - \widetilde{\varphi}(u(t)) \geq (f - \frac{du(t)}{dt}\ ,\ v-u(t)\ )$

Or $\widetilde{\varphi}(v) = 0$; d'où intégrant (8) sur $]0,T[$ il vient

$$\int_0^T \widetilde{\varphi}(u(t))dt \leq |f| \int_0^T |v-u(t)|dt + \frac{1}{2}|u_o - v|^2$$

On en déduit aisément que

(9) $\quad \frac{1}{T} \int_0^T \widetilde{\varphi}(u(t)dt \leq T|f|^2 + \frac{1}{T}|u_o - v|^2$

__Seconde estimation.__

Multipliant (7) par $\frac{du}{dt}$ et intégrant sur $[0,T]$ on obtient

sans difficultés

H. Brezis

$$\left(\int_o^T \left| \frac{du}{dt} \right|^2 dt \right)^{1/2} \leqslant |f| \sqrt{T} + \sqrt{\tilde{\varphi}(u(0))}$$

Comme $\left| \dfrac{du}{dt} \right|$ est décroissant, on a

(10) $\qquad \left| \dfrac{du}{dt} (T) \right| \leqslant |f| + \dfrac{1}{\sqrt{T}} \sqrt{\tilde{\varphi}(u(0))}$

Preuve de (6)

Soit $t > 0$ et soit $\theta \in \,]0,t\,[$. Le théorème de la moyenne et l'inéquation (9) montrent qu'il existe $t_o \in [0,\theta]$ tel que

$$\tilde{\varphi}(u(t_o)) \leqslant \theta |f|^2 + \frac{1}{\theta} |u_o - v|^2$$

L'estimation (10) appliquée sur l'intervalle $[t_o,t]$ au lieu de $[0,T]$ conduit à

$$\left| \frac{du(t)}{dt} \right| \leqslant |f| + \frac{1}{\sqrt{t-t_o}} \sqrt{\tilde{\varphi}(u(t_o))} \leqslant \left(1 + \frac{\sqrt{\theta}}{\sqrt{t-\theta}} \right) |f|$$

$+ \dfrac{1}{\sqrt{\theta(t-\theta)}} |u_o - v|$. Posant $\mathcal{E} = \dfrac{\sqrt{\theta}}{\sqrt{t-\theta}}$, on arrive au résultat.

Utilisant des techniques assez semblables, on peut résoudre le problème avec second membre.

Théorème 2. On suppose que $f \in L^2(0,T; H)$ et que $u_o \in \overline{D(A)}$. Alors il existe $u \in C(0,T;H)$ unique fonction vérifiant

(11) u est dérivable p.p. sur $]0,T[$

H. Brezis

(12) $u(t) \in D(A)$ p.p. sur $]0,T[$

(13) $\dfrac{du}{dt} \in L^2(\delta,T; H)$ $\forall 0 < \delta < T$

(14) $\varphi(u) \in C(\delta, T; H)$ $\forall 0 < \delta < T$

(15) $\dfrac{du}{dt} + (Au - f)^{\circ} = 0$ p.p. sur $]0,T[$

(16) $u(0) = u_0$

Si de plus $\varphi(u_0) < + \infty$, alors on a $\dfrac{du}{dt} \in L^2(0,T;H)$ et

$\varphi(u) \in C(0,T;H)$.

Remarque. Si A est un opérateur maximal monotone quelconque,

on savait précédemment résoudre le problème (15)-(16) avec des

hypothèses supplémentaires- i.e. $f \in C(0,T;H)$, $\dfrac{df}{dt} \in L^1(0,T;H)$

et $u_0 \in D(A)$. La solution obtenue est alors plus régulière i.e.

$u(t) \in D(A)$ pour tout $t > 0$ et $\dfrac{du}{dt} \in L^{\infty}(0,T;H)$ (cf. Kato [4]).

II. Applications aux inéquations variationnelles paraboliques.

Commencons par un bref rappel sur les inéquations varia-

tionnelles elliptiques.

Soit V un espace de Hilbert de norme $\| \ \|$ et soit V'

son dual (non identifié à V). Soit K un convexe fermé de

V et soit $L : V \longrightarrow V'$ un opérateur lineaire continu et coer

H. Brezis

cif i.e. $(Lu,u) \geqslant \alpha \|u\|^2 \quad \forall u \in V$, $\alpha > 0$.

Pour tout $f \in V'$, il existe (d'après un résultat de Stampacchia) $u \in K$ unique solution de l'inéquation.

(17) $\qquad$ $(Lu, v-u) \geqslant (f, v-u) \qquad \forall v \in K$.

On supposera dans la suite, afin de simplifier l'exposé, que $L^* = L$. Dans ce cas le problème (17) est équivalent à

$$\underset{u \in K}{\text{Min}} \quad \frac{1}{2} (Lu, u) - (f,u) \ .$$

Étant donnée une fonction $\varphi: V \longrightarrow]-\infty , +\infty]$ convexe s.c.i, $\varphi \not\equiv +\infty$, on pose

$$\partial\varphi(u) = \left\{ f \in V' \ ; \varphi(v) - \varphi(u) \geqslant (f, v-u) \ \forall v \in V \right\}$$

Avec cette notation le problème (17) peut s'ecrire

$$Lu + \partial\Psi_K(u) \ni f$$

soit

$$\partial \left[\frac{1}{2} (Lu, u) + \Psi_K(u) \right] \ni f$$

où $\Psi_K(u) = \begin{cases} 0 \ , & u \in K \\ +\infty \ , & u \notin K \ . \end{cases}$

Passons maintenant aux inéquations variationnelles <u>paraboli-ques</u>. Soit H un espace de Hilbert de norme $| \ |$ tel que

$$V \subset H = H' \subset V'$$

avec injections continues et denses.

H. Brezis

Problème. Étant donnés f et u_o , on cherche une fonction $u(t)$ telle $u(t) \in K$ p.p. et

$$(18) \begin{cases} \left(\frac{du}{dt}, v-u\right) + (Lu, v-u) \geqslant (f, v-u) \quad \forall v \in K \quad \text{p.p.} \quad \text{sur} \,]0,T[\\ \\ u(0) = u_o \end{cases}$$

Ce type de problème a été introduit pour la première fois dans Lions-Stampacchia [5] .

On se propose de montrer comment la théorie des semigroupes non linéaires peut s'appliquer à la résolution du problème (18) et conduire à une interprétation "concrète" de (18).

Comme l'opérateur $- (L + \partial \psi_K)$ engendre formellement un semigroupe de contractions dans H , on est conduit à introduire dans l'espace H l'opérateur

$$Au = (Lu + \partial \psi_K(u)) \cap H \quad \text{avec}$$

$$D(A) = \left\{ u \in K \; ; \; (Lu + \partial \psi_K(u)) \cap H \neq \emptyset \right\} ;$$

on notera que

$$\left\{ u \in K \; ; \; Lu \in H \right\} \subset D(A)$$

Il est facile de voir que A est maximal monotone dans H et d'ailleurs on a $A = \partial \phi$ (au sens du sous différentiel dans H) où

$$\varphi(u) = \begin{cases} \dfrac{1}{2}\,(Lu,\,u) & u \in K \\[2em] +\infty & u \in H\,,\ u \notin K \end{cases}$$

est une fonction convexe s.c.i. sur H .

De plus, on vérifie aisément que $\overline{D(A)} = \overline{K}$ (dans la suite,

toutes les adhérences sont à prendre au sens de H).

Le point délicat est la description explicite de $D(A)$.

Ici interviennent les théorèmes de régularité pour les inéqua-

tions variationnelles elliptiques prouvés dans Brezis Stampac-

chia [3] . Ils affirment que dans certains cas

$$D(A) = \left\{ u \in K \quad ;\quad Lu \in H \right\}$$

et par suite $Au = Lu + \partial\Psi_{\overline{K}}(u)$.

Exemple. Soit Ω un ouvert borné de frontière Γ regulière.

On prend $V = H^1_o(\Omega)$, $H = L^2(\Omega)$, $L = -\Delta$ et

$$K = \left\{ u \in H^1_o(\Omega) \ ;\quad u \geqslant \Psi \quad \text{p.p. sur } \Omega \right\}$$

avec $\Psi \in H^2(\Omega)$, $\Psi \leqslant 0$ p.p. sur Γ .

Alors $D(A) = \left\{ u \in K \,,\ Lu \in H \right\} = K \cap H^2(\Omega)$.

Appliquant à A le théorème 2 , il vient

Théorème 3. Soit $f \in L^2(0,T;H)$ et soit $u_o \in \overline{K}$. Alors il exi-

ste une fonction $u \in C(0,T;H)$ unique vérifiant

H. Brezis

(19)   u est derivable p.p. et $\dfrac{du}{dt} \in L^2(\delta,T;H)$

pour tout $0 < \delta < T$.

(20) $u \in C(\delta, T; V)$ pour tout $0 < \delta < T$

(21) $u(t) \in D(A)$ p.p. sur $]0,T[$

(22) $\left(\dfrac{du}{dt} + Lu, v{-}u\right) \geqslant (f,v{-}u)$ p.p. sur $]0,T[$, $\forall v \in K$

(23) $u(0) = u_0$

Si de plus $u_0 \in K$, alors $\dfrac{du}{dt} \in L^2(0,T;H)$ et $u \in C(0,T;V)$.

Remarque. Moyennant des hypothèses supplémentaires sur f et

u_0 , on obtient une solution plus régulière. Supposons par exem

ple $f \in C(0,T;H)$, $\dfrac{df}{dt} \in L^1(0,T;H)$ et $u_0 \in \bar{K}$, alors on a

au lieu de (19) et (21)

$\dfrac{du}{dt} \in L^\infty(\delta,T,H) \cap L^2(\delta,T;V)$ pour tout $0 < \delta < T$

et $u(t) \in D(A)$ pour tout $t > 0$.

Si l'on suppose de plus que $u_0 \in D(A)$ alors on a

$\dfrac{du}{dt} \in L^\infty(0,T;H) \cap L^2(0,T;V)$ et $u(t) \in D(A)$ pour tout $t \in [0,T]$

Interprétation de (18).

Commençons par le cas des équations différentielles ordinai

res i.e. $V = H = \mathbb{R}^n$ et supposons que $L = 0$.

H. Brezis

On a grâce à (15)

$$\begin{cases} \dfrac{du}{dt} + \left(\partial\Psi_K(u) - f\right)^{\circ} = 0 \qquad \text{p.p. sur }]0,T[\\[3mm] u(0) = u_o \end{cases}$$

Or pour tout convexe fermé C , il est immédiat que $(C-f)^{\circ} =$

$= \text{Proj}_C \, f - f$. Par ailleurs $\partial\Psi_K(u)$ est un cône, soit $\mathcal{T}_K(u)$

le cône polaire i.e. $\mathcal{T}_K(u) = \overline{\bigcup_{\lambda \geqslant 0} \lambda(K - u)}$.

On a alors $\text{Proj}_{\partial\Psi_{K(u)}} f - f = - \text{Proj}_{\mathcal{T}_K(u)} f$

D'où $\dfrac{du}{dt} = \text{Proj}_{\mathcal{T}_K(u)} f$ p.p. sur $]0,T[$.

Autrement dit, on a résolu le problème

$\dfrac{du}{dt} = f$ \qquad si $u \in \text{Int } K$

$\dfrac{du}{dt} = \text{Proj}_{\mathcal{T}_K(u)} f$ \quad si $u \in$ Frontière de K

Dans le second cas on notera que si K est un convexe fermé régulier d'interieur non vide, alors $\mathcal{T}_K(u)$ est le demi espace contenant K et déterminé par l'hyperplan tangent à K en u (fig. 1)

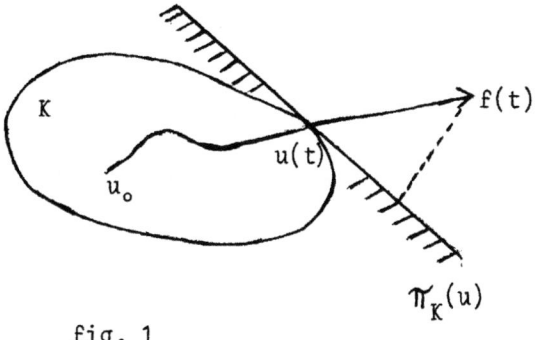

H. Brezis

fig. 1

Remarquons que ce problème peut s'ecrire $\frac{du}{dt}(t) = F(t,u(t))$; toutefois F est discontinue et donc les résultats classiques ne peuvent pas être appliqués.

Revenons maintenant au cas genéral et faisons l'hypothèse de régularité

$$D(A) = \left\{ u \in K \; ; \; Lu \in H \right\} .$$

On a alors grâce à (15)

$$\frac{du}{dt} + \left(\partial\psi_{\overline{K}}(u) + Lu - f \right)^{\circ} = 0 \quad \text{p.p. sur }]0,T[$$

Autrement dit, on a $\frac{du}{dt} = \text{Proj}_{\pi_{\overline{K}}(u)}(f - Lu)$ où

$\pi_{\overline{K}}(u) = \overline{\bigcup_{\lambda \geq 0} \lambda(K-u)}$ (les projections et adhérences sont au sens de H).

Exemple. Reprenant les mêmes notations et hypothèses qu'à l'exemple précédent il vient

H. Brezis

$$\bar{K} = \left\{ u \in L^2(\Omega) \; ; \; u \geqslant \psi \quad \text{p.p. sur} \quad \Omega \right\}$$

$$\pi_{\bar{K}}^*(u) = \left\{ v \in L^2(\Omega) \; ; \; v \geqslant 0 \quad \text{p.p. sur} \quad [u = \psi] \right\}$$

$$\text{Proj}_{\pi_{\bar{K}}^*(u)} \; g = \begin{cases} g & \text{sur} \quad [u > \psi] \\ \\ \text{Max} \left\{ g, 0 \right\} & \text{sur} \quad [u = \psi] \end{cases}$$

L'interprétation de l'inequation (18) est donc $u(x,t) \geqslant \psi(x)$

$$\frac{du}{dt} = \begin{cases} \Delta u + f & \text{sur} \quad [u > \psi] \\ \\ \text{Max} \left\{ \Delta u + f, 0 \right\} = \text{Max} \left\{ \Delta \psi + f, 0 \right\} & \text{sur} \quad [u = \psi] \end{cases}$$

$u(x,0) = u_0(x)$.

Par ailleurs le théorème 3 montre que si $f \in L^2(0,T;L^2(\Omega))$

et $u_0 \in L^2(\Omega)$ avec $u_0 \geqslant \psi$ p.p. sur Ω , alors

$u \in C(0,T; L^2(\Omega)) \cap C(\delta,T;H_0^1(\Omega)) \cap L^2(\delta,T;H^2(\Omega))$

$\dfrac{du}{dt} \in L^2(\delta,T \; ; \; L^2(\Omega))$ pour tout $0 < \delta < T$.

On trouvera dans [1] d'autres résultats de régularité (en

particulier dans les espaces L^p) pour ce problème.

H. Brezis

RÉFERENCES

[1] H. Brezis. Problèmes unilatéraux (Thèse), Paris 1970 (à paraitre).

[2] H. Brezis. Proprietés régularisantes de certains semi-groupes non linéaires (à paraitre).

[3] H. Brezis-G. Stampacchia. Sur la régularité de la solution d'inéquations elliptiques. Bull. Soc. Math. Fr. 96 (1958) p. 153-180.

[4] T. Kato. Accretive operators and non linear evolution equations in Banach spaces , Proc. Symp. Non linear Funct. Anal. Chicago A.M.S. 1968 .

[5] J.L. Lions-G. Stampacchia. Variational inequalities, Comm. pure and appl. Math. 20 (1967) p. 493-519.

CENTRO INTERNAZIONALE MATEMATICO ESTIVO

(C. I. M. E.)

F. E. BROWDER

NORMAL SOLVABILITY AND EXISTENCE THEOREMS FOR NONLINEAR

MAPPINGS IN BANACH SPACES

Corso tenuto a Varenna dal 20 al 29 Agosto 1970

NORMAL SOLVABILITY AND EXISTENCE THEOREMS FOR NONLINEAR MAPPINGS IN BANACH SPACES

by

Felix E. Browder

(University of Chicago)

Introduction: Let X be a topological space, Y a Banach space, f a mapping of X into Y. The mapping f is said to be normally solvable if $f(X)$ is closed in Y. In the preceding paper (Browder [6]), we have described how the theory of normal solvability combines this hypothesis with infinitesimal assumptions upon the mapping f to obtain sufficient conditions for a given element y of Y to lie in $f(X)$.

In our present discussion, we shall sharpen the results obtained in [6] and apply them to obtain some new existence theorems for a general family of φ - accretive mappings in the sense of Browder [2]. The chief point of the sharpening lies in abandoning the imposition of hypotheses upon the

F. E. Browder

asymptotic direction set $D_x(f)$ as in $[6]$ and using direct hypotheses upon the local structure of the mapping f . Though the symptotic direction set $D_x(f)$ can be defined for mappings f which are continuous rather than differentiable in any sense (or have no continuity properties at all), it still carries too much of the natural structure of the closure of the range of the differential df_x when the latter exists to make it useful in the most general case.

Let us begin with the simplest result in this direction, which is an extremely simple reformulation of Theorem 1 of $[5]$:

Theorem 1: Let X be a topological space, Y a Banach space, f a mapping of X into Y with $f(X)$ closed in Y . Let y be a given point in Y , and suppose that there exist $r > 0$ and $p < 1$ such that the following conditions hold:

(1) $f^{-1}(B_r(y)) \neq \emptyset$.

(2) For each x in $f^{-1}(B_r(y))$, there exist sequences $\{u_j\}$ in X and $\{\xi_j\}$ in the non-negative reals such that $f(u_j) \longrightarrow$ $\longrightarrow f(x)$ in Y , while

$$\| \xi_j^{-1}(f(u_j)-f(x)) - (y-f(x))\| \leq p \| y - f(x)\|$$

for each j .

F. E. Browder

<u>Then</u>: y <u>lies in</u> $f(X)$.

The result of the present Theorem 1 differs from that of Theorem 1 of $[6]$ in that we make no assumption that an infinite subsequence of the sequence

$$(\| f(u_j) - f(x)\|)^{-1}(f(u_j) - f(x))$$

converges to an element w of $D_x(f)$ as in the result given in $[6]$.

We obtain results under weaker hypotheses upon the local behaviour of the mapping f by strengthening our assumptions on the Banach space Y .

<u>Theorem 2</u>: <u>Let X be a topological space , Y a Banach space whose conjugate space Y^* is uniformly convex . Let f be a mapping of X into Y such that $f(X)$ is closed in Y . Suppose that there exist $\delta > 0$ such that for each x in X, there exists a dense subset R_x of the unit sphere in Y^* such that the following property holds:</u>

(P) <u>For each y^* in R_x and for the given x in X</u> ,

F. E. Browder

there exists an infinite sequence $\{u_j\}$ in X with $f(u_j) \longrightarrow$
$\longrightarrow f(x)$, $f(u_j) \neq f(x)$ for any j , such that

$$(y^*, f(u_j) - f(x)) \geq \delta \|f(u_j) - f(x)\|$$

for each $j \geq 1$.

   Then: $f(X) = Y$.

As an application of Theorem 2 , we derive the following
new result on a general class of φ - accretive mappings ([2]):

Theorem 3: Let X and Y be Banach spaces with Y^*
uniformly convex   f a mapping of X into Y with f satisfying
a Lipschitz condition on each bounded subset of X . Suppose
that there exists a mapping φ of X into Y^* such that
$\|\varphi(v)\|_{Y^*} \leq \|v\|$, $\varphi(\xi v) = \xi\varphi(v)$ for each $\xi \geq 0$, v in X ,
and with the image of φ dense in Y^* . Suppose that for each
x and u in X , and for a fixed constant $c > 0$, we have

$$(\varphi(x-u), f(x)-f(u)) \geq c \|x - u\|^2 .$$

   Then: $f(X) = Y$.

Theorem 3 provides the beginning of a connecting theory
of φ - accretive mappings which would link the theory of strongly

F. E. Browder

monotone mappings for which $Y = X^*$ with the theory of accretiv

mappings for which $Y = X$. Since the basic methods for the

theories of monotone and accretive mappings are fundamentally

different (as opposed to the results which are very similar in

character), it is of fundamental importance in linking these

two theories to obtain a new methodology from which results on

φ- accretive mappings can be derived. It is our hope that the

further development of the ideas of the theory of normally

solvable mappings using refinements of the geometrical arguments

in Banach spaces which we have been discussing may lead to the

creation of such a methodology.

Let us complete the introductory discussion by stating one

further Theorem which constitutes the appropriate form of

Theorem 2 when we are concerned only with the question of whether

a given element y of Y lies in $f(X)$.

Theorem 4: Let X be a topological space, Y a Banach

space whose conjugate space Y^* is uniformly convex. Let f

be a mapping of X into Y such that $f(X)$ is closed in Y .

Let y be a given point of Y , and suppose that there exist

F. E. Browder

$r > 0$ and $\delta > 0$ such that the following conditions hold:

(1) $f^{-1}(B_r(y))$ is non-empty

(2) For each x in $f^{-1}(B_r(y))$, there exists a sequence $\{u_j\}$ in X such that $f(u_j) \neq f(x)$ for each j, while $f(u_j) \longrightarrow f(x)$, and

$$(J(y-f(x)), f(u_j) - f(x)) \geq \delta \| y-f(x) \| \cdot \| f(u_j)-f(x) \|$$

(where J is the duality mapping of Y into Y^*).

Then: y lies in $f(X)$.

We recall that by the (normalized) duality mapping J of Y into Y^*, we mean the uniquely defined mapping J such that for each y in Y, $J(y)$ is the unique element of Y^* satisfying the two conditions:

$$(J(y), y) = \| y \|_Y^2 \; ; \quad \| J(y) \|_{Y^*} = \| y \|_Y .$$

Since Y^* is uniformly convex, J is continuous from Y to Y^* and indeed uniformly continuous on bounded subsets of Y. Indeed, this latter property is equivalent to Y^* being uniformly convex. (See [2] for the appropriate discussion).

Section 1: We now begin the proof of the results stated

F E. Browder

above:

Proof of Theorem 1: This is almost identical with the proc
of Theorem 1 of $[6]$, except for the concluding argument leadin
to a contradiction. Let $S = f(X)$. We proceed as in the proof
of Theorem 1 of $[6]$ and find and element s_0 of the set
$S \cap (s + C)$ such that for a given $\delta_1 > 0$, we have

$$(s_0 + C) \cap S \cap B_{\delta_1}(s_0) = \{s_0\} .$$

We now derive a contradiction (which is based ultimately
upon the original assumption that $d_0 = \text{dist}(y, f(X)) > 0$). By
the present hypothesis, we have the sequence $s_j = f(u_j)$ in
S which are all distinct from $s_0 = f(x)$, where as in the
original prood

$$\| y - f(x) \| \le r .$$

For each term of this sequence, we have

$$\| \xi_j^{-1}(f(u_j) - f(x)) - (y-f(x)) \| \le p \| y-f(x) \| ,$$

while $f(u_j) \longrightarrow f(w)$ as $j \longrightarrow \infty$. It follows that $\xi_j \longrightarrow 0$.
Since $p < q$, it follows that for $\epsilon > 0$, sufficiently small,

$$s + \xi_j^{-1}(f(u_j) - f(x)) \in B ,$$

F. E. Browder

and hence

$$\xi_j^{-1}(f(u_j) - f(x)) \in C .$$

Since C is a cone,

$$f(u_j) - f(x) \in C$$

for each j , i.e. $s_j = f(u_j)$ lies for all sufficiently large j in the intersection

$$(s_0 + C) \cap S \cap B_{\delta_1}(s_0)$$

while each of the s_j is distinct from s_0 . This contradicts the basic choice of s_0 , and this contradiction yields the conclusion of the Theorem.

$$q.e.d.$$

For the proofs of Theorems 2, 3, and 4 , we shall show first that Theorem 3 follows from Theorem 2 , and then that Theorem 2 follows from Theorem 4 . We complete the discussion by then giving the detailed proof of Theorem 4 .

Proof of Theorem 3 from Theorem 2: For each x and u in X , we have the inequality

$$(\varphi(x-u), f(x) - f(u)) \geq c \| x-u \|^2 .$$

F. E. Browder

Since

$$\|\varphi(x - u)\| \leq \|x - u\|$$

by the hypothesis on φ , it follows that

$$c\|x - u\|^2 \leq \|f(x) - f(u)\| \cdot \|x - u\|,$$

and for $x - u \neq 0$, we have

$$c_{_n}\|x - u\| \leq \|f(x) - f(u)\|.$$

It follows that f is one-to-one, f^{-1} is continuous (and indeed Lipschitzian) from $f(X)$ to Y , and since f is - continuous (being Lipschitzian on bounded sets), it follows that $f(X)$ is a closed subset of Y .

In order to apply Theorem 2 , we choose the dense subset of the unit sphere R_x to be the intersection of the dense range of φ with the unit sphere in Y^* . For each y^* in R_x , we may choose h in X such that $\varphi(h) = y^*$. For a given element x of X , let

$$u_t = x + th, \ (t > 0).$$

Then, by the assumption of strong φ -accretiveness on f , we have

$$(\varphi(th), f(u_t) - f(x)) \geq ct^2\|h\|^2 ,$$

or, since $\varphi(th) = t\varphi(h)$ for $t > 0$, we obtain

$$(\varphi(h), f(u_t) - f(x)) \geq ct\|h\|^2.$$

We may assume $t \leq 1$. Then all the u_t lie in a bounded set, and f satisfies a Lipschitz condition on this bounded set with constant M. Hence

$$\|f(u_t) - f(x)\| \leq M\|u_t - x\| \leq Mt\|h\|.$$

Moreover,

$$1 = \|y^*\| = \|\varphi(h)\| \leq \|h\|.$$

Finally, we have

$$(y^*, f(u_t) - f(x)) \geq cM^{-1}\|f(u_t) - f(x)\|,$$

for all $t > 0$.

Thus all the hypotheses of Theorem 2 are satisfied and the proof of Theorem 3 is complete.

Remark: We have applied a version of Theorem 2 which is slightly stronger that the one stated above since $\delta_x = cM^{-1}$ is only bounded on bounded sets. This stronger version also follows from Theorem 4, as we see below.

F. E. Browder

<u>Proof of Theorem 2 from Theorem 4</u>: Since it suffices to prove that each y in Y lies in $f(X)$ in order to prove Theorem 2 , we need only verify that the hypothesis of Theorem 2 implies the hypothesis of Theorem 4 for a given y in Y . By hypothesis,

$$(y^*, f(u_j) - f(x)) \geq \delta \| f(u_j)-f(x) \|,$$

for each y^* in the set $R(x)$. Since R_x is dense in the unit sphere in Y^* , we may choose y^* such that

$$\| y^* - (\|y - f(x)\|)^{-1} J(y-f(x)) \| \leq \frac{\delta}{2} \ .$$

Then, we obtain

$$(J(y-f(x)), f(u_j)-f(x)) \geq \frac{\delta}{2} \| y - f(x) \| \cdot \| f(u_j)-f(x) \|$$

for the same sequence $\{u_j\}$. Applying Theorem 4 , we see that y lies in $f(X)$.   q.e.d.

<u>Remark</u>: For a given y in Y , we need only apply Theorem for all x in X such that $\| f(x) - y \| \leq r$. In the application to the proof of Theorem 3 , f^{-1} maps bounded sets into bounded sets. Hence, in that application, we needed the uniformity of the constant δ only on bounded sets in x .

F. E. Browder

We now turn to the proof of Theorem 4 .

Proof of Theorem 4: Since Y^* is assumed to be uniformly
convex, the duality mapping J of Y into Y^* is uniformly
continuous on all balls in Y . We choose in particular a large
ball about y of radius $\geq 2r$. We may then find $\gamma > 0$ such
that for $\|v\| \leq \gamma$, v in Y , we have

$$\| J(z + v) - J(z)\| \leq \frac{\delta}{2} d_0$$

for all z in Y with $\|z\| \leq 2r$. Choose ϵ such that
$\epsilon > 0$ and

$$(2\epsilon + \epsilon^2)d_0^2 < \frac{\gamma \delta}{2}$$

where we suppose that

$$d_0 = \text{dist}(y, \ f(X)) > 0 \ , \ (\epsilon d_0 + \gamma) + d_0 < \gamma.$$

For this given $\epsilon > 0$, we may find a point s in $S = f(X)$
such that

$$\| y - s \| \leq (1 + \epsilon)d_0 \ .$$

Let C be the cone given by

$$C = \left\{ y \mid (y^*,y) \geq \delta \|y\| \right\}$$

F. E. Browder

where

$$y^* = (\|y - s\|)^{-1} J(y-s).$$

By its construction, C is a narrow cone in the sense of Definition 2 of $[6]$, and hence we may apply Lemma 2 of $[6]$ with respect to the cone C on bounded closed subsets of Y .

Consider a point of the set $(s+C)$ of the form

$$v = s+y, \quad y = \xi w, \quad \tfrac{\gamma}{2} \le \xi \le \gamma, \quad \|w\| = 1 .$$

Using the fact that the duality mapping J is the subgradient of the convex function $\frac{1}{2}\|y\|^2$, we see that

$$\| y-s \|^2 \ge \|y - (s+\xi w)\|^2 + 2(J(y-s-\xi w),\xi w),$$

i.e.

$$\| y-(s+\xi w)\|^2 \le (1+\epsilon)^2 d_o^2 - 2(J(y-s),\xi w) + \xi\| J(y-s) - J(y-s-\xi w)\| .$$

Since $\|\xi w\| = \xi \le \gamma$, we know that

$$\| J(y - s) - J(y - s - \xi w)\| \le \frac{\delta d_o}{2} .$$

On the other hand,

$$(J(y-s),\xi w) \ge \delta\xi\| y - s\| \ge \delta d_o \xi .$$

Hence

$$(1+\epsilon)^2 d_o^2 - 2(J(y-s), \xi w) + \xi\| J(y-s) - J(y-s-\xi w)\|$$

$$\le d_o^2 + [d_o^2(2\epsilon + \epsilon^2) - 2\delta d_o\xi + \delta d_o\xi] \le d_o^2 + [d_o^2(2\epsilon+\epsilon^2) - \delta d_o\xi$$

for $\xi \geq \frac{\gamma}{2}$, since

$$d_0^2(2\epsilon + \epsilon^2) - \delta d_0 \xi \leq d_0^2(2\epsilon + \epsilon^2) - \frac{\delta d_0 \gamma}{2} < 0 .$$

By the preceding calculation, it follows that if we set

$$K_0 = (s + C) \cap B_\gamma(s), \quad S_0 = S \cap K_0 ,$$

then for all s_0 in S_0 , we have $\| y - s_0 \| \geq d_0$, and consequently

$$\| s_0 - s \| < \frac{\gamma}{2} .$$

Choose ζ with $0 < \zeta < \frac{\gamma}{2}$. Then for any s_0 in S_0 ,

$$(s_0 + C) \cap B_\zeta(s_0) \subset K_0 ,$$

and hence

$$(s_0 + C) \cap S \cap B_\zeta(s_0) = (s_0 + C) \cap S_0 \cap B_\zeta(s_0) .$$

We may apply Lemma 2 of [6] and find a point s_0 of S_0 such that

$$(s_0 + C) \cap S_0 = \{ s_0 \} .$$

For this choice, we see by the above that

$$(s_0 + C) \cap S \cap B_\zeta(s_0) = \{ s_0 \} .$$

Moreover, $\| y - s_0 \| \leq d_0 + \epsilon d_0 + \gamma \leq r$.

F. E. Browder

By the hypothesis of Theorem 4 , we may find a sequence $\left\{ s_j = f(u_j) \right\}$ in $S = f(X)$ such that s_j differs from $s_0 = f(x)$ for each j , s_j converges to s_0 as $j \longrightarrow \infty$, while for all j ,

$$(y^*, s_j - s_0) \geq \delta \| s_j - s_0 \|$$

Hence

$$(s_j - s_0) \in C ,$$

i.e.

$$s_j \in (s_0 + C) \cap S \cap B_\xi(s_0), \quad (s_j \neq s_0),$$

for all sufficiently large j . This contradicts the choice of s_0 , and this contradiction establishes the conclusion of Theorem 4 .

q.e.d.

F. E. Browder

Bibliography

[1] E. Bishop and R.R. Phelps, The support points of convex sets, Convexity, Proc. Symposia in Pure Math., Vol. 7, Amer. Math. Soc., (1963), 27-35 .

[2] F.E. Browder, Nonlinear operators and nonlinear equations of evolution in Banach spaces, Nonlinear Functional Analysis, Proc. Symposia in Pure Math., Amer.Math. Soc., (to appear).

[3] _____, On the Fredholm alternative for nonlinear operators, Bull. Amer. Math. Soc., July 1970 .

[4] _____, Normal solvability and the Fredholm alternative for mappings into infinite dimensional manifolds, Jour. Func. Anal., (to appear).

[5] _____, Normal solvability for nonlinear mappings into Banach spaces, Bull. Amer. Math. Soc., (to appear).

[6] _____, Normal solvability for nonlinear mappings and the geometry of Banach space, CIME Symposium on Nonlinear Analysis, (1970) (to appear).

[7] S.I. Pohozhayev, On the normal solvability of nonlinear operators, Dokladi Akad. Nauk SSSR, 184 (1969), 40-43 .

./.

F. E. Browder

[8] S.I. Pohozhayev, On nonlinear operators having weakly close images and quasilinear elliptic equations, Mat. Sbornik, 78
-    (1969), 237 - 259 .

[9] _____, Normal solvability of nonlinear mappings in uniformly convex Banach spaces, Funct. Anal. and Appl., 3 (1969), 80-84.

CENTRO INTERNAZIONALE MATEMATICO ESTIVO

(C. I. M. E.)

F. E. BROWDER

NORMAL SOLVABILITY FOR NONLINEAR MAPPINGS AND THE
GEOMETRY OF BANACH SPACES

Corso tenuto a Varenna dal 20 al 29 Agosto 1970

NORMAL SOLVABILITY FOR NONLINEAR MAPPINGS AND THE GEOMETRY OF BANACH SPACES

by

Felix E. Browder

(University of Chicago)

Introduction: Let X and Y be Banach spaces, L a continuous linear mapping of X into Y. A fundamental principle of linear functional analysis originating with Hausdorff [7] states that if the range $R(L)$ of the linear mapping L is closed in the Banach space Y, then the range $R(L)$ can be characterized as $N(L^*)^{\perp}$, the annihilator in Y of the nullspace $N(L^*)$ of the adjoint mapping L^*. A mapping L having these properties is said to be <u>normally solvable</u>.

Our object in the present discussion is to consider the generalization of this property to nonlinear mappings into Banach spaces, a generalization based upon the study of the geometrical properties of closed subsets of Banach spaces.

F. E. Browder

Investigations in this direction were begun in the papers of
Pohozhayev ([10], [11], [12]) and sharpened in recent papers of
the writer (Browder [3], [4], [5]). In particular, we shall
develop in detail the results sketched in the writer's recent
note [5] .

We consider a mapping f of the topological space X
into the Banach space Y . (Manifolds modelled on Banach spaces
were treated in the results of the writer's paper [4] , but we
shall not consider them explicitly here). Such a mapping f is
said to be normally solvable in the present discussion if its
image f(X) is a closed subset of Y in the ordinary strong
topology of Y . For mappings f having this property, we shall
obtain results involving sufficient criteria for an element y
of Y to lie in f(X) in terms of local properties of the
mapping f , namely the asymptotic direction set $D_x(f)$ of f
at various points x of X :

Definition 1 : Let X be a topological space, Y a Banach
space, f a mapping of X into Y . For a given point x of
X , the asymptotic direction set $D_x(f)$ of f at x is the

subset of Y given by

$$D_x(f) = \bigcap_{\varepsilon > 0} cl(\{y \epsilon y \quad Y, \ y = \xi(f(u) - f(x)), \ \xi \geq 0, \ u \epsilon X,$$

$$\| f(u) - f(x) \| < \varepsilon \}) ,$$

where cl(.) denotes the closure in the strong topology of Y .

Our basic result is the following:

Theorem 1 : Let X be a topological space, Y a Banach space, f a mapping of X into Y such that f(X) is closed in Y . Let y be a given point in Y , and for r > 0, let $B_r(y)$ be the closed ball of radius r about y in Y . Suppose that there exists r > 0 and p < 1 such that $f^{-1}(B_r(y))$ is non-empty, while for each x in $f^{-1}(B_r(y))$,

$$\text{dist}(y - f(x), \ D_x(f)) \leq p \| y - f(x) \| .$$

Then y lies in f(X) .

(In the above result, $\text{dist}(y - f(x), \ D_x(f))$ denotes the distance of the point (y - f(x)) from the set $D_x(f)$ in Y). If D' is a subset of $D_x(f)$, then $\text{dist}(y - f(x), \ D_x(f))$

F. E. Browder

is no larger than dist$(y - f(x), D')$. To see the significance
of Theorem 1 in a more specialized context in which there exists
the possibility of computing a useful subset of $D_x(f)$, we
consider the following result:

Proposition 1: Let X be a locally convex topological
vector space, Y a Banach space, f a mapping of X into Y
which is once Gateaux differentiable at a given point x of X
with a differential df_x which is a continuous linear mapping
of X into Y . Let L denote df_x , with L^* being the
adjoint mapping of Y^* into X^* , (where X^* and Y^* are the
dual spaces of X and Y , respectively). Let R(L) denote
the range of L , $N(L^*)$ the nullspace of L^* in Y^* , $(N(L^*))^{\perp}$
its annihilator in Y .

Then:
$$D_x(f) \supset cl(R(L)) = (N(L^*))^{\perp}.$$

Proof of Proposition 1 : Let w be an element of cl(R(L)).
To show that w lies in $D_x(f)$, it suffices for each given
ε , $\delta > 0$, to show that there exists u in X with
$$\| f(u) - f(x) \| < \varepsilon$$

F. E. Browder

such that for some $\xi \geq 0$,

$$\| w - \xi(f(u)-f(x)) \| < \delta .$$

Since the map f is once Gateaux differentiable at x , it is continuous at x . Hence there exists a neighborhood U of x in X such that for u in U ,

$$\| f(u) - f(x) \| < \epsilon .$$

Hence, it suffices to obtain the desired point u from U .

Since w lies in $cl(R(L))$, there exists a point v in $R(L)$ such that

$$\| w - v \| < \frac{\delta}{2} .$$

Let h be a point of X such that $L(h) = v$. By the definition of the Gateaux derivative

$$\lim_{\gamma \to 0+} \gamma^{-1}(f(x + \gamma h) - f(x)) = L(h) = v .$$

Hence we may choose $\gamma > 0$ so small that $(x+\gamma h)$ lies in U , while

$$\| \gamma^{-1}(f(x+\gamma h) - f(x)) - v \| < \frac{\delta}{2} .$$

Hence

$$\| w - \gamma^{-1}(f(u) - f(x)) \| < \delta$$

with $u = (x + \gamma h)$.

q.e.d.

F. E. Browder

If we assume the hypothesis of Theorem 1 for each point y of Y , we obtain the following global result:

Theorem 2 : Let X be a topological space, Y a Banach space , f a mapping of X into Y such that $f(X)$ is closed in Y . Suppose that for each point y in Y , there exists $r(y) > 0$ and $p(y) < 1$ such that $f^{-1}(B_{r(y)}(y)) \neq \emptyset$, while for each x in $f^{-1}(B_{r(y)}(x))$,

$$\text{dist}(y - f(x), D_x(f)) \leq p(y) \| y - f(x) \| .$$

Then: $f(X)$ is the whole of Y .

Theorem 2 obviously follows from Theorem 1 . Using Proposition we can immediately derive some corollaries of both results:

Corollary 1 to Theorem 1 : Let X be a locally convex topological vector space, Y a Banach space, f a once Gateaux differentiable mapping of X into Y with $f(X)$ closed in Y . Let y be a given element of Y , and suppose that there exist $r > 0$ and $p < 1$ such that for all x in $f^{-1}(B_r(y)) \neq \emptyset$, we have

$$\| y - f(x) + N(df_x^*)^\perp \|_{Y/N(df_x^*)} \leq p \| y - f(x) \| .$$

F. E. Browder

Then: y <u>lies in</u> f(**X**).

Corollary 2 to Theorem 1: <u>Let</u> X <u>be a locally convex</u> <u>topological vector space,</u> Y <u>a Banach space,</u> f <u>a once Gateaux</u> <u>differentiable mapping of</u> X <u>into</u> Y <u>with</u> f(X) <u>closed in</u> Y . <u>Let</u> y <u>be an element of</u> Y , <u>and suppose that</u> $f^{-1}(B_r(y)) \neq \emptyset$ <u>for a given</u> $r > 0$. <u>Suppose that for each</u> x <u>in</u> $f^{-1}(B_r(y))$, <u>we have</u>

$$(y - f(x)) \in cl(R(df_x)) = (N(df_x^*)) .$$

Then: y <u>lies in</u> f(X) .

Corollary 2 to Theorem 1 is obviously a special case of Corollary 1 with $p = 0$. Hence, it suffices to prove Corollary to Theorem 1 . The latter follows, however, from Theorem 1 and Proposition 1 since we know that the subspace $Y_o = (N(df_x^*)) =$ $= cl(R(df_x))$ is contained in $D_x(f)$ for each x in X , that

$$dist(y - f(x), D_x(f)) \leq dist(y - f(x), Y_o),$$

while for a closed subspace Y_o of Y ,

$$dist(y - f(x), Y_o) = \| y - f(x) + Y_o \|_{Y/Y} .$$

F. E. Browder

Corollary 1 to Theorem 2: Let X be a locally convex topological vector space, Y a Banach space, f a once Gateaux differentiable mapping of X into Y with $f(X)$ closed in Y. Suppose that for each y in Y, there exists $r(y) > 0$ and $p(y) < 1$ such that $f^{-1}(B_{r(y)}(y)) \neq \emptyset$, while for all x in $f^{-1}(B_{r(y)}(y))$,

$$\text{dist}(y - f(x), \text{cl}(R(df_x)) \leq p(y)\|y - f(x)\|.$$

Then: $f(X) = Y$.

Corollary 2 to Theorem 2: Let X be a locally convex topological vector space, Y a Banach space, f a once Gateaux differentiable mapping of X into Y with $f(X)$ closed in Y. Suppose that for each x in X,

$$N(df_x^*) = \{0\}.$$

Then: $f(X) = Y$.

Corollary 1 to Theorem 2 follows immediately from Theorem 2 and Proposition 1, while Corollary 2 is a special case of Corollary 1.

The preceding results have been announced in Browder [5]

F. E. Browder

with a sketch of the proof, and we give the proof in detail in Section 1 of the present paper. Let us note the connection with previous results in the literature. In the special case in which Y is reflexive, $r = dist(y,f(X))$, and $f(X)$ satisfies the stringent condition of being closed in the weak topology of Y , Corollary 2 to Theorem 1 was proved by Pohozhayev [10] , [11] For the special case of Y uniformly convex with $f(X)$ closed in Y , Corollary 2 to Theorem 2 was proved by Pohozhayev [12] . Under the assumption that Y is a general Banach space and with $f(X)$ closed in Y , it was shown in Browder [3] that $f(X) = Y$ provided that $D_x(f) = Y$ for all x in X , which is essentially Theorem 2 with $p(y) = 0$ and $r(y) = + \infty$ for all y in Y

The result of [3] was carried over in Browder [4] to the proof that $f(X) = Y$ for Y a connected manifold modelled on an infinite dimensional Banach space B provided that $f(X)$ is closed in Y and that $D_x(f) = T_{f(x)}(Y)$ for all x in X - N , where the exceptional set N has the property that $f(N) \neq f(X)$ while $f(N)$ is a locally compact, closed subset of Y . Results on the openness of such mappings f are obtained

F. E. Browder

in [4] under the more stringent hypothesis that f is locally
a closed mapping while for each y in Y , $f^{-1}(y)$ is compact
and totally disconnected.

Remark 1 : We have altered the definition of $D_x(f)$ from
that given in [4] in two senses: First, by considering it as a
subset of Y rather than of $Y - \{0\}$, which amounts only to
the fact that under our present definitions, $D_x(f)$ always
contains the element 0 of Y ;   Second, and more essentially,
in the terminology of [4] , our present $D_x(f)$ is essentially
what we might previously have written as $D_{f(x)}(f(X))$. In the
earlier definition,

$$D_x(f) = \bigcap_U cl(\{y|y \in Y, y = \xi(f(u)-f(x)), \xi \geq 0, u \in U\})$$

where U ranges over the neighborhoods of x in X . Under
our present definition, for continuous f $D_x(f)$ is bigger
than before (or at least as large) and the force of the hypotheses
is correspondingly not increased.

Remark 2 : We have made an essential shift in the meaning
of the term normally solvable as compared with the above-cited

F. E. Browder

papers of Pohozhayev.

In the Pohozhayev papers, the term <u>normally solvable</u> refers not to the hypothesis that $f(X)$ is closed in Y , but rather to the conclusions on the properties of the once differentiable mapping f under the hypotheses of the given paper. Thus in [10] and [11] , where $f(X)$ is assumed to be weakly closed in Y and it follows that for each y there exists a nearest point $f(x_0)$ in $f(X)$, f is said to be normally solvable if y lies in $f(X)$ provided that $(y - f(x_0))$ lies in $(N(df_{x_0}^*))^\perp$. Similarly, in [12], f is said to be normally solvable provided that for any y in Y , y lies in $f(X)$ provided that the following two conditions are satisfied:

(a) There exists a sequence $\{y_n\}$ converging to y such that for each n , there exists x_n such that $\|y_n - f(x_n)\| = $ $= \operatorname{dist}(y_n, f(X))$.

(b) For each n ,

$$y_n - f(x_n) \text{ lies in } (N(df_{x_n}^*))^\perp .$$

It is the present writer's view that the increasing complexity of these definitions makes the usage introduced in

[5] and the present paper more sensible, i.e. using the term
normally solvable to refer to the hypotheses rather than the
conclusions of our Theorems.

Remark 3: There is a special case in which the result of
Corollary 2 of Theorem 2 follows more directly from known results,
namely the case in which X and Y are Banach spaces, and f
is a once continuously Frechet differentiable mapping such that
df_x has closed range for each x in X. In that case, if
$N(df_x^*) = \{0\}$, it follows that $Y = cl(R(df_x))$, i.e. $Y = R(df_x)$.
It follows from the results of Graves [6] that f is an open
mapping, and since we then know $f(X)$ is both open and closed
in Y, $f(X)$ must coincide with Y. This line of argument
works in particular if f is a Fredholm mapping in the sense
of Smale [13], i.e. df_x for each x in X is a Fredholm map
in the linear sense from X to Y. In particular, most of the
applications given by Pohozhayev in [11] are Fredholm maps for
which these remarks apply.

Remark 4 : The connection of the arguments and conclusions
of the proof of Theorem 1 with the circle of ideas of the Bishop-

F. E. Browder

Phelps Theorem on the density of support points for a closed

convex bounded subset of a Banach space make it clear, as

described in detail in [4] , that the only reasonable

assumption under which Theorem 1 is true is that Y is a Banach

space. In particular, Theorem 1 is not true nor is Theorem 2 if

Y is an arbitrary Frechet space or an incomplete normed space

(or even a pre-Hilbert space which is not complete).

Section 1 : We now turn to the detailed proof of Theorem 1

and its connection with the geometry of Banach spaces. To see

the point of the general argument which we apply, let us begin

by considering a simpler line of argument which applies only

under the restrictive hypothesis that there exists a point x_o

in X such that

$$\| y - f(x_o) \| = dist(y, f(X)) = d_o > 0 ,$$

and there by extends the arguments applied by Pohozhayev in [10],

[11], and [12] . Let B_o be the open ball of radius d_o about

y in Y , and consider any element w of Y of the form

$$w = \xi(b - s_o)$$

with b in B_o and $0 \le \xi \le 1$. Then for $s_o = f(x_o)$, we have

$$\| y - (s_0 + w) \| = \| \xi (v - b) + (1-\xi)(y-s_0) \|$$

$$\leq \xi \| y - b \| + (1-\xi) \| y - s_0 \| < d_0 \; .$$

Since $d_0 = \text{dist}(y, f(X))$, it follows that no such point $(s_c + w)$ can lie in $f(X)$, and in particular $D_{x_0}(f)$ is disjoint from the open set $B_0 - s_0$. Under the hypothesis of Theorem 1, however, $D_{x_0}(f)$ does include a point of $B_0 - s_0$. This contradiction ensures that $i_0 = 0$ and that y lies in $f(X)$.

We cannot apply this elementary argument in the general case since it is not true in general for a closed subset S of Y that the minimum distance from y to S is actually taken on by $\| y - s_0 \|$ for some point s_0 of S. We must therefore replace this relatively coarse argument by a subtler argument using the properties of cones in Banach spaces.

The basic tool of this subtler argument is the following:

Proposition 2 : Let Y be a Banach space, S_0 a closed bounded subset of Y, C a closed cone with vertex at the origin in Y which is generated by a closed, bounded, convex subset F of Y which does not include 0.

Then there exists a point s_0 of S_0 such that

$$(s_0 + C) \cap S_0 = \{s_0\} .$$

We remark that C is said to be generated by F if

$$C = \{y | y \in Y , y = \xi f, \xi \geq 0 , f \in F\}.$$

Proposition 2 is an extension of the basic argument used by Bishop-Phelps [1] for the study of the existence and density of supporto points of closed, bounded, convex subsets in a Banach space.

For the proof of Proposition 2 , we need another definition and two simple lemmas. We shall employ our standard notation for the pairing between a Banach space Y and its conjugate space Y^* in which (y^*,y) denotes the value of the linear functional y^* at the point y of Y .

Definition 2: Let Y be a Banach space, C a closed cone with vertex at 0 in Y . Then C is said to be a narrow cone in Y if there exists y^* in Y^* such that $\|y\| \leq (y^*,y)$ for all y in C .

F. E. Browder

We note that the class of <u>narrow cones</u> coincides in the terminology of Chapter 1 of Krasnoselski [8] with the class of <u>cones which admit plastering,</u> and that the following result is noted in Chapter 1 of [8] :

<u>Lemma</u> 1 : <u>Let</u> Y <u>be a Banach space,</u> C <u>a closed cone with vertex at the origin in</u> Y <u>such that</u> C <u>is generated by a closed bounded convex subset</u> F <u>of</u> Y <u>which does not include</u> O. <u>Then</u> C <u>is a narrow cone in</u> Y .

<u>Proof of Lemma</u> 1 : Since O does not lie in the closed convex set F , we may apply the standard separation theorems for convex sets in Y and obtain a continuous linear functional u^* on Y such that

$$0 = (u^*, 0) < c_0 = \inf_{v \in F} (u^*, v) .$$

By hypothesis, F is bounded, and therefore there exists $M > 0$ such that $\|v\| \leq M$ for all v in F . For each y in C , $y = \xi v$ for some $\xi \geq 0$ and v in F . Hence

$$(u^*, y) = \xi(u^*, v) \geq \xi c_0 .$$

F. E. Browder

On the other hand,

$$\|y\| = \xi \|v\| \le \xi M .$$

Hence

$$(u^*,y) \ge c_o M^{-1} \|y\|$$

for all y in C . Setting

$$y^* = c_o^{-1} M u^* ,$$

we obtain the conclusion of the Lemma. q.e.d.

Lemma 2: Let Y be a Banach space , S_o a bounded closed subset of Y , C a narrow cone in Y . Then there exists an element s_o of S_o such that

$$(s_o + C) \cap S_o = \{s_o\} .$$

Proof of Lemma 2: We introduce the standard partial opering on S_o induced by the cone C by letting $y_1 \ge y_2$ if and only if $y_1 - y_2 \in C$. With respect to this ordering, the conclusion of Lemma 2 is equivalent to the statement that there exists a maximal element s_o in S_o . By Zorn's Lemma, the existence of such a maximal element s_o will follow if for each subset S'

of S_0 which is totally ordered with respect to the given
ordering, there exists an upper bound in S_0 .

Let F be a finite subset of S' . Then F can be represented
in the form $\{y_1, y_2, \ldots, y_s\}$ such that $y_j \leq y_{j+1}$ for each j ,
($1 \leq j \leq s-1$). By hypothesis, C is a narrow cone. Hence, there
exists y^* in Y^* such that $\|y\| \leq (y^*, y)$ for all y in C .
In particular, we have

$$\| y_{j+1} - y_j \| \leq (y^*, y_{j+1} - y_j).$$

If we sum with respect to j , we obtain

$$\sum_{j=1}^{s-1} \| y_{j+1} - y_j \| \leq (y^*, x_s - x_1) \leq 2M \|y^*\| ,$$

where M is the upper bound of the norms of elements in S_0 .

We note that the bound on the right hand side of the last
inequality is independent of the choice of the finite subset F
of S' . Hence, for each $\delta > 0$, there exists a finite subset
F_δ of S' such that for any points y and y_1 in $S' - F_0$,
we have $\|y - y_1\| < \delta$. In particular, S' is countable at
most. Since the case in which S' is finite is trivial and since
S' is countable and totally ordered, we may find an ascending

sequence $\{y_j\}$ in S' with $y_j \leq y_{j+1}$ for each j such that

for each y in S' , $y \leq x_j$ for at least one value of j ,

(and hence for all larger indices). For this sequence and any

integer N , it follows from the argument of the preceding

paragraph that

$$\sum_{j=1}^{N} \| y_{j+1} - y_j \| \leq 2M \| y^* \| = M_1 ,$$

with M_1 independent of N . Hence, the infinite series

$$\sum_{j=1} \| y_{j+1} - y_j \| < \infty ,$$

and since Y is a Banach space, it follows that y_j converges

strongly in Y to some element y_0 . Since S_0 is closed,

y_0 lies in S_0 . For each y in S' , $y \leq x_j$ for all

sufficiently large j , and it follows that $y \leq y_0$. Hence y_0

is an upper bound for S' in S_0 , the inductive condition of

Zorn's Lemma is satisfied, and S_0 has a maximal element with

respect to the given ordering. Thereby the proof of Lemma 2 is

complete.

Proof of Proposition 2: By Lemma 1, C is a narrow cone. By Lemma 2 , there exists s_0 in S_0 such that

$$(s_0 + C) \cap S_0 = \{s_0\} \ .$$

q.e.d.

We can now proceed to the proof of Theorem 1 proper.

Proof of Theorem 1 : Let $S = f(X)$, where by hypothesis, S is a closed subset of the Banach space Y . Suppose that y does not lie in $f(X)$, i.e. $d_0 = \text{dist}(y, f(X)) > 0$. We shall deduce a contradiction.

For a given $\epsilon > 0$, which we shall specify later in the proof, we may choose a point s in S such that

$$d = \|y - s\| \le (1 + \epsilon)d_0 \ .$$

(If there exists a point s in S with $\|y - s\| = d_0$, we choose such an s in S and let $\epsilon = 0$. In this case, the remainder of the proof reduces to the argument at the opening of Section 1 .) By hypothesis, there exists $p < 1$ such that for every x in $f^{-1}(B_r(y))$ with $r \ge d_0$, there exists w in $D_x(f)$ such that if $\xi = \|y - f(x)\|$, then

$$\| w - (y - f(x)) \| \leq p \ .$$

We choose another constant q such that

$$0 \leq p < q < 1 \ .$$

Let B be the closed ball of radius $r = qd_o$ about the point y in Y. Then B is a closed, bounded, convex subset of Y which does not include the point s. Let C be the closed cone in Y which is generated by $F = B - s$. By Lemma 1 C is a narrow cone in Y and we may apply Proposition 2 to C.

Let K be the convex closure of the point $\{s\}$ and the closed ball B. Then K is a closed, bounded, convex subset of Y, and if u is any point of K, u may be written in the form

$$u = (1-t)s + tz$$

with $0 \leq t \leq 1$, and z a point of B. Let

$$S_o = S \cap K \ .$$

Then S_o is a closed, bounded subset of Y. If u lies in S_o, then

$$d_o \leq \| u - y \| \leq (1-t) \| s - y \| + t \| z-y \| \leq (1-t)(1+\epsilon)d_o + tqd_o \ .$$

Hence, since $d_o > 0$, it follows that

(1) $$t \leq \epsilon(\epsilon + (1-q))^{-1} .$$

We apply Proposition 2 to the closed bounded set S_o and the cone C . Then there exists a point s_o in S_o such that

$$(s_o + C) \cap S_o = \{s_o\} .$$

Since s_o lies in S_o ,

$$s_o = (1-t)s + tz$$

with z in B and t satisfying the inequality (1) above. For any element y of C , we can write y in the form

$$y = \xi(z_1 - s)$$

with $\xi \geq 0$, z_1 in B . Suppose that $y \neq 0$, and that

$$v = s + y \in S .$$

Then

$$v = (1-t)s + tz + \xi(z_1 - s) = (1-t-\xi)s + tz + \xi z_1$$

$$= (1-t-\xi)s + (t+\xi)[t(\iota + \xi)^{-1} z + \xi(t + \xi)^{-1}z_1] .$$

Suppose that

$$\xi \leq (1-q)(\epsilon + (1-q))^{-1} = \delta ,$$

F. E. Browder

where $\delta > 0$. Then

$$(t + \xi) \leq t + \delta \leq 1 ,$$

and it follows that v is a convex linear combination of s and the point

$$t(t + \xi)^{-1} z + \xi(t + \xi)^{-1} z_1$$

which lies in B since B is convex and z and z_1 are elements of B . Hence for $\xi \leq \delta$, v lies in K , and since v is assumed to be a point of S , v must lie in $S \cap K = S_0$. Since $S_0 \cap (s_0 + C) = \{s_0\}$, this is impossible. Hence, if $(s_0 + C)$ intersects S , it is only at point $(s_0 + y)$ with $y = \xi(z_1 - s)$ and $\xi > \delta$.

For any such y , however, we know that

$$\| y \| = \xi \| z - s \| > \delta \ \text{dist}(s, B) = \delta_1 > 0 .$$

Thus,

$$(s_0 + C) \cap S \cap B_{\delta_1}(s_0) = \{s_0\}.$$

Since s_0 lies in $S = f(X)$, there exists a point x in X such that $s_0 = f(X)$. We assert that $D_x(f)$ includes no points of the interior of the cone C . Indeed, suppose that

$$w \in D_x(f) \cap \text{Int}(C) .$$

F. E. Browder

Then there exists a sequence $\{u_j\}$ in X such that $f(u_j)$ converges to $f(x) = s_0$ and a sequence $\{\xi_j\}$ of non-negative real numbers such that

$$\xi_j^{-1}(f(u_j) - s_0) \longrightarrow w .$$

Since $\|f(u_j) - s_0\| \longrightarrow 0$, it follows that $\xi_j \longrightarrow 0$ as $j \longrightarrow \infty$. Hence

$$\|f(u_j) - (s_0 + \xi_j w)\| \leq r_j \xi_j$$

with $r_j \longrightarrow 0$. For j sufficiently large, since w lies in the interior of C , it follows that

$$w + \xi_j^{-1}\left[f(u_j) - {}_j w - s_0\right]$$

lies in C , i.e.

$$\xi_j^{-1}\left[f(u_j - s_0\right] \in C .$$

Hence for j sufficiently large,

$$f(u_j) \in (s_0 + C) \cap S \cap B_{\delta_1}(s_0)$$

which is a contradiction to the fact that $f(u_j)$ must differ from s_0 since $w \neq 0$.

We finally assert that if we choose $\epsilon > 0$ sufficiently small, then we may assert that $\|y - f(x)\| = |y - s_0| \leq r$, and

F. E. Browder

that $D_x(f) \cap \text{Int}(C)$ is non-empty, which yields a contradiction. For the first, we note that

$$\| y - s_0 \| \leq \| y - s \| + \| s - s_0 \| ,$$

while

$$(s - s_0) = t(s - z)$$

with t satisfying the inequality (1) written above. Hence

$$\| y - f(x) \| \leq (1+\varepsilon)d_0 + \varepsilon(\varepsilon + (1-q))^{-1}(1+\varepsilon+q)d_0 \leq (1+\varepsilon s)d_0$$

with s independent of ε for fixed q. If we choose $\varepsilon > 0$, sufficiently small, we may guarantee that

$$d_0 + \varepsilon s \, d_0 \leq r ,$$

since $r > d_0$. (If $r = d_0$, $\varepsilon = 0$ and the choice of s and s_0 coincide). Hence x lies in $f^{-1}(B_r(y))$.

By the hypothesis, finally, there exists w in $D_x(f)$ such that

$$\| w - (y - s_0) \| \leq p \| y - s_0 \| ,$$

i.e.

$$\| (s_0 + w) - y \| \leq p(1+\varepsilon s)d_0 .$$

We may choose $\varepsilon > 0$ so small that

$$p(1 + \varepsilon s) < q .$$

F. E. Browder

Then $s_0 + w$ lies in the interior of B , i.e. w lies in $D_x(f) \cap \text{Int}(C)$.

This contradiction to the initial assumption that $d_0 > 0$ establishes the conclusion of the Theorem.

q.e.d.

F. E. Browder

Bibliography

[1] E. Bishop and R.R. Phelps, The support points of convex sets, Convexity, Proc. Symposia in Pure Math., Vol. 7, Amer. Math. Socity, (1963), 27 - 35 .

[2] F.E. Browder, Nonlinear operators and nonlinear equations of evolution in Banach spaces, Nonlinear Functional Analysis, Proc. Symposia in Pure Math., Amer. Math. Soc., (to appear).

[3] _____, On the Fredholm alternative for nonlinear operators, Bull. Amer. Math. Soc., July 1970 .

[4] _____, Normal solvability and the Fredholm alternative for mappings into infinite dimensional manifolds, Jour. of Functional Anal., (to appear).

[5] _____, Normal solvability for nonlinear mappings into Banach spaces, Bull. Amer. Math. Soc., (to appear).

[6] L.M. Graves, Some mapping theorems, Duke Math. Jour., 17 (1950) 111-114 .

[7] F. Hausdorff, Zur Theorie der linearen metrischen Raume, Jour. Reine Angew. Math., 167 (1932), 294-311.

[8] M.A. Krasnoselski, Positive solutions of operator equations, Moscow, 1962 (trans. into English, P. Noordhoff Ltd., 1964).

./.

F. E. Browder

[9] R.R. Phelps, Support cones and their generalizations, <u>Convexity</u>, Proc. Symposia in Pure Math., Vol. 7, Amer. Math. Soc., (1963), 393-401 .

[10] S.I. Pohozhayev, On the normal solvability of nonlinear operators, Dokladi Akad. Nauk SSSR, 184 (1969), 40-43 .

[11] _____, On nonlinear operators having weakly closed images and quasilinear elliptic equations, Mat. Sbornik, 78 (1969), 237-259 .

[12] _____, Normal solvability of nonlinear mappings in uniformly convex Banach spaces, Funct. Anal. and Appl., 3 (1969), 80-84 .

[13] J. Smale, An infinite dimensional generalization of Sard's theorem, Amer. Jour. Math., 87 (1965), 861-866.

CENTRO INTERNAZIONALE MATEMATICO ESTIVO

(C. I. M. E.)

J. EELLS and K. D. ELWORTHY

WIENER INTEGRATION ON CERTAIN MANIFOLDS

Corso tenuto a Varenna dal 20 al 29 Agosto 1970

WIENER INTEGRATION ON CERTAIN MANIFOLDS

by

J. EELLS and K. D. ELWORTHY

(University of Warwick)

Introduction.

Here we describe briefly various topological and differentias
properties of Wiener integration.

First of all, we consider Wiener measure on certain manifold
of continuous path on Riemannian manifolds, noting how it is
related 1) to the topology of path spaces, 2) to various
fibre structures, and 3) to the differential geometry of the
path manifolds.

Secondly, we construct abstract Wiener measures on certain
Banach manifolds modelled on abstract Wiener spaces (in the sense
of Gross), using a transformation of integral formula of Kuo and
the theory of Fredholm structures. We compare abstract and
classical Wiener integration in special cases. Finally, we discuss
integral formulas representing the Brouwer degree of certain

J. Eells and K. D. Elworthy

proper Fredholm maps between Banach manifolds.

Our understanding of infinite dimensional integration has been aided by our contacts with J.L. Kelley, H.-H. Kuo, and P. Štefan. We gratefully record our appreciation here.

2. Wiener measure.

(A) Let M be a compact oriented Riemannian manifold. If Δ denotes the Laplace operator on M (actually, we could use any self-adjoint positive elliptic second order linear differential operator), then the heat operator $L = \Delta - \partial/\partial t$ has a unique fundamental solution h with the following characteristic properties:

1) $h : (M \times M - \text{Diag}) \times \mathbb{R} \ (\geq 0) \longrightarrow \mathbb{R} \ (\geq 0)$ is smooth; we emphasize that h has non-negative values.

2) $h(x, \xi ; t) = h(\xi, x, t)$ for $t \geq 0$.

3) $L_x h = 0 = L_\xi h$; here L_x is L operating on the first variable of h .

J. Eells and K. D. Elworthy

4) For each $\emptyset : M \times [t_0, t_1) \longrightarrow \mathbb{R}$ of class $C^2 \times C^1$,

$$\emptyset(x,t) =$$

$$\int_{t_0}^{t} d\tau \int_M h(x,y; \ t-\tau) L\emptyset(y,\tau) dy + \int_M h(x,y; \ t-t_0)\emptyset(y,t_0) dy$$

5) $\int_M h(x,y;t) dy = 1$ for all $x \in M$, $t \geq 0$.

6) Chapman - Kolmogoroff identity :

$$h(x,y; \ s+t) = \int_M h(x,z; \ s)h(z,y;t) dz$$

7) $h(x,y; \ t) \leq$ const. $t^{-\alpha} r(x,y)^{-n+2}$ $(0 < \alpha < 1)$

Proofs can be found in [25] , [33] , [14] .

Example. We can construct fundamental solutions on certain non-compact complete Riemannian manifolds. For instance, if M is m - dimensional Euclidean space E , then h is given by

8) $h_0(x,y;t) = (2\pi t)^{-m/2} \ \exp(- 1|x-y|^2/2t)$.

(B) Let us fix a point $a \in M$ and let $C_a(M)$ denote the space of continuous path $x : [0,1] \longrightarrow M$ with $x(0) = a$. With the topology of compact convergence $C_a(M)$ is a complete

J. Eells and K. D. Elworthy

separable metrizable space, and a smooth Banach manifold [8]

model   on the Banach space $C_o(E)$.

For each $\underline{t} = (t_1, \ldots, t_n)$ such that $0 < t_1 < \ldots < t_n \leq 1$

we define the evaluation map $\rho_{\underline{t}} : C_a(M) \longrightarrow M^n$ by $\rho_{\underline{t}}(x) =$

$= (x(t_1), \ldots, x(t_n))$.

Then (as is well known and elementary) <u>the σ - algebra</u>

<u>of the Borel subsets of</u> $C_a(M)$ <u>is generated by the fibred sets</u>

$\rho_{\underline{t}}^{-1}(S)$, where $S \subset M^n$ is a Borel subset of $M^n = M \times \ldots \times M$

(n copies).

Define $w_a(\rho_{\underline{t}}^{-1}(S)) =$

$$\int \ldots \int_S h(a, m_1; t_1) h(m_1, m_2; t_2 - t_1) \ldots h(m_{n-1}, m_n; t_{n-1}) dm_1 \ldots dm_n .$$

If $\underline{s} \subset \underline{t}$, and $n(\underline{t})$ denotes the cardinality of $\underline{t}$, then

the Chapman–Kolmogoroff identity insures that

$$w_a(\rho_{\underline{s}}^{-1}(S)) = w_a(\rho_{\underline{t}}^{-1}(\pi^{-1}(S))), \quad \text{where} \quad \pi$$

denotes the indicated projection

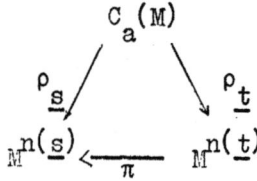

J. Eells and K. D. Elworthy

Wiener showed (see[31], [32] , [20] in case $M = \mathbb{R}$) - and the proof works in the present context, for properties 1) - 7) of h are just what is needed - that w_a is a Borel measure on $C_a(M)$ with $w_a(C_a(M)) = 1$.

From general principles concerning Borel measures on complete separable metric spaces,

1) w_a is a regular measure:

$$w_a(S) = \sup \left\{ w_a(C) : C \subset S \text{ and } C \text{ closed} \right\},$$
$$= \inf \left\{ w_a(U) : S \subset U \text{ and } U \text{ open} \right\}.$$

2) For every $\varepsilon > 0$ there is a compact subset $K \subset C_a(M)$ such that $w_a(C_a(M) - K) < \varepsilon$.

3) If $C \subset C_a(M)$ is a closed set whith $w_a(C) = 0$, then C is meager (= of first Baire category).

(C) More generally, if A , B are closed submanifolds of M and $C_{AB}(M) = \left\{ x : (I,0,1) \longrightarrow (M;A,B): x \text{ is continuous} \right\}$, then again $C_{AB}(M)$ is a smooth Banach manifold.

We have Wiener measure w_{AB} on $C_{AB}(M)$ defined by

J. Eells and K. D. Elworthy

$$w_{AB}(\rho_{\underline{t}}^{-1}(S)) =$$

$$\int_A \int_S \int_B h(a,m_1;t_1)h(m_1,m_2;t_2-t_1)\ldots h(m_{n-1},m_n;t_n-t_{n-1})h(m_n,b;1-t_n)$$

$$dadm_1\ldots dm_n db \ .$$

Then

$$w_{aB}(C_{AB}(M)) = \int_A \int_B h(a,b;1)dadb;$$

in particular ,

$$w_{MM}(C_{MM}(M)) = \text{volume } (M) \ .$$

w_{AB} is related to the locally trivial fibration

$$\rho\colon C_{AB}(M) \longrightarrow A \times B \quad (\text{defined by } \rho(x)=(x(0),x(1)))$$

in the following sense:

1) $$w_{AB}(\rho_{\underline{t}}^{-1}(S)) = \int_A \int_B w_{ab}[\rho_{\underline{t}}^{-1}(S)\cap C_{ab}(M)] dadb \ ;$$

2) $$\rho_* w_{AB} = h(a,b;1)dadb \ .$$

J. Eells and K. D. Elworthy

3. A characterization of Wiener measure.

(A) We have seen that w_{AB} respects both the topological and fibre structures of $C_{AB}(M)$. We next want to·show how, in the case of $A = a$, $B = M$, it relates to the Wiener measure w_0 on the Euclidean tangent space $M(a)$.

(B) Let $P_a(M)$ be the space of paths on M starting at a , of Sobolev class L_1^2 (i.e., paths x which are absolutely continuous and whose tangent vectors are Lebesgue square summable). Then $P_a(M) \subset C_a(M)$ as a set. With its L_1^2 - topology $P_a(M)$ is a smooth Hilbertian manifold [8] . Letting D/dt denote covariant differentiation in M along the path $x \in P_a(M)$, we obtain a complete smooth Riemannian structure on $P_a(M)$ through the inner product

$$\langle u, v \rangle_x = \int_0^1 \left\langle \frac{Du(t)}{dt} , \frac{Dv(t)}{dt} \right\rangle_{x(t)} dt .$$

If $\tau_0^s(x)$ denotes parallel transport in M along x from $x(s)$ to $x(0)$, then

$$\vartheta(x)t = \int_0^t \tau_0^s(x) \, x'(s) ds = - \int_0^t \tau_0^s(x) dx(s)$$

J. Eells and K. D. Elworthy

defines Cartan's development [6] , [21] . An equivalent

description of ϑ is the following:

Let τ denote the canonical parallelism of $P_a(M)$, which

translates every tangent vector field along x back to a path

of vectors in $M(a)$. If $E: P_a(M) \longrightarrow \mathbb{R}$ denotes the energy

function

$$E(x) = \frac{1}{2} \int_0^1 |x'(t)|^2_{x(t)} dt$$

and ∇E its gradient with respect to our Riemannian structure

on $P_a(M)$, then

$$\vartheta = \tau \circ \nabla E :$$

$$\begin{array}{ccc} TP_a(M) & \xrightarrow{\ \tau\ } & P_o(M(a)) \\ & & \\ \nabla E \nwarrow & \searrow & \nearrow \vartheta \\ & P_a(M) & \end{array}$$

Here $TP_a(M)$ denotes the tangent vector bundle of $P_a(M)$.

The development determines a diffeomorphism

$$\vartheta : P_a(M) \longrightarrow P_o(M(a)) ,$$

where $M(a)$ denotes the Euclidean tangent space to M at a .

J. Eells and K. D. Elworthy

(C) $P_a(M)$ has Wiener w_a - measure zero, by a theorem of Paley-Wiener-Zygmund [27] . However, techniques of stochastic integration (as developed by Itô [19] ; see also [14]) permit an extension of $\vartheta \colon \tilde{C}_a(M) \longrightarrow \tilde{C}_o(M(a))$, where $\tilde{C}_a(M)$ is a subset of $C_a(M)'$ of w_a - measure 1 .

Now the manifold $C_a(M)$ is modelled on the Banach space $C_o(M(a))$; that model has the Euclidean Wiener measure w_o , constructed using h_o in the formula 8).

The basic relationship between it and w_a is expressed by the formula

(1)
$$\vartheta * w_a = w_o .$$

More precisely, for any Borel subset $S \subset C_a(M)$,

$$w_a(S) = w_o(\vartheta (S \cap \tilde{C}_a(M)) .$$

The formula (1) could be considered as a transformation of integral formula.

Remark. In [14] Gangolli offered an alternative construction of the fundamental solution of the heat operator, based on stochastic integral equations. The identity (1) above should

J. Eells and K. D. Elworthy

be viewed as a reformulation and reinterpretation of Gangolli's approach.

4 - <u>Wiener measures on abstract manifolds.</u>

(A) There is an abstract version of Wiener integration due to Gross [17] which can be applied to produce a Wiener-type measure on any separable Banach space. We shall describe some of the results of Gross and Kuo [23] , and show how to carry over to infinite dimensional manifolds the theory of smooth measures or densities. In particular, <u>if</u> X <u>is a metrizable</u> C^h <u>manifold modelled on a Banach space</u> E <u>with separable</u> <u>dual</u> E^* <u>then we can construct abstract Wiener measures on</u> X . These will be determined essentially by choice of abstract Wiener space (i,H,E) on the model E , together with a choice of Fredholm map f: $X \longrightarrow E$ (see § 5 C below).

It seems possible that much of integration theory of finite dimensional manifolds (e.g. harmonic integrals; see [16,18]) has some analogue in an infinite dimensional context.

J. Eells and K. D. Elworthy

(B) Consider a separable infinite dimensional Hilbert space H together with a continuous linear inclusion $i : H \longrightarrow E$ of H onto a dense subset of a Banach space E . The norm of H will be denoted by $|\ \ |$, and that of E by $\|\ \ \|$. There is a naturally induced inclusion $j = i^* : E^* \longrightarrow H$ of the dual space of E onto a dense subspace of H . The maps i , j will be considered as identifications; the inner product $\langle - , - \rangle : H \times H \longrightarrow \mathbb{R}$ of H then extends over $E^* \times E$ to the natural pairing, also written $\langle -,- \rangle$.

A cylinder set of E is a subset of the form

$$C(S; \, l_1, \ldots, l_n) = \left\{ x \in E : (\langle l_1, x \rangle), \ldots, \langle l_n, x \rangle \in S \right\}$$

where $l_1, \ldots, l_n$ are in E^* and S is a Borel subset of $\mathbb{R}^n$. Any cylinder set of H may be written in the form

$$P^{-1}(S) = \left\{ x \in H : Px \in S \right\}$$

where P is an (orthogonal) projection with finite dimensional range, and S is a Borel subset of that range. The Gaussian measure on H (with variance parameter $t > 0$) assigns to $P^{-1}(S)$ the measure

J. Eells and K. D. Elworthy

$$(2) \qquad \mu_t(P^{-1}(S)) = (2\pi t)^{-n/2} \int_S \exp(-|x|^2/2t)\,dx$$

where n is the dimension of PH and dx is Lebesgue measure

on PH .

The norm of E is said to be measurable <u>with respect to</u>

i: H $\longrightarrow$ E if for any $\epsilon > 0$ there exists a projection P_0 of

finite rank such that for any other such projection P with $P \perp P_0$

we have

$$\mu_1(\{x: \| Px \| < \epsilon\}) > 1 - \epsilon .$$

If so, the triple (i, H, E) is called an <u>abstract Wiener space</u>

[17] .

For any cylinder set D in E define $p_t(D) = \mu_t(D \cap H)$.

If (i, H, E) is an abstract Wiener space then p_t extends

uniquely [17] to a Borel measure p_t on E . This is the <u>Wiener</u>

<u>measure</u> of (i, H, E) with variance parameter t .

(C) We will call a C^r map f: U $\longrightarrow$ E on an open subset U

of E <u>admissible</u> if $f(x) = x + \alpha(x)$, where

1. $\alpha(U) \subset E^*$ and $\alpha : U \longrightarrow E^*$ is C^1 ;

J. Eells and K. D. Elworthy

2. the restriction of the differential $D\alpha(x) \mid H \in L_1(H)$ for

all $x \in U$, and the map $x \longrightarrow D\alpha(x) \mid H$ is continuous from

U into $L_1(H)$.

Here $L_1(H)$ denotes the space of trace class operators

$T: H \longrightarrow H$ with their trace class norm.

The set of admissible C^r diffeomorphisms between open subsets

of E forms a pseudo-group W^r . A stronger version (using a

somewhat broader concept of admissible map) of the following

transformation of integral formula is proved by Kuo in his thesis

[23] :

Let (i, H, E) be an abstract Wiener space with E^* separable.

Suppose that $\varphi : U \longrightarrow V$ is an admissible diffeomorphism between

open subsets of E , and define

$$g_t(\varphi, -): U \longrightarrow \mathbb{R}$$

by

$$g_t(\varphi, x) = \det(D\varphi(x) \mid H) \exp \left\{ [-2\langle \varphi(x)-x, x\rangle -|\varphi(x)-x|^2]/2t \right\} .$$

Then for any Wiener measurable function f on V which makes

either side of (3) exist, the other side exists and there is equality:

J. Eells and K. D. Elworthy

(3) $\qquad \int_V f(y) \, dp_t(y) = \int_U f \circ \varphi(x) \mid g_t(\varphi,x) \mid dp_t(x)$.

(Kuo assumes that the norm $\| \ \|$ on E can be chosen to be C^1 on $E - \{0\}$, but this is equivalent to the separability of E^* [28]).

(D) For an abstract Wiener space (i,H,E) define a W^r - __structure__ X_W on a metrizable C^r Banach manifold X to be a C^r atlas $\{(\varphi_i, U_i)\}$ of charts $\varphi_i : U_i \longrightarrow E$ for X , maximal with respect to the property that $\varphi_i \circ \varphi_j^{-1} \in W^r$ when defined (c.f. Kuo [23]).

Since the inclusion $i: H \longrightarrow E$ in an abstract Wiener space is always compact, $\underline{a}$ W^r - __structure is a special kind of Fredholm__ __structure on__ X [13] . Fix $t > 0$; given such a structure, for each i , j define

$$g_{ij} = g_{ij}^t \colon U_i \cap U_j \longrightarrow GL(\mathbb{R})$$

by $\quad g_{ij}(x) = g_t(\varphi_j \varphi_i^{-1}, \varphi_i(x)) \ g_t(\varphi_j \varphi_i^{-1}, \varphi_i(x) =$

$= \exp\{\frac{1}{2t} [-2<\varphi_j(x) - \varphi_i(x), \varphi_i(x)> - |\varphi_j(x) - \varphi_i(x)|^2]\} \times \det(D\varphi_j \varphi_i^{-1}(\varphi_i(x)|H)$.

The family $\{g_{ij}\}_{i,j}$ then form the transition functions for a line bundle over X [30] . This will be denoted by $W_t(X_w)$ and called the bundle of Wiener densities over X_w (with variance parameter t); the sections of $W_t(X_w)$ will be called Wiener densities on X_w .

The existence of a never zero Wiener density on X_w means that the underlying Fredholm structure of X_w is orientable [13] . Then X_w can be given an orientation X_w^o i.e. a maximal W^r-atlas $\{(\varphi_i, U_i)\}_i$ such that each $g_{ij}(x) > 0$. A density ξ on X_w^o can be called positive if each of its representatives $\xi_i: \varphi_i(U_i) \longrightarrow \mathbb{R}$ in the trivializations determined by the charts of X_w^o is a positi function. Just as in the finite dimensional situation, when E^* is separable such a positive density ξ determines a positive Borel measure $\mu(\xi)$ on X by setting

$$\mu(\xi)(V) = \int_{\varphi_i(V)} \xi_i(x) \, dp_t(x)$$

for any open set V in U_i and any (φ_i, U_i) in X_w^o . By Kuo's Theorem 4C this is independent of the choice of (φ_i, U_i) . Since any density λ (recall that it is fixed) is a multiple of such a positive density ξ , say $\lambda(x) = f(x) \xi(x)$ for some function $f: X \rightarrow \mathbb{R}$, we can define the integral of λ over a Borel set S

J. Eells and K. D. Elworthy

of X by

$$\int_S \lambda = \int_S f d\mu(\xi) \ .$$

If X_w is not orientable the corresponding procedure can be followed by considering the bundle of absolute Wiener densities defined by the transition maps $\{|g_{ij}|\}_{i,j}$.

Otherwise said:

There is (by the construction above) a unique Wiener measure class associated to any W^r - structure on X (two Borel measures being equivalent if each is absolutely continuous with respect to the other).

(E) Examples. The natural inclusion $i\colon P_0(\mathbb{R}^n) \longrightarrow C_0(\mathbb{R}^n)$ defined an abstract Wiener space $(i, P_0(\mathbb{R}^n), C_0(\mathbb{R}^n))$, and the abstract Wiener measure p_1 on $C_0(\mathbb{R}^n)$ agrees with the classical Euclidean Wiener measure w_0 (see [15]). Since the dual space $C_0(\mathbb{R}^n)^*$ is not separable, Kuo's theorem does not apply immediately. However, if we let $\Lambda_0^\alpha(\mathbb{R}^n)$ denote the closure of $P_0(\mathbb{R}^n)$ in the norm

$$\| x \|_\alpha = \sup \{|x(s)-x(t)|/| s-t|^\alpha : s = t\} \text{ for } 0 < \alpha < 1/2 \ ,$$

the natural inclusion i_α makes $(i_\alpha, P_0(\mathbb{R}^n), \Lambda_0^\alpha(\mathbb{R}^n))$ an abstract

J. Eells and K. D. Elworthy

Wiener space $[15]$, and the inclusion $\Lambda_0^\alpha(\mathbb{R}^n) \longrightarrow C_0(\mathbb{R}^n)$

determines an equivalence of the resulting abstract measures

for each parameter $t > 0$. In particular, $\Lambda_0^\alpha(\mathbb{R}^n)$ has full

Wiener measure in $C_0(\mathbb{R}^n)$. The space $\Lambda_0^\alpha = \Lambda_0^\alpha(\mathbb{R})$ is topological

linearly isomorphic to the Banach space c_0 of sequences

tending to zero $[7]$, so $(\Lambda_0^\alpha)^*$ is linearly isomorphic to l_1

and hence separable. Thus Kuo's theorem applies.

For a submanifold B of codimension p in $\mathbb{R}^n$ set

$\Lambda_{oB}^\alpha = \Lambda_0^\alpha(\mathbb{R}^n) \cap C_{oB}(\mathbb{R}^n)$. It is then a smooth submanifold of

$\Lambda_0^\alpha(\mathbb{R}^n)$ of codimension p . Let $\pi^p: \mathbb{R}^n \longrightarrow \mathbb{R}^p$ be the projection

on the last p coordinates, and define the projection

$$\Pi_p : \Lambda_0^\alpha(\mathbb{R}^n) \longrightarrow \Lambda_{0\ \mathbb{R}^{n-p}}^\alpha$$

by
$$\Pi_p(f)(t) = f(t) - t\, \pi^p f(1).$$

Then (see 5A below) $\Pi_p | \Lambda_{0\ B}^\alpha : \Lambda_{0\ B}^\alpha \longrightarrow \Lambda_{0\ \mathbb{R}^{n-p}}^\alpha$ is a C^∞

Fredholm map of index zero and so (see 5C below) induces a

W^{oo} - structure on $\Lambda_{0\ B}^\alpha$ modelled on the abstract Wiener space

$(i_\alpha, P_{0\ \mathbb{R}^{n-p}}, \Lambda_{0\ \mathbb{R}^{n-p}}^\alpha)$.

From the view-point of differential topology the spaces Λ_0

J. Eells and K. D. Elworthy

are considerably more tractable than C_o ; for instance, they
have an equivalent norm of class C^∞ off the origin (whereas
C_o does not have an equivalent C^1 - norm). The same holds for
the analogous spaces $\Lambda^{p+\alpha}(M,\mathbb{R}^n)$ of differentiable maps
$M \to \mathbb{R}^n$ defined on a compact manifold [1] . It therefore seems
particularly sensible to look for naturally defined Wiener
measures on these spaces - or more generally on the corresponding
manifolds $\Lambda^{p+\alpha}(M,N)$ of maps $f: M \to N$ of manifolds.

It is not clear whether the path manifolds $\Lambda_a^\alpha(M)$ carry
a W^∞- structure whose measure class contains the Wiener measure
$w_a | \Lambda_a^\alpha(M)$ constructed in § 2 .

5 - Degree.

(A) Let X, Y be metrizable smooth manifolds modelled on
a separable Banach space. A __Fredholm map__ $\varphi: X \to Y$ __of index__
0 (a Φ_o - __map__, for short) is a smooth map such that each
differential $D\varphi(x) : X(x) \to Y(\varphi(x))$ is a Fredholm operator
with kernel and cokernel of the same finite dimension. Smale

J. Eells and K. D. Elworthy

[29] has defined a Brouwer degree mod 2 for proper Φ_0 - maps; and, by introducing additional structure (precisely, oriented Fredholm structures) on X, Y, Elworthy - Tromba [12] and Mukherjea [26] have developed a theory of the degree $d_\varphi \in \mathbb{Z}$ for such maps. This degree is a global form of the Leray-Schauder degree [24] .

(B) It seems likely (although we have not verified all details in this generality) that for path manifolds we can define d_φ for proper Φ_0 - maps $\varphi: C_{AB}(M) \longrightarrow C_{CD}(N)$ by an integral formula

(4)
$$d_\varphi = \int_{C_{AB}(M)} J_\varphi(\chi) \, dw_{AB}(\chi)$$

for a suitable function J_φ ; and show that (with suitable orientability assumptions, and assuming that $C_{AB}(M)$ and $C_{CD}(N)$ are connected) if $J_\varphi \geq 0$ and is not identically o, then φ is surjective. The relation (4) would be a consequence of a transformation of integral formula

(5)
$$\int_{\varphi(S)} \beta(y) \, dw_{CD}(y) = \int_S \beta(\varphi(\chi)) J_\varphi(\chi) \, dw_{AB}(\chi)$$

J. Eells and K. D. Elworthy

for measurable subsets $S \subset C_{AB}(M)$ and integrable functions β on $C_{CD}(N)$.

Example. If ϕ is defined on $C_{oR}(\mathbb{R})$ by

$$\phi(x)t = x(t) + \int_0^1 K(t,s)x(s)ds, \text{ where } K \text{ in } C^2 \text{ on } I \times I,$$

then its Fredholm determinant

$$D_\phi = \sum_{\nu=0}^{\infty} \frac{1}{\nu!} \int_0^1 \cdots \int_0^1 \det(K(s_i,s_j)_{1 \le i, j \le} ds_1 \cdots ds \quad ;$$

and

$$|D_\phi|^{-1} = \int_{C_{\mathbb{R} \mathbb{R}}(\mathbb{R})} \exp(-\bar\phi(u)) \, dw_o(u), \quad \text{where}$$

$$\bar\phi(u) = \int_0^1 \left[\frac{d}{dt} \int_0^1 K(t,s)u(s)ds \right]^2 dt + 2 \int_0^1 \frac{\partial K(t,s)}{\partial t} u(s)ds \, du(t) \; .$$

(This is due to Cameron-Martin [2]).

More generally, we consider the non-linear operator

$$\phi(x)t = x(t) + \int_0^1 K(t,s,x(s))ds \; .$$

With certain conditions on the kernel K we have $[3,4,5]$

$$\int_{\varphi(S)} \beta(y)dw_o(y) = \int_S \beta(\varphi(x))J_\varphi(x)dw_o(x)$$

J. Eells and K. D. Elworthy

for any measurable subset $S \subset C_{\text{oIR}}(\mathbb{R})$ and integrable function β

where

$$J_{\varphi}(x) = |D_{\varphi}(x)| \exp(-\check{\Phi}(x)$$

and in the expression for $\check{\Phi}$ we have $K(t,s,x(s))$ in place of $K(t,s)x(s)$.

(C) let (i,H,E) be an abstract Wiener space with E^* separable, a metrizable C^1 and X Banach manifold modelled on E. It is known [11] that <u>there exist (proper)</u> $\check{\Phi}_0$ - <u>maps</u> $f : X \longrightarrow E$. The method of Proposition 2D of [9] can be used to obtain:

<u>A</u> $\check{\Phi}_0$ - <u>map</u> $F: X \longrightarrow E$ <u>of class</u> $C^r(r \geq 1)$ <u>induces on</u> X <u>a unique</u> W^r - <u>structure</u> $\{X,f,i\}$ <u>modelled on</u> (i,H,E) <u>in whose</u> <u>charts</u> f <u>is represented by admissible maps</u>. The $\check{\Phi}_0$ - map f is said to be <u>orientable</u> if $\{X,f,i\}$ is orientable in the sense of § 4D .

Let (φ_i,U_i) be a chart of $\{X,f,i\}$; then $f \circ \varphi_i^{-1}$: $\varphi_i(U_i) \longrightarrow E$ is admissible, so that $g_t(f \circ \varphi_i^{-1},-) : \varphi_i(U_i) \longrightarrow \mathbb{R}$ is defined. <u>The family</u> $\{\xi_i^f\}_i$, <u>with</u> $\xi_i^f(x) = g_t(f \circ \varphi_i^{-1},x)$,

J. Eells and K. D. Elworthy

determines a Wiener density ξ^f on $\{X,f,i\}$. The method of

Smale [29] in his version of the Morse-Sand theorem can be

used to show that the set of critical values of f has p_t -

measure zero in E . Together with Kuo's theorem this can be

used to establish the following result:

Let X be a separable metrizable C^r - manifold $(r \geq 2)$

and $f: X \longrightarrow E$ a proper $\check{\phi}_0$ - map of class C^r . Assume that

f is orientable. With suitable choice of orientations, if ξ^f

is integrable over X , then

$$d_f = \int_X \xi^f \ .$$

The integral with exist if the number of points in $f^{-1}(y)$

is bounded as y ranges over the set of regular values of f .

In particular, if

$$0 < \left| \int_X \xi^f \right| < \infty$$

then f is surjectiv

J. Eells and K. D. Elworthy

References

[1] Bonic R., Frampton J., and Tromba A., "Λ - manifolds", J. Fun. Analysis 3 (1969), 310 - 320.

[2] Cameron R.H. and Martin W.T., "Transformation of Wiener integrals under a general class of linear transformations T.A.M.S. 58 (1945), 184 - 219 .

[3] _____, "The transformation of Wiener integrals by non-linear transformations", T.A.M.S. 66 (1949), 253 - 283 .

[4] _____, "Nonlinear integral equations", Annals of Math. 51 (1950), 629 - 642.

[5] Cameron R.H. and Shapiro J.M., "Nonlinear integral equations", Annals of Math. 62 (1955) 472 - 497 .

[6] Cartan E., "La methode de repère mobile, la theorie des groupes continus et les espaces généralisés", Paris 1935.

[7] Ciesielski Z., "On the isomorphism of the spaces H^{α} and m ", Bull. Acad. Polon. Sci. 8 (4) (1960), 217 - 222 .

[8] Eells J, "On the geometry of function spaces", Symp. Inter. de Top. Alg. Mexico 1956 (1958), 303 - 308 .

J. Eells and K. D. Elworthy

[9] Eells J. and Elworthy K.D., "Open embeddings of certain
Banach manifolds", Annals of Math. 91 (1970), 465-485.

[10] _____, "On Fredholm manifolds", Proc.
Int. Congress Nice 1970.

[11] Elworthy K.D., "Embeddings, isotopy and stability of Banach
manifolds", to appear.

[12] _____ and Tromba A.J., "Degree theory on Banach
manifolds", Proc. Symp. Non-linear Functional Analysis
(Chicago), A.M.S. 1968.

[13] _____, Fredholm maps and differential
structure on Banach manifolds", Summer Institute on
Global Analysis (Berkeley) A.M.S. 1968 .

[14] Gangolli R., "On the construction of certain diffusion on a
differentiable manifold", Z. Wahrscheinlichkeitstheorie
2 (1964), 406 - 419 .

[15] Gross L., "Measurable functions on Hilbert space", T.A.M.S.
105 (1962), 372 - 390 .

[16] _____, "Harmonic analysis on Hilbert space", Mem. A.M.S.
46 (1963), 62 pp.

[17] _____, "Abstract Wiener spaces", Proc. Fifth Berkeley
Symp. in Math. Stat. and Probability 1965/66, 31 - 42.

J. Eells and K. D. Elworthy

[18] Gross L., "Potential theory on Hilbert space", J. Fun. Analysis 1 (2) (1967), 123 - 181.

[19] Itô K., "Stochastic differential equations on a differentiable manifold", Nagoya Math. J. 1 (1950), 35 - 47 .

[20] Ito K. and McKean H.P., "Diffusion processes and their sampl path", Grundlehren vol. 125. Springer.

[21] Kobayashi S., "Theory of connections", Annali di Mat. 43 (1957), 119 - 194.

[22] Koval'chik I.M.,"The Wiener Integral", Russ. Math. Surveys 18 (1963), 97 - 134 .

[23] Kuo H.H., "Integration theory on infinite dimensional manifolds" , to appear.

[24] Leray J. and Schauder J., "Topologie et equations fonctionnelles", Ann. E.N.S. 51 (1934), 45 - 78 .

[25] Milgram A. and Rosenbloom P., "Harmonic forms and heat conduction I", P.N.A.S. 37 (1951), 180 - 184 .

[26] Mukherjea K.K., "Cohomology theory for Banach manifolds", J. Math. and Mech. 19 (1970) 731 - 744.

[27] Paley R.E.A.C., Wiener N, and Zygmund A., "Notes on random functions", Math. Z. 37 (1963), 647 - 668.

J. Eells and K. D. Elworthy

$\begin{bmatrix} 28 \end{bmatrix}$ Restrepo G., "Differentiable norms in Banach spaces", Bull.
A.M.S. 70 (1964), 413 - 414 .

$\begin{bmatrix} 29 \end{bmatrix}$ Smale S., "An infinite dimensional version of Sard's theorem",
Am. J. Math. 87 (1965), 861 - 866.

$\begin{bmatrix} 30 \end{bmatrix}$ Steenrod N., "The topology of fibre bundles", Princeton
Univ. Press 1951.

$\begin{bmatrix} 31 \end{bmatrix}$ Wiener N., "The average value of a functional", Proc. London
Math. Soc. 2 (22) (1924), 454 - 467.

$\begin{bmatrix} 32 \end{bmatrix}$ _____, "Generalized harmonic analysis", Acta Math. 55
(1930), 117 - 258 .

$\begin{bmatrix} 33 \end{bmatrix}$ Yoshida K., "On the fundamental solution of the parabolic
equation in a Riemannian space" , Osaka Math. J. 5(1)
(1953), 65 - 74 . See also K. Îto, "The fundamental
solutions of the parabolic differential equations in
differentiable manifold", Osaka Math. J. 5(1953),
75 - 92 .

University of Warwick

CENTRO INTERNAZIONALE MATEMATICO ESTIVO

(C. I. M. E.)

W. H. FLEMING

NONLINEAR PARTIAL DIFFERENTIAL EQUATIONS -

-PROBABILISTIC AND GAME THEORETIC METHODS

Corso tenuto a Varenna dal 20 al 29 Agosto 1970

Course of Lectures - CIME, Varenna, Italy, August 1970

Nonlinear Partial Differential Equations -
Probabilistic and Game Theoretic Methods

Wendell H. Fleming

Department of Mathematics
Brown University
Providence, Rhode Island

and

Center for Dynamical Systems
Division of Applied Mathematics
Brown University
Providence, Rhode Island

Contents

W. H. Fleming

1. The Cauchy Problem for First Order and Second Order Parabolic Semilinear Equations

These lectures are concerned with first order partial differential equations of the form

$$(1.1) \qquad \varphi_s + F(s,x,\varphi,\varphi_x) = 0, \qquad s \leq T$$

and with second order equations of the form

$$(1.2) \qquad \varphi_s + \sum_{i,j=1}^{n} a_{ij}(s,x)\varphi_{x_i x_j} + F(s,x,\varphi,\varphi_x) = 0, \qquad s \leq T.$$

Here s denotes a time-like variable, $x = (x_1,\ldots,x_n)$ an n-tuple of space-like variables, and $\varphi_x = (\varphi_{x_1},\ldots,\varphi_{x_n})$ the gradient in these variables. The characteristic values of the symmetric matrices $(a_{ij}) = a$ are bounded below by a positive constant. Thus, equation (1.2) is semilinear and uniformly parabolic. Both equations (1.1) and (1.2) are to be considered with the Cauchy data

$$(1.3) \qquad \varphi(T,x) = \Phi(x).$$

Equations of the form (1.1) arise as the Hamilton-Jacobi equations of problems in calculus of variations and optimal control theory. As these problems are usually formulated, $F = F(s,x,\varphi_x)$.

W. H. Fleming

Moreover, F is necessarily a concave (or convex) function of its third argument.

The Isaacs equation associated with a differential game also has the form (1.1); see §3. In this case F need not have the concavity property in question.

Equations of the form (1.2) arise in considering stochastic optimal control problems and stochastic differential games. In these problems the differential equations describing the evolution of the states of the optimal control problem, or differential game, contain an additive "white noise" term. See §'s 2,5.

There are many other sources of equations of the form (1.1) or (1.2). For instance, when $n = 1$ the equation $v_s + F(s,x,v)_x = 0$ of a nonlinear conservation law [16],[21] can be reduced to the form (1.1). In population models of logistic growth together with diffusion [19] and with branching [20] equations of the form (1.2) appear, with $F = F(\varphi)$. A probabilistic representation for such solutions in terms of branching diffusion processes has been given [11], [26]. An implicit probabilistic representation of solutions when (1.2) is possibly degenerate parabolic, and results about generalized solutions, are given in [32], [33], in case only φ appears nonlinearly in (1.2).

Our objective is to study equations (1.1) and (1.2) using methods from probability, control theory, and differential

W. H. Fleming

games. Given F, we therefore need to find a control problem for which (1.1) is the Hamilton-Jacobi equation, or a differential game for which (1.1) is the Isaacs equation. This can be done in many ways, two of which are indicated in §6.

In particular, we are interested in results which hold approximately when (1.2) is nearly of first order. For simplicity, let

$$a_{ij} = \varepsilon \delta_{ij}, \quad i,j = 1,\ldots,n,$$

where δ_{ij} is the Kronecker symbol. Also, for simplicity, let $F = F(s,x,\varphi_x)$. Let φ^ε be the solution of (1.2) - (1.3). Thus

(1.4) $\qquad \varphi_s^\varepsilon + \varepsilon \triangle_x \varphi^\varepsilon + F(s,x,\varphi_x^\varepsilon) = 0, \qquad s \leq T$

with $\varphi^\varepsilon = \Phi$ when $s = T$. By representing $\varphi^\varepsilon(s,x)$ as the value of a stochastic differential game, one finds (§6) that φ^ε tends uniformly to a limit φ^0 as $\varepsilon \to 0$, under rather general assumptions on F and Φ. The function φ^0 is a generalized solution to (1.1), in the sense of §2. Under a different set of assumptions, including a strict concavity condition on F, stronger results are obtained (§7). These include the convergence of φ_x^ε to φ_x^0 at "regular" points, and asymptotic expansions in powers of ε of φ^ε and φ_x^ε. Such expansions are valid in

W. H. Fleming

regions made up of regular, nonconjugate points (s,x). In such

regions φ^o is a smooth function.

The convergence of φ^ε_x to φ^o_x as $\varepsilon \to 0$ has also been

studied by the method of finite differences [21]. Our methods

give results which are stronger in several respects. Probabilistic

methods similar in spirit, though quite different in detail, have

been used by Donsker and by Varadhan [29] to study the same question.

For another probabilistic approach to nonlinear parabolic equa-

tions see McKean [18].

Our methods seem to apply to first order equations in

divergence form

$$(1.5) \qquad v_s + \sum_{i=1}^{n} F_i(s,x,v)_{x_i} = 0,$$

only when $n = 1$. For $n = 1$, (1.5) becomes (1.1) by the sub-

stitution $v = \varphi_x$. Rather complete results about generalized

solutions of (1.5) are known [14], [31], especially for the case

$F_i = F_i(v)$.

Unfortunately, methods of probability and optimization

have shed no light on the question of generalized solutions of

the Cauchy problem for a system of first order nonlinear equations.

Recent work on this problem (by other methods) suggests that such

systems are inherently much more difficult than a single nonlinear

equation. See [9], [27].

W. H. Fleming

2. Optimal Control Models, of Pontryagin Type

The following minimization problem, which we call of Pontryagin type, has been much studied. Consider a system of n ordinary differential equations, written in vector form as

(System equations)

$$(2.1) \qquad \dot{\xi} = f(t,\xi(t),u(t)), \qquad s \leq t \leq T,$$

(Initial data)

$$(2.2) \qquad \xi(s) = x.$$

Here $\xi(t)$ is a vector in n dimensional space R^n denoting the state of the system at time t, and $u(t)$ a vector denoting the control applied at time t. We require that

(Control constraints)

$$(2.3) \qquad u(t) \in K,$$

where K is a given closed convex subset of some R^d. No restriction is placed on the final state $\xi(T)$ in the present discussion. Let

(Performance criterion)

$$(2.4) \qquad J(s,x; u) = \int_s^T L(t,\xi(t),u(t))dt + \Phi[\xi(T)].$$

W. H. Fleming

The problem is to find a control function u minimizing J.

Pontryagin's principle gives a necessary condition for a minimum

[17], [22]. If there are no control constraints $(K = R^d)$, this

problem is of a type considered in calculus of variations. If,

moreover, $\dot{\xi} = u$, $K = R^n$, then the problem is of the simplest type

in calculus of variations.

Let, given s,x, and $p \in R^n$,

$$(2.5) \qquad F(s,x,p) = \min_{y \in K} [L(s,x,y) + pf(s,x,y)].$$

Then (1.1) is the Hamilton-Jacobi equation associated with the

control problem. From its definition, F is concave in the vari-

able p. A function φ is called a generalized solution if φ

is locally lipschitzian and satisfies (1.1) for almost all (s,x).

Let

$$(2.6) \qquad \varphi^o(s,x) = \inf_u J(s,x;\ u).$$

It turns out that, under the assumptions in §6 (for the case of

one controller), φ^o is a generalized solution. Clearly, φ^o

satisfies the Cauchy data (1.3).

If u^o is optimal (for (s,x) as initial data) and

ξ^o is the corresponding solution of (2.1), then

W. H. Fleming

$$(2.7) \qquad \varphi^{o}(s,x) = \int_{s}^{T} L(t,\xi^{o}(t),u^{o}(t))dt + \Phi[\xi^{o}(T)].$$

This gives an integral representation for the generalized solution φ^{o}; it agrees with the formula for classical solutions obtained from the method of characteristics. Besides (2.7) it is useful to have integral representations for certain partial derivatives of φ^{o}, in particular, for $\varphi^{o}_{x_{i}}$. See §7.

Let us now describe a stochastic version of the Pontryagin model, obtained by adding a certain "noise" term in the system equations (2.1) and taking expected values in (2.4). Let w denote an m dimensional brownian motion process (also called a Wiener process), and as system equations take the stochastic differential equations

$$(2.1') \qquad d\xi = f(t,\xi(t),u(t))dt + \sigma(t,\xi(t))dw.$$

Now, ξ,u are stochastic process on the time interval $[s,T]$. The initial data (2.2) and control constraints (2.3) are as before. We allow u to be any process (with values in K) which does not anticipate future increments of the w process [4, p. 262]. In effect, this means that the controller can use at time t any information about past states, controls, or noise. Let

$$(2.4') \qquad J'(s,x; u) = E \int_{s}^{T} L(t,\xi(t),u(t))dt + E\Phi[\xi(T)],$$

where E denotes expected value. The problem is to minimize J'.

Again, let F be as in (2.5) and let

(2.8) $$a = \frac{1}{2} \sigma\sigma^*,$$

where * denotes matrix transpose. Suppose that the elements a_{ij} are bounded, Lipschitz, and that the characteristic values of $a(s,x)$ are bounded below by $c > 0$. Also, make the assumptions in §4 on L, f, Φ, K. Then (1.2) - (1.3) has a unique bounded solution φ, and all partial derivatives appearing in (1.2) are Holder continuous functions.

It can be shown that, corresponding to (2.6),

(2.6') $$\varphi(s,x) = \inf_{u} J'(s,x; u).$$

See [4, p. 263]. It can also be shown that an optimal nonanticipative u^{o} exists. In place of (2.7) we have

(2.7') $$\varphi(s,x) = E \int_{s}^{T} L(t,\xi^{o}(t),u^{o}(t))dt + E\Phi[\xi^{o}(T)],$$

where ξ^{o} is the corresponding solution of (2.1').

From the nature of the control problem it is plausible that the state $\xi(t)$ at time t contains all relevant information about the past up to t. By a <u>control policy</u> let us mean a

W. H. Fleming

function $Y(s,x)$ with values in K. If the controller uses a control policy, then in (2.1') one puts

$$(2.9) \qquad u(t) = Y(t, \xi(t)).$$

If Y is bounded and Lipschitz, then the usual Ito conditions hold and guarantee that the solution of (2.1') with initial data (2.2) is well-defined. Even if Y is merely bounded and measurable, the solution of (2.1') - (2.2) is well-defined since we have assumed uniform ellipticity of the operator $\sum a_{ij} \partial^2/\partial x_i \partial x_j$. See Stroock-Varadhan [28].

One can show that

$$\varphi(s,x) = \inf_Y J'(s,x; Y),$$

where $J'(s,x; Y) = J'(s,x; u)$, when u is as in (2.9). An optimal policy Y^o is found from the condition

$$L(s,x,y) + \varphi_x(s,x)f(s,x,y) = \min \text{ on } K \text{ when } y = Y^o(s,x).$$

The policy Y^o is defined almost everywhere, and need not be continuous. If ξ^o, u^o are as in (2.1'), (2.9) with $Y = Y^o$, then u^o is an optimal nonanticipative process.

W. H. Fleming

Remark 1. The uniform ellipticity assumption above has recently been weakened [25], in a way which includes assumptions on σ (hence on a) more often encountered in applications.

Remark 2. In the above problems, $F = F(s,x,\varphi_x)$. To obtain a Hamilton-Jacobi equation in which φ appears in (1.1) one can take

$$(2.10) \qquad J(s,x; u) = \int_s^T e^{\theta(t)} L(t,\xi(t),u(t))dt + e^{\theta(T)}\Phi[\xi(T)],$$

where

$$\theta(t) = g(t,\xi(t),u(t)).$$

In this case

$$(2.11) \qquad F(s,x,\varphi,p) = \min_{y \in K} [L+pf+\varphi g],$$

L,f,g being evaluated at (s,x,y). If, in the stochastic problem we take for J' the expected value of the right side of (2.10), then the corresponding partial differential equation is (1.2).

3. Differential Games

We being with a formal description of such games. The

W. H. Fleming

idea is that a differential game is a control problem of the type
in §2, in which there are two controllers with opposing interests.
Both controllers have perfect information about the past states
and control choices. As in §2 let us suppose that the game is
played on a fixed time interval [s,T]. The variable stopping
time case introduces further difficulties, which we mention
briefly in §4.

(System equations)

(3.1) $\dot{\xi} = f(t,\xi(t),u(t),v(t))$, $s \leq t \leq T$.

(Initial data)

(3.2) $\xi(s) = x$.

(Control constraints)

(3.3) $u(t) \in K_1, \quad v(t) \in K_2$.

(Performance criterion (also called payoff)

(3.4) $J(s,x; u,v) = \int_{s}^{T} L(t,\xi(t),u(t),v(t))dt + \Phi[\xi(T)]$.

Controller I wishes to choose u to maximize J; and controller
II to choose v to minimize J. Since both take into account
the information available, one could let $u(t)$ and $v(t)$ be

W. H. Fleming

obtained from functionals on past states and controls $\xi(r)$, $u(r),v(r)$ for $s \le r \le t$. Another possibility is to use control policies:

$$u(t) = Y(t,\xi(t))$$
$$v(t) = Z(t,\xi(t)).$$

Intuitively, the current state $\xi(t)$ contains all relevant information for further play of the game. Hence, it should suffice to use control policies.

Unfortunately, if one admits sufficiently large classes of functionals on the past or control policies (including discontinuous Y,Z), one does not know that the system equations (3.1) can be solved. This analytical difficulty can be circumvented in ways indicated in §4.

Another difficulty is that to obtain a saddle point for a differential game, control policies which involve mixing may be needed. To avoid this difficulty we shall deal with what are called minorant and majorant games. Roughly speaking in the minorant game controller II has a slight information advantage, while in the majorant game his opponent has a corresponding advantage. Let

(3.5) $$F(s,x,p) = \max_{y \in K_1} \min_{z \in K_2} (L+pf)$$

$$(3.5^*) \qquad F^*(s,x,p) = \min_{z \in K_2} \ \max_{y \in K_1} \ (L + pf).$$

With the minorant game is associated a value function φ^o, which turns out to be a generalized solution of (1.1) with this choice for F. Similarly, there is a majorant value function φ^{o*}, which is a generalized solution when F is replaced by F^*. This is explained in §'s 4,5.

The condition $F = F^*$ is called Isaacs' minimax condition [12, p. 35]. It implies $\varphi^o = \varphi^{o*}$, and therefore, that the differential game has a saddle point without mixing. A proof is outlined in §5.

For a more complete introduction to the subject of differential games we refer to [1], [7], [9], [12], [23], [24].

4. The Value Function

Let us suppose that the control sets K_1, K_2 are compact and convex. Moreover, we suppose that f, L, Φ are of class C^1 and are bounded together with their gradients f_x, L_x, Φ_x in the variables $x = (x_1, \ldots, x_n)$.

One method to circumvent the analytical difficulty mentioned in §3 is to discretize time. For the discretized problem moves occur at times $t_j = j\Delta_n$, where $\Delta_n = 2^{-n}T$, $n = 1, 2, \ldots$. Here we take $0 \leq s \leq T$, $s = t_i$ a dyadic rational times T. Let us give controller II a slight information advantage.

At time t_j, controller I chooses a control vector u_j knowing the initial state $x = \xi_i$ and all choices u_k, v_k for $i \leq k < j$. Controller II then chooses v_j knowing in addition u_j. The states obey the difference analogue of (3.1):

$$(3.1_n) \qquad \xi_{j+1} = \xi_j + \Delta_n f(t_j, \xi_j, u_j, v_j),$$

with $\xi_i = x$; and the performance criterion (or payoff) is:

$$(3.4_n) \qquad J_n = \sum_{j=i}^{\ell-1} \Delta_n L(t_j, \xi_j, u_j, v_j) + \Phi[\xi_\ell],$$

where $t_\ell = T$. Define φ_n by the nonlinear difference equation

$$(4.1_n) \qquad \varphi_n(s, x) = \max_{y \in K_1} \min_{z \in K_2} [\Delta_n L(s, x, y, z) + \varphi_n(s + \Delta_n, X)],$$

$$X = x + \Delta_n f(s, x, y, z),$$

with $\varphi_n(T, x) = \Phi(x)$. From the theory of positional games with finitely many moves, for each n and initial data (s, x) the time-discrete game has a saddle point. Its value is $\varphi_n(s, x)$.

Theorem 4.1. The function φ_n tends to a limit φ^o as $n \to \infty$, uniformly for $0 \leq s \leq T$, $x \in R^n$.

See §5 for an outline of a proof. If controller I

W. H. Fleming

instead of II has the information advantage above, then one has

in (4.1_n) min max. Let be resulting value function be φ_n^*. In a

similar way, φ_n^* tends to a limit φ^{o*} uniformly. Since $\varphi_n \leq \varphi_n^*$,

we have $\varphi^o \leq \varphi^{o*}$. If $\varphi^o(s,x) = \varphi^{o*}(s,x)$, then the number

$\varphi^o(s,x)$ is called the value of the differential game for initial

data (s,x). Otherwise, $\varphi^o(s,x)$, $\varphi^{o*}(s,x)$ are called the minorant

and majorant values, respectively.

Theorem 4.2. Let F be as in (3.5). The function φ^o in

Theorem 4.1 is lipschitzian and satisfies (1.1) for almost all

(s,x), $0 \leq s \leq T$, $x \in R^n$. Similarly, φ^{o*} satisfies (1.1) in the

same sense with F replaced by F^*.

This result was proved in [3, p. 1005]. It states that

φ^o is a generalized solution of Isaacs' equation, with the Cauchy

data $\varphi^o = \Phi$ when $s = T$. If φ^o happens to be C^1 everywhere,

then φ^o is just the classical solution of the Cauchy problem

given by the method of characteristics. In many cases, the method

of characteristics for solving differential games gives a function

$\varphi(s,x)$ which is C^1 in certain regions, but not everywhere. The

number $\varphi(s,x)$ is often called in the literature the value. See

[1], [12], [23]. A general theorem relating φ and φ^o does not

seem to have been proved.

If one allows games of variable, rather than fixed

duration, then the value function may be discontinuous. This is shown, for instance, by Isaacs' homicidal chauffeur example [12]. One has in such problems boundary data, rather than Cauchy data, for (1.1). Some partial results appear in [21, §VII], [8].

A more elegant way of defining a value was used by Varaiya and Lin [30] and by Friedman [7], [8]. It is as follows. Given n, define t_j as above. On the interval $[t_{j-1}, t_j)$ controller I chooses $u(t)$ knowing the initial state $x = \xi_i$ and $u(r), v(r)$ for $s \leq r < t_{j-1}$. Controller II then chooses $v(t)$ on $[t_{j-1}, t_j)$ knowing in addition $u(t)$ on $[t_{j-1}, t_j)$. The system equations are (3.1) and the performance criterion (3.4). For each n, there is a value function $\psi_n(s,x)$. It is easy to show than ψ_n is nondecreasing in n, and hence

$$\psi^o = \lim_{n \to \infty} \psi_n$$

exists. Similarly, let ψ_n^* be the value when controller I has the information advantage (ψ_n^* is nonincreasing in n, and $\psi_n \leq \psi_n^*$). Let

$$\psi^{o*} = \lim_{n \to \infty} \psi_n^*.$$

If $\psi^o(s,x) = \psi^{o*}(s,x)$, then this number is the value in the sense of [7], [30].

Lemma. Let $L(s,x,y,z)$ be concave in y and convex in z; let $f(s,x,y,z)$ be bilinear in (y,z). Then $\varphi^o \geq \psi^o$, $\varphi^{o*} \leq \psi^{o*}$.

This lemma can be proved by adapting reasoning in [21, §'s III, IV]. From the lemma, the two numbers $\psi^o(s,x)$, $\varphi^o(s,x)$ are equal if there is a value in the sense of [7], [30].

It was shown in [8] that $\psi^o = \psi^{o*}$, under the further restriction that f and L both separate into the sum of a function of (s,x,y) and a function of (s,x,z). Moreover, ψ^o is a generalized solution of (1.1). A corresponding result, for a different type of performance criterion, is given in [30].

5. Some Stochastic Difference Games

For $\varepsilon > 0$ consider the stochastic difference equations

$$(3.1_n^\varepsilon) \qquad \xi_{j+1}^\varepsilon = \xi_j^\varepsilon + \Delta_n f(t_j, \xi_j^\varepsilon, u_j^\varepsilon, v_j^\varepsilon) + (2\varepsilon\Delta_n)^{1/2} \eta_j,$$

with initial conditions $\xi_i^\varepsilon = x$, $t_i = s$, and $\eta_i, \eta_{i+1}, \ldots, \eta_{\ell-1}$ independent gaussian random variables each with 0 and variance 1. (The time-continuous analogue, which we do not consider, would be

$$d\xi^\varepsilon = f(t, \xi^\varepsilon, u^\varepsilon, v^\varepsilon)dt + (2\varepsilon)^{1/2} dw,$$

where w is an n dimensional brownian motion.) As performance

W. H. Fleming

criterion take

$$(3.4_n^{\mathcal{E}}) \qquad J_n^{\mathcal{E}} = E\{ \sum_{j=i}^{\ell-1} \Delta_n L(t_j, \xi_j^{\mathcal{E}}, u_j^{\mathcal{E}}, v_j^{\mathcal{E}}) + \Phi[\xi_\ell^{\mathcal{E}}] \}.$$

The controls $u_j^{\mathcal{E}}, v_j^{\mathcal{E}}$ are chosen with the same information rule as in §4. This defines a stochastic difference game. The value function $\varphi_n^{\mathcal{E}}$ satisfies the nonlinear difference equation

$$(4.1_n^{\mathcal{E}}) \qquad \varphi_n^{\mathcal{E}}(s,x) = \max_{y \epsilon K_1} \min_{z \epsilon K_2} \int_{R_n} [\Delta_n L(s,x,y,z) + \varphi_n^{\mathcal{E}}(s+\Delta_n,X')]G(\eta)d\eta,$$

$$X' = x + \Delta_n f(s,x,y,z) + (2\mathcal{E}\Delta_n)^{1/2}\eta,$$

$$G(\eta) = (2\pi)^{-n/2}\exp(-\eta^2/2).$$

Also, let $\varphi^{\mathcal{E}}$ be the unique bounded solution of (1.4) with the Cauchy data (1.3). Here F is as in (3.5).

Lemma 1. For fixed $\mathcal{E} > 0$, $\varphi_n^{\mathcal{E}} \to \varphi^{\mathcal{E}}$ as $n \to \infty$, uniformly for $0 \leqq s \leqq T, x \in R^n$.

The proof of Lemma 1 is a straightforward application of $(4.1_n^{\mathcal{E}})$, Taylor's formula, and Holder continuity of the partial derivatives $\varphi_s^{\mathcal{E}}, \varphi_{x_i}^{\mathcal{E}}, \varphi_{x_i x_j}^{\mathcal{E}}$. See [3, pp. 993-994].

Lemma 2. For fixed n, $\varphi_n^{\mathcal{E}} - \varphi_n$ tends to 0 as $\mathcal{E} \to 0$. The convergence is uniform with respect to x,s,n.

Let us indicate the proof. Suppose that in (3.1_n) and $(3.1_n^{\mathcal{E}})$, $u_j^{\mathcal{E}} = u_j$, $v_j^{\mathcal{E}} = v_j$; thus both controllers make the same choices for the deterministic and stochastic difference games. By an elementary estimate for stochastic difference equations,

$$E \max_j |\xi_j^{\mathcal{E}} - \xi_j|^2 \leq C\mathcal{E}T$$

for suitable C. This implies, for suitable M,

$$|J_n^{\mathcal{E}} - J_n| \leq M(\mathcal{E}T)^{1/2},$$

independent of s,x,n, and the control choices. Then $|\varphi_n^{\mathcal{E}} - \varphi_n| \leq M(\mathcal{E}T)^{1/2}$, whence the lemma. For details see [2II, p. 205] or [3, p. 1000].

Lemma 3. There exists M_1 (not depending on $\mathcal{E}$ and n) such that $|\varphi_n^{\mathcal{E}}(s,x)| \leq M_1$, and

(5.1) $|\varphi_n^{\mathcal{E}}(s',x') - \varphi_n^{\mathcal{E}}(s,x)| \leq M_1[|s'-s|^{1/2} + |x'-x|]$

for $0 \leq s, s' \leq T$, $x,x' \in R^n$.

The proof uses again standard estimates for stochastic difference equations. For instance, let $s' = s$. Let $\xi_n^{\mathcal{E}}, \xi_n^{\mathcal{E}'}$ denote the solutions of $(3.1_n^{\mathcal{E}})$ for respective initial data x,x'. Let $J_n^{\mathcal{E}}, J_n^{\mathcal{E}'}$ denote the respective performance criteria, when both controllers use the same control choices $u_j^{\mathcal{E}}, v_j^{\mathcal{E}}$, for both sets $(s,x),(s,x')$ of initial data. Then

$$E \max_{j} |\xi_j^{\mathcal{E}'} - \xi_j^{\mathcal{E}}|^2 \leq C_1 |x' - x|^2,$$

$$|J_n^{\mathcal{E}'} - J_n^{\mathcal{E}}| \leq M_1 |x' - x|$$

for suitable C_1, M_1. The latter inequality implies (5.1) when $s = s'$. A complete proof appears in [3, pp. 995-996].

From these lemmas we obtain:

Theorem 5.1. As $n \to \infty$, $\mathcal{E} \to 0$, $\varphi_n, \varphi_n^{\mathcal{E}}, \varphi^{\mathcal{E}}$ all tend to the same limit φ^o, uniformly for $0 \leq s \leq T$, $x \in R^n$.

In a similar way, by exchanging the information advantage we can define $\varphi_n^{\mathcal{E}*}, \varphi^{\mathcal{E}*}$, where $\varphi^{\mathcal{E}*}$ satisfies (1.4) with F replaced by F^*. Then $\varphi_n^*, \varphi_n^{\mathcal{E}*}, \varphi^{\mathcal{E}*}$ tend to the same limit φ^{o*}. If the minimax condition $F = F^*$ holds, then $\varphi^{\mathcal{E}} = \varphi^{\mathcal{E}*}$. Hence by Theorem 5.1, $\varphi^o = \varphi^{o*}$ if the minimax condition holds.

6. An Application

Given $F(s,x,p)$ let us discuss convergence of $\varphi^{\mathcal{E}}$ in (1.3) - (1.4) to a generalized solution φ^o of (1.1). [See [3] for the more general case $F = F(s,x,\varphi,p)$, treated by introducing an exponential factor like (2.11).] We do this under two

W. H. Fleming

different sets of assumptions on F. In both instances we assume

that Φ is C^1 with Φ, Φ_x bounded.

In the first instance we apply Theorems 5.1 and 4.2

about differential games. To do so we need to find f, L, K_1, K_2

such that F satisfies (3.5). Let us assume that F is C^1 and

that, for some positive B,

(6.1) $\qquad |F(s,x,0)| \leqq B, \quad |F_x| \leqq B(1+|p|), \quad |F_p| \leqq B.$

An F satisfying these inequalities is sometimes called "mildly

nonlinear". For suitable m, c, let

$$K_1 = \{q \in R^n : |q| \leqq m\}$$

$$K_2 = \{\lambda \in R^n : |\lambda| \leqq c\}$$

$$L(s,x,q,\lambda) = \frac{F(s,x,q)}{1+|q|^2} - \lambda \cdot q$$

$$f(s,x,q,\lambda) = \frac{F(s,x,q)q}{1+|q|^2} + \lambda.$$

<u>Lemma.</u> <u>For suitable</u> m, c

$$F(s,x,p) = \max_{|q| \leqq m} \min_{|\lambda| \leqq c} [L+pf].$$

A proof is given in [3, p. 999]. From the lemma and

W. H. Fleming

Theorems 5.1, 4.2, φ^{ε} tends as $\varepsilon \to 0$ uniformly to a generalized solution φ^0 of (1.1).

In particular, suppose that $F = F(p)$. From the maximum principle, $|\varphi_x^{\varepsilon}| \le M$ if $|\Phi_x| \le M$, where Φ is the Cauchy data (1.3). Hence it is no restriction to assume that (6.1) holds in this case if F is C^1. Thus φ^{ε} tends uniformly to φ^0 as $\varepsilon \to 0$.

In place of (6.1) let us now consider another set of assumptions:

(6.2)   (a) $F \in C^3$, $F(s,x,p) \le C$, $|F(s,x,0)| \le C_o$,

$$\lim_{|p| \to \infty} \frac{F(s,x,p)}{|p|} = -\infty;$$

(b) $|F_x| \le c'(F-pF_p) + c''$;

(c) $|F_p| \le R(|p|)$. For each $r > 0$ there exists v_r such that $|F_p| \le r$ implies $|p| \le v_r$.

(d) $C_1(|p|)|v|^2 \le -vF_{pp}v \le C_2(|p|)|v|^2$ for all $v \in R^n$, where $C_1(v)$, $C_2(v)$ are positive and respectively nonincreasing, nondecreasing in v.

These assumptions permit faster growth of $|F|$ as $|p| \to \infty$ than (6.1). On the other hand, (d) is a strong concavity condition in p not imposed in (6.1). Consider the following optimal control problem, which is, in fact, of the simplest type in calculus of variations. Let $\dot{\xi} = u$, $K = R^n$, and

$$(6.3) \qquad L(s,x,y) = \max_{p \in R^n} [F - py].$$

Then

$$(6.4) \qquad F(s,x,p) = \min_{y \in R^n} [L + py].$$

Formulas (6.3) and (6.4) express the duality between the function F concave in p, and L convex in y. At the points p,y where max and min occur, we have

$$y = F_p, \quad p = -L_y.$$

These are the classical Legendre transformation and its inverse. See [5].

According to (2.5) and (6.4), the Hamilton-Jacobi equation for this calculus of variations problem is (1.1). For the corresponding stochastic problem the system equations (2.1') become

$$(6.5) \qquad d\xi = u(t)dt + (2\varepsilon)^{1/2}dw,$$

W. H. Fleming

when we put $\sigma = (2\varepsilon)^{1/2}$ (identity). We apply the discussion in §2. The function φ^o in (2.6) is a generalized solution of (1.1) and (1.3) while the function φ^ε in (2.6') is the unique bounded solution of (1.4) and (1.3). Using assumptions (6.2)(b)(c) it can be shown that there are a priori bounds $|\varphi_x^\varepsilon| \le M$, $|u^\varepsilon| \le C$, where u^ε is an optimal control process. It is then not difficult to show that φ^ε tends uniformly to φ^o as $\varepsilon \to 0$. See [5, p. 525, 527]. Stronger results will be stated in the next section.

In case $F = F(p)$, the behavior of F for large $|p|$ is unimportant (as noted above). Suppose that the matrices $F_{pp}(p)$ are negative definite. In (6.3), $L = L(y)$; it suffices in (2.6) to use constant controls, $u(t) \equiv y$. Then

$$(6.6) \qquad \varphi^o(s,x) = \inf_y \{(T-s)L(y) + \Phi[x + (T-s)y\}.$$

Formulas for generalized solutions equivalent to (6.6) appear often in the literature; see for instance [16]. If $F(p)$ is neither concave nor convex in p, then might ask whether $\varphi^o(s,x)$ is expressible in a similar way taking max min among constant controls, $u(t) \equiv y$, $v(t) \equiv z$. (This would be simpler than representing $\varphi^o(s,x)$ as the value of a differential game as done above.) The answer to this question is generally "no". However, in certain cases the answer is "yes", for instance if φ^o is C^2 and $T - s$ is sufficiently small [13].

W. H. Fleming

7. More Precise Results

Let us again assume (6.2), and give more precise state-ments regarding the sense in which φ^ε and φ^o are close for small ε. We consider two kinds of results: (i) convergence of φ_x^ε to φ_x^o, almost everywhere in x for fixed s; (ii) asymptotic expansions of $\varphi^\varepsilon, \varphi_x^\varepsilon$ in powers of ε.

Definition. A point (s,x) is <u>regular</u> if the problem

$$\int_s^T L(t,\xi(t),\dot{\xi}(t))dt + \Phi[\xi(T)] = \min$$

with initial data $\xi(s) = x$ has a unique solution ξ^o.

We recall that this minimum problem is precisely the one introduced at the end of §6, and (1.1) is its Hamilton-Jacobi equation. The minimum value $\varphi^o(s,x)$ defines a generalized solution.

Theorem (Kuznetzov-Šiškin). The generalized solution φ^o is differentiable at (s,x) if and only if (s,x) is a regular point.

A proof is given in [5, pp. 520-521].

Theorem 7.1. As $\varepsilon \to 0$, $\varphi_x^\varepsilon(s,x) \to \varphi_x^o(s,x)$ at every regular point (s,x).

W. H. Fleming

The proof is based on the formulas

$$\varphi_x^{\mathcal{E}}(s,x) = E \int_s^T L_x(t,\xi^{\mathcal{E}},u^{\mathcal{E}})dt + E\Phi_x[\xi^{\mathcal{E}}(T)],$$

$$\varphi_x^o(s,x) = \int_s^T L_x(t,\xi^o,u^o)dt + \Phi_x[\xi^o(T)],$$

where $u^{\mathcal{E}}$ is an optimal stochastic control process for initial data (s,x), $\xi^{\mathcal{E}}$ as in (6.5) with $u = u^{\mathcal{E}}$, $\overset{\bullet o}{\xi} = u^o$. It can be shown that $u^{\mathcal{E}}$ is uniformly bounded and tends in mean square to u^o, with probability 1. Moreover, $\xi^{\mathcal{E}}$ tends uniformly with probability 1 to ξ^o. For details see [5].

For the next result, let Φ, F be C^∞ and F satisfy (6.2) as above. Let N denote a set open relative to the half space $s \leq T$.

Definition. N is a region of strong regularity if:

    (1)   every $(s,x) \in N$ is regular; and

    (2)   for each $(s,x) \in N$, the unique optimal ξ^o with initial point (s,x) has no conjugate points and $(t,\xi^o(t)) \in N$ for $s \leq t \leq T$.

By the method of characteristics, φ^o is C^∞ in N.

Theorem 7.2. Let N be a region of strong regularity. Then for any positive integer ℓ,

W. H. Fleming

$$\varphi^{\varepsilon} = \varphi^{0} + \varepsilon\theta_1 + \varepsilon^2\theta_2 + \cdots + \varepsilon^{\ell}\theta_{\ell} + o(\varepsilon^{\ell})$$

$$\varphi_x^{\varepsilon} = \varphi_x^{0} + \varepsilon(\theta_1)_x + \varepsilon^2(\theta_2)_x + \cdots + \varepsilon^{\ell}(\theta_{\ell})_x + o(\varepsilon^{\ell}),$$

where $\varepsilon^{-\ell}o(\varepsilon^{\ell}) \to 0$ as $\varepsilon \to 0$, uniformly for (s,x) in any compact set $N' \subset N$. The coefficient θ_j satisfies the linear first order equation obtained by formally differentiating (1.4) j times with respect to ε and setting $\varepsilon = 0$.

This theorem is proved in [6,§7].

W. H. Fleming

References

[1] BERKOVITZ, L.D., A survey of differential games, Math. Theory of Control (eds. A.V. Balakrishnan and L.W. Neustadt), Academic Press, Inc., 1967, pp. 342-372.

[2] FLEMING, W.H., The convergence problem for differential games, I. J. Math. Anal. Appl., Vol. 3 (1961), pp. 102-116. II. Advances in Game Theory (eds. M. Dresker, L.S. Shapley, A.W. Tucker), Princeton University Press, 1964, pp. 195-210.

[3] FLEMING, W.H., The Cauchy problem for degenerate parabolic equations, J. Math. and Mech., Vol. 13 (1964), pp. 987-1008.

[4] FLEMING, W.H., Duality and apriori estimates in markovian optimization problems, J. Math. Anal. Appl., Vol. 16 (1966), pp. 254-279; Erratum, J. Math. Anal. Appl., Vol. 19, p. 204.

[5] FLEMING, W.H., The Cauchy problem for a nonlinear first-order partial differential equation, J. Diff. Eq., Vol. 5 (1969), pp. 515-530.

[6] FLEMING, W.H., Stochastic control for small noise intensities, SIAM J. Control (submitted).

[7] FRIEDMAN, A., Differential Games, Academic Press (to appear).

[8] FRIEDMAN, A., Existence of value and saddle points in differential games of survival, J. Diff. Eq., Vol. 7 (1970), pp. 111-125.

[9] GLIMM, J. and LAX, P.D., Decay of solutions of systems of hyperbolic conservation laws, Courant Inst. of Math. Sci. report, April 1969.

[10] HO, Y.C., Differential games, dynamic optimization, and generalized control theory, J. Optimization Theory Appl. (to appear in 1970).

[11] IKEDA, N., NAGASAWA, M., and WATANABE, S., Branching
Markov Process, J. Math. Kyoto University, Vol. 8 (1968),
pp. 233-278, 365-410; Vol. 9 (1969), pp. 95-160.

[12] ISAACS, R., Differential Games, John Wiley, 1965.

[13] KRUZHKOV, S.N., On minimax representations of solutions of
nonlinear equations of first order, Funtzional'nii Analiz:
Prilozh., Vol. 3, (1969), pp. 57-66 (in Russian).

[14] KRUZHKOV, S.N., Quasilinear equations of first order in
many independent variables, Mat. Sbornik, Vol. 81 (123)(1970),
pp. 228-255 (in Russian).

[15] KRYLOV, N.V., Minimax type equations in the theory of elliptic
and parabolic equations in the plane, Mat. Sbornik, Vol. 81
(123)(1970), pp. 3-23 (in Russian).

[16] LAX, P.D., Hyperbolic systems of conservation laws II, Comm.
Pure Appl. Math., Vol. 10 (1957), pp. 537-566.

[17] LEE, E.B. and MARKUS, L., Foundations of Optimal Control
Theory, John Wiley, 1967.

[18] MCKEAN, H.P., JR., A class of Markov processes associated
with nonlinear parabolic equations, Proc. Nat. Acad. Sci.
USA, Vol. 56 (1966), pp. 1907-1911.

[19] MONTROLL, E.W., On nonlinear processes involving population
growth and diffusion, J. Applied Probability, Vol. 4 (1967),
pp. 281-290.

[20] MOYAL, J., Multiplicative population processes, J. Applied
Probability, Vol. 1 (1964), pp. 267-283.

[21] OLEINIK, O.A., Discontinuous solutions of non-linear differential equations, Uspekhi Mat. Nauk (NS), Vol. 12 (1957),
No. 3, pp. 3-73; American Math. Soc. Transl. Ser. 2, No. 26,
pp. 95-172.

[22] PONTRYAGIN, L.S., BOLTYANSKII, V.G., GAMKRELIDZE, R.V., and MISHCHENKO, E.F., The Mathematical Theory of Optimal Processes, John Wiley, New York, 1962 [English Translation].

[23] PONTRYAGIN, L.S., On the theory of differential games, Russian, Math. Surveys, Vol. 21 (1966), pp. 193-246.

[24] PONTRYAGIN, L.S., Linear differential games 1,2, Soviet Math. Dokl., Vol. 8 (1967), No. 3, pp. 769-771; No. 4, pp. 910-912.

[25] RISHEL, R., Weak solutions of the partial differential equation of dynamic programming, Bell Labs. Tech. Memo. MM-70-4165-2.

[26] SAWYER, S., A formula for semigroups, with an application to branching diffusion processes (to appear in Trans. of A.M.S.)

[27] SMOLLER, J.A. and JOHNSON, J., Global solutions for an extended class of hyperbolic systems, Arch. Rat. Mech. Anal., Vol. 32 (1969), pp. 169-189.

[28] STROOCK, D.W. and VARADHAN, S.R.S., Diffusion process with continuous coefficients, Comm. Pure Appl. Math., Vol. 22 (1969), pp. 345-400, 479-530.

[29] VARADHAN, S.R.S., Asymptotic probabilities and differential equations, Comm. Pure Appl. Math., Vol. 19 (1966), pp. 261-286.

[30] VARAIYA, P. and LIN, J., Existence of saddle points in differential games, SIAM J. on Control, Vol. 7 (1969), pp. 142-157.

[31] VOL'PERT, A.I., The spaces BV and quasilinear equations, Math. USSR-Sbornik, Vol. 2 (1967), pp. 225-267.

[32] FREIDLIN, M.I., Quasilinear parabolic equations and measures in function spaces, Functional Analysis and Its Applications, Vol. 1 (1967), pp. 234-240.

[33] NISIO, M., On stochastic differential equations associated with certain quasilinear parabolic equations (to appear).

CENTRO INTERNAZIONALE MATEMATICO ESTIVO

(C. I. M. E.)

C. FOIAŞ

SOLUTIONS STATISTIQUES DES ÉQUATIONS D'ÉVOLUTIONS

NON LINÉAIRES

Corso tenuto a Varenna dal 20 al 29 Agosto 1970

SOLUTIONS STATISTIQUES DES ÉQUATIONS D'ÉVOLUTIONS NON LINÉAIRES

par

C. FOIAȘ

(Académie des Sciences - Bucarest)

Une des plus belle théorie classique est sans doute celle
des invariants intégraux des systemes differentiels autonomes,
qui rélie ces systèmes aux équations aux dérivées partielles de
premier ordre et à la théorie de la mesure. L'étude d'un des
modèles mathématiques pour la turbulence (HOPF [1]) conduit à
une sorte de théorie des invariants intégraux du système de
Navier-Stokes (PRODI [2], [3] , FOIAS [1]) ouvrant ainsi la
voie vers une variante fonctionnelle, loin d'ètre achevée, de
la théorie classique. Le but de ces leçons est de servir
d'introduction stimulante dans cette variante fonctionnelle.
Pour cela nous ne chercherons pas la plus grande généralité,
mais étudierons seulement des equations d'évolution proches à
la forme fonctionnelle des équations de Navier-Stokes (voir
LIONS [1] , Ch. I, § 6) dans un domaine borné, à frontière
assez régulière. La plupart de ces exposés s'appuit sur un

C. Foiaş

travail en preparation, concernant les équations de Navier-Stokes, de G. PRODI et de l'auteur.

1. Préliminaires.

a. On supposera le lecteur familiarisé avec la théorie élémentaire des espaces de Banach et Hilbert (convergence faible, forte, opérateurs compacts, autoadjoints, etc.; par exemple RIESZ-NAGY [1]), à la théorie de l'integration des fonctions vectorielles (en particulier la représentation des fonctionnelles sur des espaces L^p de telles fonctions; voir DINCULEANU [1]) et aux méthodes fonctionnelles linéaires dans les problèmes aux limites telles qu'on les trouvent dans LIONS-MAGENES [1] , Ch. I, et LIONS [1] , Ch. I.

b. On considèrera dans la suite un espace de Hilbert reél H , un opérateur autoadjoint $A \geq 0$ dont l'inverse existe, est partout défini et est compact, donc ayant un système complet de vecteurs propres. Ainsi nous pouvons choisir une base orthonormale $\{w_n\}_{n=1}^{\infty}$ de H telle que $Aw_n = a_n w_n (n=1,2,\ldots)$

C. Foiaş

et $0 < a_1 \leq a_2 \leq \ldots$. Pour $\alpha > 0$, on désignera par H^α le domaine de l'opérateur $A^{\alpha/2}$ normé par $|u|_\alpha = |A^{\alpha/2}u|$ ou $|u| = (u,u)^{1/2}$, (u,v) désignant le produit scalaire dans H. On pose aussi $\|u\| = |u|_1$ et $((u,v)) = (A^{1/2}u, A^{1/2}v)$ pour $u, v \in H^1$. Puisque H^α est dense dans H on peut plonger H dans le dual $H^{-\alpha}$ de H ; par cette identification la valeur de $h \in H$ (considerée comme fonctionelle sur H^α) en $u \in H^\alpha$ est $(h,u) = (u,h)$; par un abus de notation on désignera par (v,u) aussi la valeur de $v \in H^{-\alpha}$ en $u \in H^\alpha$.

c.   On se donnera de plus un opérateur bilinéaire continu $B(u,v)$ de $H^1 \times H^1$ dans H^{-1} qui se prolonge en un opérateur continu (désigné de même manière) de $H \times H^1$ et $H^1 \times H$ dans H^{-2}. En outre on suppose que B vérifie aussi la condition

(1.1) $(B(u,v),v) = 0$ pour tous $u,v \in H^1$.

d.   Si X est un espace de Banach (en particulier si $X = H^\alpha$, $-\infty < \alpha < \infty$), alors on désignera par $L^p(0,T;X)$ $(1 \leq p \leq \infty)$, resp. $C^k(0,T;X)$ $(k=0,1,2,\ldots)$, l'espace L^p, resp. de classe C^k, des fonctions définies sur l'intervalle $[0,T]$ à valeurs dans X.

C. Foiaş

2. Équations d'évolution de type de Navier-Stokes.

a. L'équation d'évolution qu'on étudiera dans la suite
sera

(2.1) $\qquad u'(t) + Au(t) + B(u(t), u(t)) = f(t)$

(dans $(0,T)$) avec $u(0) = u_0$ où le membre droit $f(.)$
appartient à $L^2(0,T; H^{-1})$ et est fixé, tandisque u_0 est la
donnée initielle, arbitraire dans H .

Par définition, une solution de (2.1) sera, une fonction

(2.2) $\qquad u(.) \in L^\infty(0,T;H) \cap L^2(0,T;H^1)$

telle que

$$(2.3) \quad -\int_0^T (u(t),v'(t))dt + \int_0^T \left[((u(t),v(t))) + (B(u(t),u(t)),v(t))\right]dt =$$

$$= \int_0^T (f(t),v(t))dt + (u_0,v(0))$$

pour toute fonction

(2.4) $\qquad v(.) \in C^1(0,T;H) \cap C^0(0,T;H^1),$

à support compact dans $[0,T)$. On remarque que pour toute telle

C. Foiaş

$v(.)$, qui de plus appartient aussi à $L^2(0,T; H^2)$, on a

$$\int_0^T (u(t),v'(t))dt = \int_0^T (U(t),v(t))dt + (u_0,v(0))$$

ou $U(.)$ est une fonction de $L^2(0,T;H^{-2})$ dependant seulement de $u(.)$. Il en découle facilement que $u(.)$ est, presque partout dans $(0,T)$, fortement dérivable dans H^{-2} et que

$$(2.5) \qquad u'(t) \in L^2(0,T; H^{-2}) .$$

b. Il est instructif de montrer que le système de Navier-Stokes avec le problème mixte homogène est de type (2.1). Rappelons qu'il s'agit du problème

$$(2.6) \quad \begin{cases} u'_t - \Delta u + (u.\text{grad})u = -\text{grad } p + f, \quad \text{div } u = 0 \text{ dans } \Omega, \\ (u/t=0) = u_0 \text{ (la donnée initiale)}, \quad (u/\partial\Omega) = 0 , \end{cases}$$

où u est une fonction à valeur dans R^n définie dans $[0,T] \times \Omega$, Ω étant un ouvert de R^n .

L'étude fonctionnel de (2.6) se fait de la manière suivante (voir LIONS [1] , Ch. I, § 6):

C. Foiaş

On pose

$$H^1 = \left\{ u: u \in (H_o^1(\Omega))^n,\ \text{div } u = 0 \right\}$$ (1)

et

$$H = \left\{ u: u \in (L^2(\Omega))^n,\ \text{div } u = 0 \right\}.$$

Alors il existe un seul opérateur autoadjoint A dans H tel que H^1 soit le domaine de $A^{1/2}$ et que

$$(A^{1/2}u,\ A^{1/2}u) = \sum_{i,j}^n \int_\Omega \left(\frac{\partial u_i}{\partial x_j}\right)^2 dx \quad \text{pour tout } u \in H^1;$$

ici pour $u, v \in H$, (u,v) désigne le produit scalaire usuel

$$\sum_{i=1}^n \int_\Omega u_i\, v_i\, dx.$$

En supposant Ω borné, la compacité de l'injection $H^1 \longrightarrow H$ (conséquence de celle de $H_o^1(\Omega)$ dans $L^2(\Omega)$) assure que A^{-1} est compact. Si on suppose que la frontière $\partial\Omega$ de Ω est de classe C^2, alors on a de plus que le domaine de A coincide avec

$$H^1 \cap (H^2(\Omega))^n,$$

C. Foiaş

la norme $|Au|$ étant equivalente à la norme

$$\sum_{i=1}^{n} \|u_i\|_{H^2(\Omega)}$$

(Ceci est une conséquence d'un théorème profond de

Cattabriga - Solonnikov - Yudovic; voir CATTABRIGA [1]).

Pour toutes les fonctions u, v, w aux valeurs dans R^n

définies dans Ω posons

(2.7) $b(u,v,w) = \sum_{i,j=1}^{n} \int_{\Omega} u_i \frac{\partial v_j}{\partial x_i} w_j \, dx$

chaque fois que les integrales ont un sens. Remarquons qu'en

vertu de l'inégalité de Hölder

(2.8) $|b(u,v,w)| \leq \left(\sum_{i=1}^{n} \|u_i\|_{L^4(\Omega)}^{4} \right)^{1/4} \|v\| \left(\sum_{i=1}^{n} \|w_i\|_{L^4(\Omega)}^{4} \right)$.

En tenant compte du théorème de Sobolev qui donne, en

particulier, l'injection continue de

(2.9) $H^1(\Omega)$ $\begin{cases} \text{dans} & L^{\frac{2n}{n-2}}(\Omega) \text{ si } n \geq 3 \text{ et} \\ \\ \text{dans} & L^q(\Omega) \text{ pour tout } 1 \leq q \leq \infty \text{ si } n=2, \end{cases}$

C. Foiaş

on déduit de (2.8) que si $n \le 4$ alors

(2.10) $$|b(u,v,w)| \le C \|u\| \|v\| \|w\| .$$

Ainsi

(2.11) $$b(u,v,w) = (B(u,v),w),$$

ou $B(.,.)$ est continue de $H^1 \times H^1$ dans H^{-1} ; la condition (1.1)
se vérifie aussitôt par une intégration par parts et un passage
à la limite. Enfin, toujours en supposant $n \le 4$ et en utilisant
le théorème de Sobolev plusieurs fois, nous avons (avec des
constantes diverses)

$$|b(u,v,w)| = |b(u,v,w)| = c_1 \left(\sum_{i=1}^{n} \|u_i\|_{L^4(\Omega)} \right) \left(\sum_{i,j=1}^{n} \left\| \frac{\partial w_j}{\partial x_i} \right\|_{L^4(\Omega)} \right) |v|$$

$$= c_2 \left(\sum_{i=1} \|u_i\|_{H^1(\Omega)} \right) \left(\sum_{i=1} \|w_i\|_{H^2(\Omega)} \right) |v| \le c_3 \|u\| . |v| . |Aw|,$$

ce qui prouve que l'application B définie par (2.11) jouit
des propriétés exigées. Dépuis LERAY [1] , une solution (faible)
du problème (2.6) est, par définition, une solution de
l'équation (2.1) avec le choix particulier de A et B indiqué
ci-dessus, qui de plus vérifie l'inégalité de l'énergie

C. Foiaş

$$(2.12) \qquad \frac{1}{2} \left| u(t) \right|^2 + \int_0^t \left\| u(\tau) \right\|^2 \, d\tau \leq \frac{1}{2} \left| u_0 \right|^2 + \int_0^t (f(\tau), u(\tau)) d\tau$$

presque partout dans $(0, T)$.

3. Solutions individuelles.

a. Le but de ce paragraphe est de démontrer l'éxistence des solutions dans le sens du § 2 de (2.1); pour différencier ces solutions des solutions statistiques de (2.1), qui feront l'objet principal de ces exposés, nous les appelerons solutions individuelles.

Proposition 1. Pour tout $u_0 \in H$, il existe au moins une solution individuelle de (2.1) vérifiant (2.12) presque partout dans $(0, T)$.

Démonstration. Soit $u_n(t)$ la solution du système différentiel ordinaire

C. Foiaş

(3.1) $\qquad u'_n + AP_n u_n + P_n B(u_n, u_n) = P_n f$, $u_n(0) = P_n u_0$,

où P_n désigne la projection orthogonale de H sur l'espace engendré par $w_1, w_2, ..., w_n$. On obtient aisément

(3.2) $\qquad \dfrac{1}{2} \dfrac{d}{dt} |u_n|^2 + \|u_n\|^2 = (f, u_n) \le |f|_{-1} \|u_n\|$,

d'où il résulte aussitôt

(3.3) $|u_n(t)|^2 + \displaystyle\int_0^t \|u_n(\tau)\|^2 \, d\tau \le |u_0|^2 + \int_0^t |f(\tau)|_{-1}^2 \, d\tau$, $t \in (0, T)$.

Par conséquent la suite $\left\{ u_n(.) \right\}_{n=1}^{\infty}$ est bornée dans $L^2(0, T; H^1)$ et $L^\infty(0, T; H)$, donc on peut extraire une sous-suite $\left\{ u_{n_j}(.) \right\}$ faiblement convergente dans $L^2(0, T; H^1)$ et convergeant aussi dans la topologie $\sigma(L^\infty(0, T; H), L^1(0, T; H))$ de $L^\infty(0, T; H)$. D'autre part

$|(u'_n, v)| \le |u_n| |Av| + |f|_{-1} \|v\| + |(B(u_n, u_n), P_n v)| \le$

$\qquad\qquad = c_1 |Av| + c_2 \|u_n\| . |Av|$

d'où

C. Foiaş

$$\left|u_n'(t)\right|_{-2} \leq c_1 + c_2 \left\|u_n(t)\right\|, \quad t \in (0,T),$$

de manière que $\left\{u_n'(.)\right\}$ est une suite bornée dans $L^2(0,T;H^{-2})$.

On utilise maintenant le suivant lemme de compacité (LIONS [ ], Ch. I, Th. 5):

Si pour trois espaces de Banach reflexifs B_0, B, B_1 , on a $B_0 \subset B \subset B_1$, la première inclusion étant compacte et la deuxième continue, alors l'application identique de l'espace

$$\left\{v(.) \, : \, v(.) \in L^{p_0}(0,T;B_0), \ v'(.) \in L^{p_1}(0,T;B_1)\right\}$$

(normé de manière evidente) dans $L^p(0,T;B)$ est compacte (où $1 < p_0$, p, $p_1 < \infty$).

On applique ce lemme pour $B_0 = H^1$, $B = H$, $B_1 = H^{-2}$ et on en conclut en passant peut-être à une nouvelle sous-suite) que $\left\{u_{n_j}(.)\right\}$ est fortement convergente dans $L^2(0,T;H)$; en faisant un nouveau choix, on peut aussi supposer que $\left\{u_{n_j}(t)\right\}$ converge (fortement) dans H pour presque tout $t \in (0,T)$. Soit $u(.)$ la fonction limite. (2.12) s'obtient en intégrant (3.2) et en passant à la limite, l'inégalité provenant du fait

C. Foias

que la convergence faible dans $L^2(0,T;H^1)$ entraîne celle dans $L^2(0,t;H^1)$ (pour tout $t \in (0,T)$) et y décroit les normes. En intégrant (3.1) on obtient aisément

$$- \int_0^T r'(t)(u_n(t),w_m)dt + \int_0^T r(t)\left[((u_n(t),w_m))+(B(u_n(t),u_n(t)),w_m)\right]dt$$

$$= r(0)(u_0,w_m) + \int_0^T r(t)(f(t),w_m)dt$$

quels que soient $m \leq n$ et $r(.)$ de classe C^1 à valeurs réelles et à support compact dans $[0,T)$. Pour déduire

$$(3.4) \quad - \int_0^T r'(t)(u(t);w_m)dt +$$

$$+ \int_0^T r(t)\left[((u(t),w_m)) + (B(u(t),u(t)),w_m)\right]dt =$$

$$= r(0)(u_0,w_m) + \int_0^T r(t)(f(t),w_m)dt$$

on doit prouver seulement que

$$b_j = \int_0^T r(t)(B(u_{n_j}(t), u_{n_j}(t)),w_m)dt \longrightarrow$$

$$\longrightarrow \int_0^T r(t)(B(u(t),u(t)),w_m)dt = b .$$

C. Foias

Comme $w_m \in H^2$ et $r(.)$ est borné on a

$$|b_j - b| \leq c_1 \left[\int_0^T |u_{n_j}(t) - u(t)| \, \|u_{n_j}(t)\| \, dt + \right.$$

$$\left. + \int_0^T \|u(t)\| \, |u_{n_j}(t) - u(t)| \, dt \right] \leq$$

$$= c_1 \left[\left(\int_0^T \|u_{n_j}(t)\|^2 dt \right)^{1/2} + \left(\int_0^T \|u(t)\|^2 dt \right)^{1/2} \right] \left(\int_0^T |u_{n_j}(t) - u(t)|^2 dt \right)^{1/2}$$

d'où la convergence désirée.

Finalement toute fonction $v(.)$ utilisée dans (2.3) s'obtient comme limite dans

$$C^1(0,T;H) \cap C^0(0,T;H^1) \quad \text{(normé par la somme des normes)}$$

les sommes finies des fonctions de la forme $r(.)w_m$ envisagée ci-dessus. Ainsi (3.4) implique (2.3) ce qui finit la preuve.

b. Faisons quelques remarques utilise sur les solutions individuelles de (2.1).

(i) L'inégalité de l'énergie entraîne de manière manifeste

$$3.5) \qquad \|u(.)\|^2_{L^\infty(0,T;H)} + \|u(.)\|^2_{L^2(0,T;H^1)} \leq$$

C. Foiaş

$$|u_o|^2 + \|f(.)\|^2_{L^2(0,T;H^{-1})} \ .$$

(ii) <u>Toute solution</u> $u(.)$ <u>de</u> (2.1) <u>est, comme fonction</u> <u>à valeurs dans</u> H^{-2}, <u>absolument continue et sa dérivée</u> (prise dans la topologie de H^{-2}) <u>appartient à</u> $L^2(0,T;H^{-2})$ <u>et</u> <u>vérifie</u>

$$(3.6) \quad \|u'(.)\|_{L^2(0,T;H^{-2})} \leq c_1(1+\|u(.)\|_{L^2(0,T;H^1)})\|u(.)\|_{L^\infty(0,T;H)}$$
$$+ \|f(.)\|_{L^2(0,T;H^{-2})} \ .$$

En effet si

$$U(t) = Au(t) + B(u(t),u(t))-f(t)$$

alors la fonction $U(.)$ appartient à $L^2(0,T;H^{-2})$ et sa norme dans cet espace est majorée par le membre droit de (3.6). D'autre part, (2.3) peut s'écrire sous la forme

$$(3.7) \quad -\int_0^T (u(t),v'(t))dt + \int_0^T (U(t),v(t))dt = 0$$

au moins dès que $v(.)$ satisfait la condition supplémentaire

$$v(.) \in L^2(0,T;H^2) \ .$$

De (3.7) il vient par un argument standard que dans H^{-2}

C. Foias

on a

(3.8) $\qquad u(t) = u_0 - \int_0^t U(\tau)d\tau$ presque pour tout $t \in (0,T)$.

Ceci prouve l'assertion (ii), puisque en vertu de (3.8) on

a le droit de remplacer u(.) par le membre de (3.8) qui juit

de propriétes exigées.

(iii) <u>Toute solution</u> u(.) <u>de</u> (2.1) <u>est continue de</u> $[0,T)$

<u>à</u> H <u>muni de la topologie faible.</u>

C'est immédiat si on utilise (iii), la densité de H^2 dans

H et le fait que |u(.)| est bornée sur $[0,T)$.

4. <u>Unicité des solutions individuelles.</u>

a. Il semble bien probable que dans les conditions de la

proposition 1 il n'y a pas une solution unique; d'ailleurs il

est certainement plus facile de trouver un exemple de telle

sorte pour (2.1) que pour le problème plus concret (2.6).

D'autre part en s'inspirant de ce exemple on peut aisément

C. Foiaş

donner une condition sur le terme non linéaire B de (2.1)
pour assurer l'unicité des solutions individuelles.

A cet effet rappelons au lecteur la suivante inégalité
de LADYZENSKAYA [1]

$$(4.1) \qquad \| u \|_{L^4(\Omega)} = \sqrt[4]{2} \, |u|^{1/2} \, \| u \|^{1/2} \quad \text{pour tout} \quad u \in H_o^1(\Omega),$$

valable si la dimension de $R^n \supset \Omega$ est $n = 2$. [2] Ainsi
dans ce cas l'inégalité de Hölder donne immédiatement

$$(4.2) \quad |(B(u,v)w)| = |(B(u,w)v)| = \sqrt{2} \, |u|^{1/2} \| u \|^{1/2} |v|^{1/2} \| v \|^{1/2} \| w \|$$
$$\text{pour tous} \quad u,v,w \in H^1 .$$

Proposition 2. Supposons que B vérifie (4.2). Alors
pour tout $u_o \in H$ il existe une solution unique de (2.1) telle
que $u(0) = u_o$; de plus, elle satisfait l'équation de l'énergie

$$(4.3) \qquad \frac{1}{2} |u(t)|^2 + \int_0^t \| u(\tau) \|^2 \, d\tau = \frac{1}{2} \, |u_o|^2 + \int_0^t (f(\tau), u(\tau)) d\tau$$

pour tout $t \in [0,T]$, et est continue de $[0,T]$ dans H .

Démonstration. On remarque d'abord qu'en vertu de (4.2)

C. Foiaş

on a

$$u'(.) \in L^2(0,T;H^{-1})$$

où la dérivée est celle prise dans H^{-2} conformément à la
Proposition (ii) du paragraphe 3. Ainsi $u(.)$, comme fonction
à valeurs dans H^{-1} , est dérivable presque partout dans $(0,T)$.
En tenant compte de ce que $u(t) \in H^1$ presque partout dans
$(0,T)$ et de (1.1) nous déduissons que

$$(4.4) \qquad \frac{d}{dt} |u(t)|^2 = 2(u'(t),u(t)) \quad \text{presque partout dans } (0,T).$$

De plus $P_n u(.)$ est évidemment de classe C^1 de $(0,T)$
(car P_n est continue de H^{-2} dans H^α pour tout $\alpha \geq 0$)
dans H^1 ; il en résulte (pour $0 \leq s \leq t \leq T$)

$$\left| |P_n u(t)|^2 - |P_n u(s)|^2 \right| \leq \left(\int_0^t |u'(\tau)|_{-1}^2 \, d\tau \right)^{1/2} \left(\int_s^t \|u(\tau)\|^2 d\tau \right)^{1/2} ,$$

d'où en laissant $n \to \infty$ et en utilisant ensuite l'inégalité
ainsi obtenue on peut déduire facilement que $u(.)$ est
absolument continue sur $(0,T)$. Par conséquent on peut
intégrer (4.4) pour obtenir (4.2) . Celle-ci assure la continuité

C. Foiaș

de $|u(t)|$ en t . En vertu de (iii) § 3. b il resulte que

$u(.)$ est continue de $[0,T)$ dans H .

Le même argument que celui conduisant à (4.2) nous montre

que si $u(.)$ et $v(.)$ sont deux solutions aux données initiales

u_0 et resp. v_0 , alors $|u(.) - v(.)|^2$ est absolument

continue sur $(0,T)$ et

$$\frac{1}{2}\frac{d}{dt} |u(t)-v(t)|^2 = -\|u(t)-v(t)\|^2 - (B(u(t)-v(t),v(t)),u(t)-v(t))$$

(4.5)
$$\leq |u(t) - v(t)|^2 \|v(t)\|^2$$

presque partout sur $(0,T)$. En intégrant (4.5) il vient

(4.6) $|u(t)-v(t)| \leq |u_0-v_0| \exp\left(\int_0^t \|v(\tau)\|^2 d\tau\right)$ pour tout $t \in (0,T)$.

En particulier cette inégalité nous montre que $u_0 = v_0$

entraîne $u(.) = v(.)$, ce qui finit la preuve.

Remarque. Cette proposition est de facto, due à LIONS -

PRODI [ ] et . PRODI [ ] qui l'ont établit pour le cas du système

de Navier-Stokes en dimension $n = 2$, indiqué plus haut.

C. Foiaş

b. En vertu de la Proposition 2, si B vérifie (4.2), il existe une application S(t) de H dans H telle que si u(.) est la solution de (2.1) à donnée initiale u_0 , alors

$$u(t) = S(t)u_0 \text{ pour tout } t \in [0,T);$$

en outre en vertu de la même proposition, S(.)u est continue de [0,T) dans H , quel que soit $u \in H$.

Les inégalités (4.6) et (3.5) montrent que l'application S(t) satisfait une condition Lipschitz, notamment

(4.7) $|S(t)u-S(t)v| \leq c_0 |u-v| \exp(|v|^2)$ pour tous $u,v \in H$,

quel que soit $t \in [0,T)$, c_0 étant une constante indépendente de t,u,v . On déduit aisément que S(t)u est continue en $(t,u) \in [0,T) \times H$ dans H .

c. Il est manifeste que

$$S(0)u = u \text{ pour tout } u \in H ,$$

et que si f(.) est constante, alors on peut définir S(t) pour tout $t \geq 0$ de manière qu'on a

C. Foiaş

$$S(t_1)S(t_2) = S(t_1+t_2) \quad \text{pour tous} \quad t_1, t_2 \geq 0 .$$

Ainsi on obtient un semi-groupe d'applications non linéaires dont l'étude est loin d'être achevée même dans le cas concret du système de Navier-Stokes en dimension $n = 2$.

5. Solutions statistiques.

a. Dans des cas intéressant la physique mathématique, la donnée initiale u_0 n'est pas déterminée univoquement, mais plutôt sa distribution statistique; cela veut dire qu'on se donne une probabilité sur H , c'est-à-dire une mesure borelienne $\mu = 0$ telle que $\mu(H) = 1$ et on se demande qu'elle sera la probabilité μ_t donnant la distribution statistique des solutions individuelles dans le moment t . Dans le cas où on a le système $\{S(t) : 0 \leq t \leq T\}$ d'applications de H , la réponse s'obtient simplement en identifiant d'une façon abusive la probabilité avec la frequence relative dans un raisonnement heuristique, d'ailleurs typique pour la physique théorique: Soit N le nombre des données initiales envisagées et pour un

C. Foias

sous-ensemble borelien ω de H soit $N_t(\omega)$ le nombre des valeurs en le moment t des solutions individuelles qui partent de l'ensemble des données initiales. Alors évidemment

$$\frac{N_t(\omega)}{N} = \frac{N_o(S(t)^{-1}\omega)}{N} \quad ;$$

de cette manière on arrive à la relation

(5.1) $\mu_t(\omega) = \mu(S(t)^{-1}\omega)$ pour tout borelien $\omega \subset H$.

Dans notre cas, qui englobe le système Navier-Stokes, avec le problème (2.6), aussi si la dimension n = 3 ou 4 , on n'a plus la possibilité d'utiliser le système $\{S(t) : 0 \leq t \leq T\}$. Nous allons contourner cette difficulté en établissant une équation qui doit être vérifiée par la mesure μ_t donnée par la formule (5.1), quant celle-ci fait sens, et dont la dépendance de l'équation (2.1) ne se fait plus par l'intermédiaire du $\{S(t) : 0 \leq t \leq T\}$.

b. Dans ce but, soit $\vec{\Phi}(t,u)$ une fonctionnelle réelle définie pour $(t,u) \in [0,T) \times H$, telle que

C. Foiaş

(5.2) $\qquad \phi(t, P_m u) = \phi(t, u)$ pour tous t, u ,

et pour un certain $m = 1, 2, \ldots$. (Rappelons au lecteur que P_m est la projection orthogonale de H dans l'espace engendré par les vecteurs propres $w_1, w_2, \ldots, w_m$ de A). Supposons de plus que $\phi(., .)$ est de classe C^1 (c'est-à-dire, différentiable au sens de Fréchet de $[0, T) \times H$ dans R , à différentielle continue) et que

(5.3) $\qquad |\phi'_u(t, u)| \leq c_1$ et $|\phi'_t(t, u)| \leq c_2 + c_3 |u|$ pour tous t, u ;

c_1, c_2, c_3 sont des constantes dépendant de ϕ .

Soit de plus $u(.)$ une solution de (2.1); alors en utilisant (5.2), le fait que P_m est continue de H^{-2} dans H et la proposition (ii) du paragraphe 3, b., on déduit que $\phi(t, u(t))$ est absolument continue et que

(5.4) $\qquad \dfrac{d}{dt} \phi(t, u(t)) = \phi'_t(t, u(t)) + (\dot{u}'_u(t), \phi'_u(t, u(t)))$

presque partout dans $(0, T)$. En tenant compte de (3.5) et (3.6) nous obtenons par suite

C. Foiaş

$$\frac{d}{dt} \Phi(t, u(t)) \leq |u'(t)|_{-2} \; |P_m \Phi_u'(t, u(t))|_2 + c_2 + c_3 |u| \leq$$

$$\leq |u'(t)|_{-2} \; c_1 a_m + c_4 + c_5 |u_o| \leq$$

$$\leq c_6 + c_7 |u(t)| + c_8 |u(t)| \cdot \|u(t)\| + c_5 \|u_o\| \leq$$

$$\leq c_9 + c_{10} |u_o| + c_{11} \|u(t)\| + c_{12} |u_o| \cdot \|u(t)\|$$

$$\text{5.5)} \quad \int_0^T \left| \frac{d}{dt} \Phi(t, u(t)) \right| dt \leq c_{13} (1 + |u_o|)(1 + \int_0^T \|u(t)\|^2 dt)^{1/2} = c_{14}(1 + |u_o|)$$

Supposons maintenant que la mesure initiale μ vérifie

$$\text{5.6)} \qquad |u|^2 \; d\mu(u) < \infty \; ,$$

et plaçons nous dans les conditions de la Proposition 2. Alors en vertu de (3.5) et (5.6) on a

$$\text{5.7)} \quad \sup \left\{ \int |u|^2 \; d\mu_t(u) = \int |S(t)u|^2 \; d\mu(u) : 0 \leq t \leq T \right\} < \infty$$

$$\text{5.8)} \quad \int_0^T \left(\int \|u\|^2 \; d\mu_t(u) \right) dt = \int_0^T \left(\int \|S(t)u\|^2 \; d\mu(u) \right) dt < \infty \; .$$

De plus, pour toute fonction réelle $r(.)$ de classe C^1

C. Foiaş

au support compact dans $(0,T)$ on a

$$- \int_0^T r'(t)(\int \Phi(t,S(t)u))dt = \int(- \int_0^T r'(t)\Phi(t,u(t))dt)\, d\mu(u) =$$

$$= \int (\int_0^T r(t)\frac{d}{dt}\Phi(t,u(t))dt)\, d\mu(u) = \int_0^T r(t)(\int \frac{d}{dt}\Phi(t,u(t))d\mu(u))dt$$

où on a utilisé (5.5), (5.6) et le théorème de Fubini; en utilisant

encore une fois (5.5) et (5.6) et ensuite (5.4) et (2.1), il

vient

$$\frac{d}{dt}\int \Phi(t,u)d\mu_t(u) = \frac{d}{dt}\int \Phi(t,S(t)u)d\mu(u) =$$

$$= \int \frac{d}{dt}\Phi(t,u(t))d\mu(u) =$$

$$(5.9) \quad = \int \left[\Phi_t'(t,u(t)) + (u'(t),\Phi_{u(t)}'(t,u(t)))\right] d\mu(u) =$$

$$= \int \left[\Phi_t'(t,S(t)u)+(-AS(t)u-B(S(t)u,S(t)u)+f(t),\Phi_{S(t)u}'(t,S(t)u))\right] d\mu(u)=$$

$$= \int \left[\Phi_t'(t,u) + (-Au - B(u,u) + f(t),\Phi_u'(t,u))\right] d\mu_t(u) ,$$

d'où en intégrant en t on obtient finalement

$$- \int_0^T \left[\int \Phi'_t(t,u)\, d\mu_t(u) \right] dt +$$

$$(5.10) \quad + \int_0^T \left\{ \int \left[((u, \Phi'_u(t,u))) + (B(u,u), \Phi'_u(t,u)) \right] d u_t(u) \right\} dt =$$

$$= \int \Phi(0,u)\, d\mu(u) - \int_0^T \left[\int (f(t), \Phi'_u(t,u))\, d\mu_t(u) \right] dt \ ,$$

dès que Φ est en outre nulle pour t assez proche de T .

Désignons par $\mathfrak{J}_0$ la classe des fonctionnelles Φ pour
lesquelles nous venons d'établir (5.10); rappelons que
$\Phi(.,.) \in \mathfrak{J}_0$ si elle est de classe C^1 de $[0,T) \times H$ dans R ,
satisfait les conditions (5.2), (5.3) et est nulle pour t
assez proche de T . Soit $\mathfrak{J}$ la classe (qui contient $\mathfrak{J}_0$)
des fonctionnelles réelles $\Phi(.,.)$ définies sur $[0,T) \times H^1$
et satisfaisant aux conditions suivantes:

i) $\Phi'_t(t,u)$, est continue en $(t,u) \in [0,T) \times H^1$ et vérifie
la deuxième relation (5.3);

ii) $\Phi(t,u)$ est différentiable en u dans le sens suivant:
Il existe $\Phi'_u(t,u) \in H^1$ tel que

$$\frac{1}{|v|} \left| \Phi(u+v) - \Phi(u) - (\Phi'_u(t,u), v) \right| \longrightarrow 0 \quad \text{pour} \quad v \in H^1, \ |v| \longrightarrow 0$$

iii) $\Phi_u'(.,)$ est continue de $[0,T) \times H^1$ dans un $H^{-\alpha}$ (avec

un $\alpha \gtrless 0$) est $\|\Phi_u'(.,.)\|$ est bornée.

Lemme. Soit $\{\mu_t : 0 \leq t < T\}$ une famille de

probabilités sur H telle que

(5.11) $\int |u|^2 d\mu_o(u) \in L^\infty(0,T;R)$ et $\int \|u\|^2 d\mu_o(u) \in L^2(0,T;R)$.

Alors si (5.10) est vérifiée pour toute $\Phi(.,.) \in \mathcal{T}_o$, elle

est aussi vérifiée pour toute $\Phi \in \mathcal{T}$.

Démonstration. Soit $\psi \in \mathcal{T}$; on pose

$$\Phi^{(n)}(t,u) = \psi(t,P_n u)$$

pour tous t,u et n . Alors $\Phi^{(n)}(.,.) \in \mathcal{T}_o$ donc (5.10)

est vérifiée pour $\Phi = \Phi^{(n)}$ quel que soit $n=1,2,\ldots$. On

fait ensuite $n \longrightarrow \infty$ en applicant plusieurs fois le théorème

de convergence dominée de Lebesgue.

b. Par définition, une solution statistique de (2.1)

est une famille $\{\mu_t : 0 \leq t \leq T\}$ vérifiant (5.10) pour toute

C. Foiaş

fonctionelle $\Phi(.,.)\in\mathfrak{J}$ et (5.11) ; la mesure μ est la donnée initiale.

Le lemme nous enseigne que si (5.11) est vérifiée alors il suffit de vérifier (5.10) seulement pour $\Phi\in\mathfrak{J}_o$ ce qui est, comme on le verra dans la suite, plus facile. Remarquons que la définition d'une solution statistique a été choisie de la manière que, dans les conditions de la Proposition 2, une telle solution soit donnée par la formule (5.1).

Rappelons que cette remarque concerne en particulier le problème (2.6) pour le système de Navier-Stokes en dimension n = 2 .

6. Existence des solutions statistiques.

a. Le but de ce paragraphe, qui constitue le partie centrale de cette exposition, est de prouver le suivant théorème d'existence:

Proposition 3. Pour toute donné initiale μ vérifiant

C. Foiaș

(5.6), <u>il existe une solution statistique</u> $\{\mu_t: 0 \le t < T\}$ <u>de</u> (2.1) <u>qui, en outre satisfait presque partout dans</u> $(0,T)$ <u>a</u> <u>l'inégalité de l'energie suivante:</u>

$$
\begin{aligned}
\textbf{(6.1)} \quad &\frac{1}{2}\int \varphi(|u|^2)d\mu_t(u) + \int_0^t \left[\int \varphi'(|u|^2)\|u\|^2 \, d\mu_\tau(u)\right] d\tau = \\
&= \frac{1}{2}\int \varphi(|u|^2)d\mu(u) + \int_0^t \left[\int \varphi'(|u|^2)(f(t),u)d\mu_\tau(u)\right] d\tau \, ,
\end{aligned}
$$

<u>quelle que soit la fonction</u> $\varphi \ge 0$ <u>de classe</u> c^1 <u>dans</u> $[0,\infty)$ <u>telle que</u>

$$
\textbf{(6.2)} \qquad 0 \le \varphi'(x) \le c(\varphi) < \infty \qquad \underline{\text{pour tout}} \ \ x \in [0,\infty) \, .
$$

<u>Démonstration.</u> On la fera en plusieurs étapes.

1) On désigne par $S^{(n)}(t)u_0$ la valeur en t de la solution de l'équation (3.1) à valeur initiale $u_0 \in P_n H$. C'est une fonction de classe C^0 en $(t,u_0) \in [0,T] \times P_n H$. On définit une mesure borelienne $\mu^{(n)}$ dans H par la formule

$$
\textbf{(6.3)} \quad \mu^{(n)}(\omega) = \mu(P_n^{-1}(\omega \cap P_n H)) \quad \text{pour tout borelien} \ \ \omega \subset H \, .
$$

C. Foiaş

C'est une probabilité sur H , à support inclus dans $P_n H$ et telle que pour toute fonctionelle $\Phi(.)$ continue de H dans R vérifiant

$$|\Phi(u)| = c_1 \ (1 + |u|) \quad \text{pour tout} \quad u \in H$$

on a, comme conséquence de (5.6),

$$\int \Phi(u) d\mu^{(n)}(u) = \int \Phi(P_n u) d\mu(u)$$

et par suite, en utilisant encore une fois (5.6),

$$(6.4) \qquad \int \Phi(u) d\mu^{(n)}(u) \longrightarrow \int \Phi(u) d\mu(u)$$

quelle que soit la fonctionelle $\Phi(.)$ du type envisagé.

On définit ensuite la mesure $\mu_t^{(n)}$ dans H comme l'image par $S^{(n)}{}_t$ de $\mu^{(n)}$ ce qui fait sens puisque cette application (étant définie dans H , qui contient le support de $\mu^{(n)}$) est presque partout définie par rapport à $\mu^{(n)}$. On a donc

$$(6.5) \qquad \int \Phi(u) d\mu_t^{(n)}(u) = \int \Phi(S^{(n)}(t) P_n u) d\mu(u)$$

quelle que soit la fonctionelle $\Phi(.)$ du même type que celles

C. Foiaş

considerées dans (6.4).

Evidemment les mesures $\mu_t^{(n)}$ $(0 \le t < T)$ sont des probabilités dans H aux supports inclus dans $P_n H$ et $\mu_0^{(n)} = \mu^{(n)}$. De plus par une intégration de (3.3) et par l'utilisation de (6.5) on obtient facilement les suivantes estimations:

$$(6.6) \qquad \sup_n \left[\text{vrai} \sup_{0 \le t < T} \int |u|^2 \, d\mu_t^{(n)}(u) \right] < \infty ,$$

$$(6.7) \qquad \sup_n \int_0^T \left[\int \|u\|^2 \, d\mu_t^{(n)}(u) \right] dt < \infty .$$

2) Soit $C(k)$ $(k = 0,1,2)$ l'espace des fonctionnelles réelles $\Phi(.)$ continues sur H telles que

$$(6.8) \qquad \|\Phi\|_{C(k)} = \sup \left\{ \frac{|\Phi(u)|}{1+|u|^k} : u \in H \right\} < \infty ,$$

et soit $C(1,1)$ l'espace des fonctionnelles réelles $\psi(.)$ continues sur H^1 telles que

$$(6.9) \qquad \|\psi\|_{C(1,1)} = \sup \left\{ \frac{|\psi(u)|}{1+|u| \cdot \|u\|} : u \in H^1 \right\} < \infty .$$

C. Foiaş

Ce sont des espaces de Banach non séparables; en outre

$$C(2) \subset C(1,1) .$$

Il en résulte que

(6.10) $L^2(0,T;C(2)) \subset L^2(0,T;C(1,1)) \cap L^1(0,T;C(2)) .$

De plus le premier espace dans (6.10) est évidemment dense dans le dernier; ainsi toute fonctionnelle linéaire continue de $L^2(0,T;C(1,1))$ dans R , dont la restriction à $L^2(0,T;C(2))$ est continue par rapport à la norme de $L^1(0,T;C(2))$ se prolonge par continuité en une, et une seule, fonctionnelle linéaire de $L^1(0,T;C(2))$ dans R . On utilisera cette remarque tout de suite.

Commençon par prouver que si pour une fonctionnelle $\Phi(.)$ définie dans (une partie de) H on pose

(6.11) $F^{(n)}(\Phi) = \int_0^T \left[\int \Phi(u) d\mu_t^{(n)}(u) \right] dt$

chaque fois que la formule ait un sens, alors $\left\{ F^{(n)} \right\}_{n=1}^{\infty}$ est bornée dans $(L^2(0,T); C(1,1)))$ et dans $(L^1(0,T;C(2)))$. En effet pour $\Phi(.,.) \in L^1(0,T;C(2))$ on a

C. Foiaş

$$|F^{(n)}(\Phi)| \le \int_0^T \|\Phi(t,.)\|_{C(2)} \left[\int (1+|u|^2)d\mu_t^{(n)}(u)\right]dt \le$$

$$= c_1 \int_0^T \|\Phi(t,.)\|_{C(2)} dt = c_1 \|\Phi(.,.)\|_{L^1(0,T;C(2))}$$

où c_1 est une constante (independante de n) qu'on obtient de (6.6) .

Pour $\Phi(.,.) \in L^2(0,T;C(1,1))$ on a, un peu moins banalement,

$$|F^{(n)}(\Phi)| \le \int_0^T \|\Phi(t,.)\|_{C(1,1)}\left[\int (1+|u| \cdot \|u\|)d\mu_t^{(n)}(u)\right]dt \le$$

$$\le \int_0^T \|\Phi(t,.)\|_{C(1,1)}\left(\int (1+|u|^2)d_t^{(n)}(u)\right)^{1/2} \cdot$$

$$\left(\int (1+\|u\|^2)d_t^{(n)}(u)\right)^{1/2} dt \le$$

$$\le c_2 \int_0^T \|\Phi(t,.)\|_{C(1,1)}\left(\int (1+|u|^2)d\mu_t^{(n)}(u)\right)^{1/2} dt \le$$

$$\le c_2 \|\Phi(.,.)\|_{L^2(0,T;C(1,1))} \cdot \left(\int \left(\int (1+|u|^2)d\mu_t^{(n)}(u)\right)dt\right)^{1/2}$$

$$\le c_3 \|\Phi(.,.)\|_{L^2(0,T;C(1,1))}$$

C. Foiaş

ou $c_{2,3}$ sont des constantes indépendantes de n dont l'éxistence est assurée par (6.6) et (6.7).

En utilisant la compacité *-faible des boules (propres) d'un dual d'un espace de Banach on obtient une fonctionnelle $F \in (L^2(0,T;C(1,1)))'$ qui est limite *-faible de la suite $\left\{F^{(n)}\right\}_{n=1}^{\infty}$, c'est-à-dire appartenant à l'adhérence *-faible de $\left\{F^{(n)}\right\}_{n=m}^{\infty}$ pour tout m = 1,2,... . Il est manifeste que la restriction de cette fonctionnelle à $L^2(0,T;C(2))$ est continue par rapport à la norme de $L^1(0,T;C(2))$ donc se prolonge en une fonctionnelle linéaire continue de cet espace dans R ; on désignera cette fonctionnelle aussi par F . En utilisant la remarque qui suit (6.10), on peut aisément déduire que F est aussi une limite * - faible dans $(L^1(0,T;C(2)))'$ de la suite $\left\{F^{(n)}\right\}_{n=1}^{\infty}$.

3) Soit maintenant $\phi(.,.) \in \mathfrak{T}_o$; par la même méthode que celle qui nous a conduit à (5.10), nous pouvons arriver à l'équation

C. Foiaș

$$- \int_0^T \left[\int \Phi_t'(t, P_n u) \, d\mu_t^{(n)}(u) \right] dt +$$

$$(6.12) \quad + \int_0^T \left\{ \int \int \left[((u, P_n \Phi_{P_n u}'(t, P_n u))) + (B(P_n u, P_n u), P_n \Phi_{P_n u}'(t, P_n u)) \right] d\mu_t^{(n)}(u) \right\} dt =$$

$$= \int \Phi(0, P_n u) \, d\mu(u) + \int_0^T \left[\int \int (f(t), \Phi_{P_n u}'(t, P_n u) \, d\mu_t^{(n)}(u) \right] dt .$$

Mais le support de $\mu_t^{(n)}$ est contenu dans $P_n H$ et en vertu de (5.2) on a $\Phi_u'(t, u) = P_n \Phi_u'(t, u)$ pour tous t, u dès que n est assez grand.

Ainsi pour de tels n, (6.12) peut s'écrire

$$- \int_0^T \left[\int \Phi_t'(t, u) \, d\mu_t^{(n)}(u) \right] dt +$$

$$(6.13) \quad + \int_0^T \left\{ \int \int \left[((u, \Phi_u'(t, u)) + (B(u, u), \Phi_u'(t, u)) \right] d\mu_t^{(n)}(u) \right\} dt =$$

$$= \int \Phi(0, P_n u) \, d\mu(u) + \int_0^T \left[\int \int (f(t), \Phi_u'(t, u)) \, d\mu_t^{(n)}(u) \right] dt .$$

En introduisant la fonctionnelle $F^{(n)}$, (6.13) prend la forme

C. Foiaş

$$- F^{(n)}(\oint_t'(.,.)) + F^{(n)}(\ ((.,\oint_u'(.,.)))+(B(.,.),\oint_u'(.,.)) \ )$$

(6.14)
$$= \int\oint(0,P_n u)d\mu(u) + F^{(n)}(\ (f(.),\oint_u'(.,.)) \ ),$$

où en vertu du choix de $\oint(.,.)$ toutes les fonctionnelles en
(t,u) appartienent à $L^1(0,T;C(2))$, à l'exception de

$$(B(.,.),\oint_u'(.,.)) \quad \text{appartenant à} \quad L^2(0,T;C(1,1)).$$

Ainsi en vertu du point 2) on a le droit de remplacer $F^{(n)}$
dans (6.14) par F ; on en obtient

$$- F(\oint_t'(.,.)) + F(\ ((.,\oint_u'(.,.)))+(B(.,.),\oint_u'(.,.))) =$$

(6.15)
$$= \int\oint(0,u)d\mu(u) + F(\ (f(.),\oint_u'(.,.)) \ ).$$

5) Cette étape, la plus simple, consiste en l'application
directe d'un théorème profond de la théorie de la mesure,
notamment le théorème suivant (forme particulière d'un théorème
général de A . et C. IONESCU TULCEA ; pour les détails nous
renvoyons le lecteur à la monographie DINCULEANU [1] , §§ 10-13):

C. Foiaş

Soit X un espace de Banach quelconque (c'est-à-dire qui n'est supposé ni séparable ni réflexif). Alors toute fonctionelle $F \in (L^1(0,T;X))'$ admet la représentation

$$(6.16) \quad F(x(.)) = \int_0^T F_t(x(t))dt \quad \text{pour toute } x(.) \in L^1(0,T;X),$$

où $F_t \in X'$ (pour tout $t \in (0,T)$) et $\sup \left\{\|F_t\| : 0 \le t < T\right\} = \|F\|$. Ici la famille des fonctionnelles F_t peut être (et sera) choisie telle que

$$(6.17) \quad \lambda(F(x))(t) = F_t(x) \quad \text{pour tous } t \in (0,T) \text{ et } x \in X ,$$

où λ est un relèvement fixé de $L^\infty(0,T;R)$. (Rappelons que λ est un relèvement de $L^\infty(0,T;R)$ si il constitue une application linéaire multiplicative de l'algèbre des fonctions mesurables et essentiellement bornées (dans $(0,T)$, la mesure étant celle de Lebesgue) dans celle des fonctions mesurables bornées, telle que

$$(6.18) \quad \varphi \ge 0 \text{ presque partout} \implies \lambda(\varphi) \ge 0 \text{ partout;}$$

l'existence d'un tel relèvement était decouverte par VON NEUMANN

C. Foias

dépuis longtemps).

Dans notre cas, X sera $C(2)$ (qui est un espace de Banach non séparable et non réflexif) et la fonctionnelle F sera celle introduite dans le point 2) et considerée dans (6.15). On prendra la représentation (6.16) vérifiant (6.17) et on cherchera à montrer que presque partout dans $(0,T)$, les fonctionnelles F_t sont des intégrales par rapport à des probabilités sur H.

5) Commençons d'abord par remarquer que $F^{(n)} \geq 0$, c'est-à-dire $F^{(n)}(\phi) \geq 0$ si $\phi \geq 0$. Comme F est une limite *-faible de $\{F^{(n)}\}$, F est aussi ≥ 0. Ainsi

$$\int_0^T r(t) F_t(\phi) dt = F(r \otimes \phi) \geq 0$$

pour tous $r(.) \in L^1(0,T;R)$, $r(.) \geq 0$ et $\phi \in C(2)$, $\phi \geq 0$. Cela entraîne evidemment que $F_t(\phi) \geq 0$ presque partout. En utilisant (6.17) et (6.18) on déduit que

(6.19) $\quad F_t(\phi) \geq 0$ pour tout $t \in (0,T)$ et toute $\phi \in C(2)$, $\phi \geq 0$,

C. Foias

c'est-à-dire $F_t \geq 0$ pour tout t. De manière analogue, comme

$$F^{(n)}(r \otimes 1) = \int_0^1 r(t)dt \quad \text{pour tout} \quad n = 1,2,3,\ldots,$$

on déduit aussitôt

(6.20) $\qquad F_t(1) = 1$ pour tout $t \in (0,T)$.

Posons

(6.21) $\qquad b_p = \left\{ u : u \in H^1, \|u\| \leq p \right\}$ pour $p = 1,2,3,\ldots$.

C'est un ensemble compact dans H . De plus pour toute fonctionnelle Φ appartenant à $C(0)$ nous avons

$$|F^{(n)}(1 \otimes \Phi)| \leq \int_0^T \left[\int |\Phi(u)| d\mu_t^{(n)}(u)\right] dt \leq \int_0^T \left[\int_{b_p} |\Phi(u)| d\mu_t^{(n)}(u) + \right.$$

$$\left. + \int_{H \setminus b_p} |\Phi(u)| d\mu_t^{(n)}(u)\right] dt \leq$$

$$\leq \left(\max_{b_p} |\Phi| \right) T + \|\Phi\|_{C(0)} \int_0^T \mu_t^{(n)}(H \setminus b_p) dt ,$$

et comme, en vertu de (6.7),

C. Foiaş

$$\int_0^T p^2 \, \mu_t^{(n)}(H \backslash b_p) dt = \int_0^T \left[\int_{H \backslash b_p} \|u\|^2 \, d\mu_t^{(n)}(u) \right] dt =$$

$$= \int_0^T \left[\int \|u\|^2 \, d\mu_t^{(n)}(u) \right] dt \leq c$$

(où c est une constante indépendante de n), il résulte

$$\left| F^{(n)}(1 \otimes \Phi) \right| \leq \left(\max_{b_p} |\Phi| \right) T + \|\Phi\|_{C(0)} \frac{c}{p^2}$$

d'où on conclut aussitôt

(6.23) $\left| F(1 \otimes \Phi) \right| \leq \left(\max_{b_p} |\Phi| \right) T + |\Phi|_{C(0)} \frac{c}{p^2}$

quel que soit $p = 1,2,3,\ldots$ et la fonctionnelle $\Phi \in C(0)$.

Soit maintenant

(6.24) $\Phi_1 \geq \Phi_2 \geq \Phi_3 \geq \cdots \geq \Phi_r \geq \cdots \geq 0$

une suite de fonctions de $C(0)$ convergeant vers 0 (en chaque

point de H). Alors, puisque b_p est compact, en vertu du

théorème de Dini,

C. Foiaş

$$\epsilon_{p,r} = \max_{b_p} |\bar{\phi}_r| \longrightarrow 0 \quad \text{pour} \quad r \longrightarrow \infty .$$

Ainsi en tenant compte de ce que $F_t \geq 0$ (pour tout t) on déduit de (6.23), par le lemme de Fatou,

$$\int_0^T \lim_r F_t(\bar{\phi}_r) dt \leq \|\bar{\phi}_1\|_{C(0)} \frac{c}{p^2}$$

pour tout $p = 1,2,3,..$, ce qui entraîne évidemment que

(6.25) $\qquad \lim_r F_t(\bar{\phi}_r) = 0$ presque partout dans $(0,T)$.

Dans cette relation l'ensemble exceptionnel dépend de la suite (6.24).

Nous allons utiliser ce résultat incomplet pour montrer que (6.25) est vraie en dehors d'un ensemble σ de mesure de Lebesgue nulle, indépendant de la suite (6.24).

Dans ce but introduisons les fonctions

$$d_{p,q}(u) = \begin{cases} q \cdot (\text{distance de u à } b_p), & \text{si cette distance est } \leq \frac{1}{q} \\ 1 & , \text{ " " " " } > \frac{1}{q} . \end{cases}$$

C. Foiaș

Evidemment, $d_{p,q}$ est une fonction de $C(0)$. Pour p,q fixés, soit de plus

$$\varphi_r(u) = \min\left\{d_{p,q}(u), \frac{1}{p^2}\|P_r u\|^2\right\} \text{ et } \psi_r = d_{p,q} - \varphi_r .$$

Il est manifeste que φ_r et ψ_r appartienent aussi à $C(0)$. De plus

(6.26) $$\|P_r u\|^2 = \|P_{r+1} u\|^2 \le \|u\|^2 ,$$

donc φ_r est une suite croissante convergeant vers

$$\min\left\{d_{p,q}(.), \frac{1}{p^2}\|\cdot\|^2\right\} = d_{p,q}(.) ;$$

par conséquent $\left\{\psi_r\right\}$ est une suite decroissante convergeant vers 0 .

Par ce que nous avons déjà établit, il existe un ensemble $\sigma_{p,q} \subset (0,T)$ de mesure de Lebesgue nulle, tel que $\lim_r F_t(\psi_r) = 0$, c'est-à-dire

(6.27) $$\lim_r F_t(\varphi_r) = F_t(d_{p,q}) \text{ pour tout } t \notin \sigma_{p,q} .$$

D'autre part on a

C. Foiaş

$$\mathcal{G}_r(u) \leq \frac{1}{p^2} \left\| P_r u \right\|^2 \qquad \text{pour tout } u \in H \ ,$$

où la fonction $\frac{1}{p^2} \left\| P_r \cdot \right\|^2$ appartient à $C(2)$. Par conséquent,

en utilisant k fait que $F_+ \geq 0$, on a

$$F_t(\mathcal{G}_r) = \frac{1}{p^2} F_t(\| P_r \cdot \|^2) \qquad \text{pour tous } t \in (0,T) \quad \text{et} \quad r = 1,2,\ldots,$$

d'où

(6.28) $\qquad F_t(d_{p,q}) \leq \frac{1}{p^2} \lim_r F_t(\| P_r \cdot \|^2)$ pour tout $t \in (0,T) \backslash \sigma_{p,q}$.

Ici la limite, que nous désignons par $\theta(t)$, existe
(pouvant être aussi $= \infty$) en vertu de (6.19) et (6.26). Mais
la relation (6.7) donne, en vertu de (6.19),

$$F^{(n)}(1 \otimes \| P_r \cdot \|^2) \leq c \qquad \text{pour tous} \quad ,n = 1,2,\ldots \ ,$$

(c étant une constante $< \infty$). En passant au limite il vient

$$F(1 \otimes \| P_r \cdot \|^2) \leq c$$

quel que soit $r = 1,2,\ldots$. Par le théorème de Beppo-Levi on

C. Foiaş

en déduit

$$\int_0^T \theta(t)dt = \lim_r \int_0^T F_t(\|P_r \cdot \|^2)dt = \lim_r F(1 \otimes \|P_r \cdot \|^2) \le c \; ,$$

ce qui montre finalement que $\theta(.)$ est intégrable. Ainsi si

$$\sigma = \left\{ t : \theta(t) = \infty \right\} \cup \left(\bigcup_{p,q} \sigma_{p,q} \right)$$

on obtient un ensemble de mesure de Lebesgue nulle qui, compte

tenu de (6.28), juit de la suivante propriété :

$$(6.29) \quad F_t(d_{p,q}) = \frac{1}{p^2} \; \theta(t) \quad \text{et} \quad \theta(t) < \infty \quad \text{pour tout} \quad t \in (0,T) \setminus \sigma \; ,$$

quels que soient $p,q = 1,2,\dots$.

Nous allons maintenant montrer que

$$(6.30) \qquad \lim_r F_t(\Phi_r) = 0 \quad \text{pour tout} \quad t \in (0,T) \setminus \sigma \; ,$$

quelle que soit la suite (6.24).

Dans ce but soit

$$\varepsilon_{p,r} = \max_{b_p} \Phi_r$$

C. Foias

et pour tout $u_0 \in b_p$ soit $0_0 = \{ u : u - u_0 < \varepsilon_0 \}$ telle que

$$| \Phi_r(u) - \Phi_r(u_0) | < \varepsilon_{p,r} \quad \text{pour tout } u \in 0_0 .$$

Comme b_p est compact, il existe un ensemble fini de boules $0_1, 0_2, \ldots, 0_k$ qui couvrent b_p . On a donc

$$| \Phi_r(u) | < 2 \varepsilon_{p,r}$$

dans le voisinage $0_1 \cup 0_2 \cup \ldots \cup 0_k$ de b_p . Mais pour un q assez grand on a

$$\left\{ u : \text{distance de } u \text{ à } b_p \leq \frac{1}{q} \right\} \subset 0_1 \cup 0_2 \cup \ldots \cup 0_k .$$

Cette inclusion nous montre que pour un tel q la majoration suivante est valable

$$| \Phi_r(u) | \leq 2\varepsilon_{p,r} + \| \Phi_r \|_{C(0)} \, d_{p,q}(u)$$

pour tous u, r . Il s'ensuit

$$0 \leq F_t(\Phi_r) \leq 2 \, \varepsilon_{p,r} \, F_t(1) + \| \Phi_r \|_{C(0)} \, F_t(d_{p,q}),$$

d'où

C. Foiaş

$$0 \leq \lim_{r} F_t(\Phi_r) \leq \|\Phi_1\|_{C(0)} \frac{1}{p^2} \; \theta(t)$$

quels que soient $t \in (0,T) \smallsetminus \sigma$ et $p = 1,2,\dots$. Comme pour

ces t on a $\theta(t) < \infty$, en faisant $p \longrightarrow \infty$, on obtient

finalement (6.30).

De cette manière on vient de prouver que si $t \in (0,T) \smallsetminus \sigma$,

alors la restriction de F_t à l'espace $C(0)$ satisfait à la

condition de Daniell. C'est classique que dans ce cas il existe

une mesure borelienne positive (voir (6.19)) μ_t sur H telle

que

(6.31) $\qquad F_t(\Phi) = \int \Phi(u) d\mu_t(u)$ pour toute $\Phi(.) \in C(0)$.

En vertu de (6.20) cette mesure μ_t est une probabilité.

6) Le but de ce alinéa est de prouver que

(6.32) $\qquad F(\Phi(.,.)) = \int_0^T \left[\int \Phi(t,u) d\mu_t(u) \right] dt$

pour toute

(6.33) $\qquad \Phi(.,.) \in L^1(0,T;C(2))$.

Pour cela commençons par prouver que (6.31) reste vraie

C. Foias

Pour cela commençons par prouver que (6.31) reste vraie aussi pour toute de C(2) . Considérons d'abord une fonction $\Phi(.) \in C(2)$ nulle pour $|u| \leq r$ (où r est un nombre quelconque ≥ 0 ; en particulier $|.|^2$ est une telle fonction avec $r = 0$) . Alors pour toute $\varphi(.) \in L^1(0,T;R)$ et tout $n = 1,2,\ldots$, nous avons

$$F^{(n)}(\varphi \otimes \Phi) \leq \int_0^T |\varphi(t)| \left[\int_{\{u:\, |u| \geq r\}} (1+|u|^2) \frac{|\Phi(u)|}{1+|u|^2} \, d\mu_t^{(n)}(u) \right] dt \leq$$

$$\leq \| \Phi \|_{C(2)} \int_0^T |\varphi(t)| \left[\int_{\{u:\, |u| \geq r\}} (1+|u|^2) d\mu_t^{(n)}(u) \right] dt .$$

où, en vertu de (3.3),

$$\int_{\{u:\, |u| \geq r\}} (1+|u|^2) d\mu_t^{(n)}(u) = \int_{\{u:\, |S^{(n)}(t)| \geq r\}} (1+|S^{(n)}(t)P_n u|^2) d\mu(u) =$$

$$\leq \int_{\{u:\, |S^{(n)}(t)u| \geq r\}} \leq (1+|P_n u|^2 + c_1) d\mu \leq \int_{\{u:\, |u|^2 + c_1 \geq r^2\}} (1+|u|^2 + c_1) d\mu(u) =$$

$$\leq c_1' \int_{\{u:\, |u|^2 + c_1 \geq r^2\}} (1+|u|^2) d\mu(u) = \varepsilon(\frac{1}{r}) \longrightarrow 0 \quad \text{pour} \quad r \longrightarrow \infty \quad (c_1 , c_1'$$

étant des constantes indépendantes de $n = 1,2,\ldots$ ou r). Ainsi

$$\mid F^{(n)}(\varphi \otimes \bar{\Phi}) \mid \; \le \parallel \bar{\Phi} \parallel_{C(2)} \left[\int_0^T \mid \varphi(t) \mid dt \right] \epsilon(\tfrac{1}{r})$$

quel que soit $n = 1,2,\ldots$, donc

$$\mid F(\varphi \otimes \bar{\Phi}) \mid \; \le \parallel \bar{\Phi} \parallel_{C(2)} \left[\int_0^T \mid \varphi(t) \mid dt \right] \epsilon(\tfrac{1}{r}) \; .$$

Puisque cette relation est valable pour toute $\varphi(.)$ de $L^1(0,T;R)$, la représentation (6.16) entraîne

(6.34) $$\mid F_t(\bar{\Phi}) \mid \; \le \; \parallel \bar{\Phi} \parallel_{C(2)} \; \epsilon(\tfrac{1}{r})$$

presque partout dans $(0,t)$, donc, vu (6.17) et (6.18), partout dans $(0,T)$, en particulier pour $t \notin \sigma$. Soit maintenant $\bar{\Phi}$ une fonction arbitraire de $\overset{\circ}{C}(2)$ et soit, pour $r=1,2,\ldots$,

$$\varphi_r(x) = \begin{cases} 1 & \text{si} \quad 0 = x = r \; , \\ \text{linéaire} & \text{si} \quad r = x = r+1 \; , \\ 0 & \text{si} \quad r+1 = x \; . \end{cases}$$

Alors pour

$$\bar{\Phi}_r(.) = \bar{\Phi}(.) - \varphi_r(\mid . \mid) \bar{\Phi}(.)$$

C. Foiaş

on a, d'après (6.34),

$$|F_t(\Phi_r)| \leq \|\Phi_r\|_{C(2)} \, \varepsilon(\tfrac{1}{r}) \leq \|\Phi\|_{C(2)} \, \varepsilon(\tfrac{1}{r}) \longrightarrow 0 \quad \text{pour} \quad r \longrightarrow \infty \, ,$$

ce qui entraîne

$$(6.35) \qquad F_t(\varphi_r(|.|) \Phi(.)) \longrightarrow F_t(\Phi) \quad \text{pour} \quad r \longrightarrow \infty \, .$$

Or la suite $\{\varphi_r(|.|) \Phi(.)\}_{r=1}^{\infty}$ tend, en croissant, vers $\Phi(.)$ et est contenue dans $C(0)$. En utilisant (6.35) et le théorème de Beppo-Levi on déduit l'intégrabilité de $\Phi(.)$ par rapport à μ_t et la validité de (6.31) (pour $t \notin \sigma$). Enfin, la représentation (6.16) nous montre que vraiment (6.33) entraîne (6.32).

7) On montrera maintenant que la famille des mesures μ_t vérifie (5.11).

Dans ce but remarquons que $\|P_n.\|^2$ appartient à $L^1(0,T;C(2))$ et que par conséquent,

$$\int_0^T \left[\int \|P_n u\|^2 \, d\mu_t(u) \right] dt = F(1 \otimes \|P_n.\|^2) = \sup_m \overline{F^{(m)}}(1 \otimes \|P_m.\|^2) =$$

$$= \sup_m \int_0^T \left[\int \|P_n u\|^2 \, d\mu_t^{(m)}(u) \right] dt \, ,$$

C. Foiaș

qui, en vertu de (6.7) et (6.26), est majoré par une constante

c < ∞ indépendante de n = 1,2,... . Par le théorème de Beppo-

Levi on conclut

$$(6.36) \qquad \int_0^T \left[\int \| u \|^2 \, d\mu_t(u) \right] dt \leq c \; .$$

Ainsi on a obtenu la deuxième relation (5.11). Quant à la

prémière, il faut seulement remarquer que $|.|^2 \in C(2)$ et que

$$\int |u|^2 \, d\mu_t(u) = F_t(|.|^2) \leq \| |.| \|_{C(2)} \| F_t \| \leq \| F \| \; .$$

Finalement nous sommes à même de prouver que la famille

des mesures μ_t que nous avons obtenue vérifie (5.10) pour

toute $\phi \in \mathfrak{I}_0$. En vertu de (6.15) et du point 6) de notre

démonstration tout revient à prouver que si ψ est définie

par

$$\psi(t,u) = (B(u,u), \underline{\Phi}_u'(t,u))$$

où $\phi \in \mathfrak{I}_0$ est fixée, alors

$$(6.37) \qquad F(\psi) = \int_0^T \left[\int \psi(t,u) d\mu_t(u) \right] dt \; .$$

Posons

C. Foiaş

$$\psi_n(t,u) = (B(u,P_nu), \Phi_u'(t,u)) \quad \text{pour tous} \quad t \in (0,T), u \in H, n=1,2,\dots$$

Alors en tenant compte d'une des propriétés de continuité de B et de (5.2) il vient

$$|\psi_n(t,u)| \leq c_1 \|P_nu\| \cdot |u| \cdot |\Phi_u'(t,u)|_2 \leq c|u|^2$$

et ensuite que

$$\psi_n(.,.) \in L^\infty(0,T;C(2)) \subset L^1(0,T;C(2)),$$

de sorte que la relation (6.32) est vérifiée par $\Phi = \psi_n$. Il nous reste à prouver les convergences

$$(6.38) \quad F(\psi_n) \longrightarrow F(\psi) \quad \text{et} \quad \int_0^T \left[\int \psi_n(t,u)d\mu_t(u)\right] dt \longrightarrow$$

$$\longrightarrow \int_0^T \left[\int \psi(t,u)d\mu_t(u)\right] dt$$

pour $n \longrightarrow \infty$. Comme la deuxième convergence se prouve d'une manière analogue, mais plus simple, que la prémière, on prouvera seulement celle-ci. On commence par remarquer que

$$1 \otimes |(I-P_n).|\|.\| \in L^2(0,T;C(1,1)), \quad 1 \otimes |(I-P_n).|^2 \in L^2(0,T;C(2))$$

C. Foiaş

et

$$F^{(m)}(1 \otimes |(I-P_n).|.\|.\|) = \int_0^T \left[\int |(I-P_n)u|.\|u\| d\mu_t^{(m)}(u) \right] dt \leq$$

$$\leq \int_0^T \left(\int |(I-P_n)u|^2 d\mu_t^{(m)}(u) \right)^{1/2} \left(\int \|u\|^2 d\mu_t^{(m)}(u) \right)^{1/2} dt =$$

$$= c' \left(\int_0^T \left[\int |(I-P_n)u|^2 d\mu_t^{(m)}(u) \right] dt \right)^{1/2} = c'(F^{(m)}(1 \otimes |(I-P_n).|^2))^{1/2}$$

où c' est une constante indépendante de n,m (notamment
$c' = c^{1/2}$, c étant la constante de (6.36)); l'inégalité
entre les termes extrêmes ci-dessus se conserve au limite (dans
$(L^2(0,T;C(1,1)))'$ *-faible); on en obtient

$$F(1 \otimes |(I-P_n).|.\|.\|) \leq c' (F(1 \otimes |(I-P_n).|^2)).$$

Mais en vertu de (6.32) et (5.11) le deuxième terme de la
relation précedente tend vers 0 lorsque $n \rightarrow \infty$. Par suite

$$|F(\psi_n) - F(\psi)| \leq F(c'' |(I-P_n)|.\|.\|) \rightarrow 0 \quad \text{pour} \quad n \rightarrow \infty ,$$

car c'' est une constante dépendant de Φ , mais pas de n .

La preuve de l'existence d'une solution statistique à

C. Foiaş

donnée initiale μ est terminée. Le fait que la solution $\{\mu_t\}$, que nous venons de construire, satisfait l'inéquation de l'énergie (6.1) sera prouvé et étudié dans le paragraphe suivant.

7. L'inéquation de l'énergie.

a. Le but de ce alinéa sera d'esquisser la preuve de l'inéquation de l'energie (6.1-2). Soit donc $\varphi(.)$ une fonction réelle définie sand $[0,\infty)$ satisfaisant aux conditions indiquées dans l'énoncé de la Proposition 3 ; alors en vertu de (3.2) on a

$$\frac{1}{2}\frac{d}{dt}\int \varphi(|u|^2)d\mu_t^{(n)}(u) + \int \varphi'(|u|^2)\|u\|^2 \, d\mu_t^{(n)}(u) =$$

$$= \frac{1}{2}\frac{d}{dt}\int \varphi(|S^{(n)}(t)P_n u|^2)d\mu(u) + \int \varphi'(|S^{(n)}(t)P_n u|^2)\|S^{(n)}(t)P_n u\|^2 \, d\mu(u) =$$

$$= \frac{1}{2}\int \varphi'(|S^{(n)}(t)P_n u|^2)\left[\frac{1}{2}\frac{d}{dt}|S^{(n)}(t)P_n u|^2 + \|S^{(n)}(t)P_n u\|^2\right] d\mu(u) =$$

$$= \int (f(t),S^{(n)}(t)P_n u)\,\varphi'(|S^{(n)}(t)P_n u|^2)d\mu(u) =$$

$$= \int (f(t),u)\,\varphi'(|u|^2)d\mu_t^{(n)}(u),$$

C. Foiaş

d'où, en remplaçant le terme $\|u\|^2$ par $\|P_k u\|^2$ $(k = 1, 2, \ldots)$

on déduit

$$- \int_0^T r'(t)\left[\int \varphi(|u|^2)\,d\mu_t^{(n)}(u)\right]dt + \int_0^T r(t)\left[\int \varphi'(|u|^2)\|P_k u\|^2\,d\mu_t^{(n)}(u)\right]dt$$

$$\leq \int_0^T r(t)\left[\int (f(t),u)\,\varphi'(|u|^2)\,d\mu_t^{(n)}(u)\right]dt \; ,$$

quelle que soit $r(.)$ de classe C^1 à support compact dans $(0, T)$.

Dans cette relation il est permis de passer "au limite" en n ; on en obtient

$$- \int_0^T r'(t)\left[\int \varphi(|u|^2)\,d\mu_t(u)\right]dt + \int_0^T r(t)\left[\int \varphi'(|u|^2)\|P_k u\|^2\,d\mu_t(u)\right]dt \leq$$

$$\leq \int_0^T r(t)\left[\int \varphi'(|u|^2)(f(t),u)\,d\mu_t(u)\right]dt \; .$$

En faisant $k \longrightarrow \infty$ et ensuite en "intégrant", on obtient finalement (6.1).

b. Approchant la fonction $\max(0, x-r)$ (où $0 \leq r$, $x < \infty$) par des fonction du type envisagé dans (6.1) on déduit aisément

C. Foiaş

de l'inéquation de l'énergie (6.1) une estimation intéressante,

notamment la suivante

$$(7.1) \quad \int\limits_{\{u:\,|u|\geq r\}} |u|^2 d\mu_t(u) + \int_0^t \left[\int\limits_{\{u:\,|u|>r\}} \|u\|^2 d\mu_\tau(u) \right] d\tau \leq \int\limits_{\{u:\,|u|>r\}} |u|^2 d\mu(u)$$

presque partout dans $(0,T)$, valable des que $r \geq c_o$, où c_o

est une constante qui ne dépend pas de μ .

En particulier on déduit de (7.1) d'une manière évidente

la suivante propriété de la solution statistique donnée par la

Proposition 3 :

Proposition 4. Si la donnée initiale μ a le support

borné dans H , alors le support de μ_t est uniformément

borné dans H pour presque tout $t \in (0,T)$.

Pour une étude plus poussée des solutions statistiques

(surtout pour le système de Navier-Stokes) nous renvoyons aux

recherches citées dans l'introduction de G. PRODI et de l'auteur.

C. Foiaş

Notes de bas

(^{1}) Rappelons que $H_o^1(\Omega)$ est l'adhérence dans $H^1(\Omega)$ de
l'espace $C_o^\infty(\Omega)$ des fonctions de classe C^∞ au support
compact dans Ω et que

$$H^k(\Omega) = \left\{ u : \frac{\partial^\alpha u}{\partial x^\alpha} \in L^2(\Omega) \text{ pour tout ordre } |\alpha| \le k \right\},$$

la norme dans $H^k(\Omega)$ étant donnée par $\left(\sum \| \partial^\alpha u/\partial x^\alpha \|^2_{L^2(\Omega)} \right)^{1/2}$

(^{2}) Pour une démonstration élémentaire de (4.1) voir LIONS [2] ,
p. 70.

C. Foiaş

Ouvrages cités

CATTABRIGA, L.

[1] Su un problema al contorno relativo al sistema di equazioni di Stokes, Rend.Sem.Univ.Padova, 31 (1961), 1-33 .

DINCULEANU, N.

[1] Vector Mesures, Berlin (1966).

FOIAŞ, C.

[1] Ergodic problems in functional spaces related to Navier-Stokes equations, Proc.Intern.Conf.on Funct. Anal., Tokyo (1969), 290-304 .

HOPF, E.

[1] Statistical hydromechanics and functional calculus, Journ.Rat.Mech.Anal.,1 (1952), 87-123.

LADYZENSKAYA, O.A.

[1] On the solution in large of the boundary problem for Navier-Stokes equations involving two space variables, Dokl.Akad.Nauk SSSR, 123 (1958), 427-429.

C. Foias

LERAY, J.

[1] Essai sur les mouvements plans d'un liquide visqueux
que limitent des parois, Journ.Math.Pures et Appl.,
9e sér., 13 (1934), 331-418.

LIONS, J.L.

[1] Quelques méthodes de résolution des problèmes aux
limites non linéaires, Paris (1969).

LIONS, J.L. - PRODI, G.

[1] Un théorème d'existence et unicité dans les équations
de Navier-Stokes en dimension 2, C.R.Acad.Sci.Paris,
248 (1959), 3519-3521.

LIONS, J.L. - MAGENES, E.

[1] Problèmes aux limites non homogènes et applications,
vol. I, Paris (1968).

PRODI, G.

[1] Qualche risultato riguardo alle equazioni di Navier-
Stokes nel caso bidimensionale, Rend.Sem.Mat.Padova,
30 (1960), 1-15 .

C. Foiaș

[2] Teoremi ergodici per le equazioni della idrodinamica, C.I.M.E. Roma (1960).

[3] On probability measures related to the Navier-Stokes equations, Trieste (1961).

RIESZ, F. - SZ. NAGY, B.

[1] Leçons d'analyse fonctionnelle, Budapest (1953).

CENTRO INTERNAZIONALE MATEMATICO ESTIVO
(C.I.M.E.)

J. L. LIONS

QUELQUES PROBLEMÈS DE LA THEORIE DES EQUATIONS
NON LINÉAIRES D'ÉVOLUTION

Corso tenuto a Varenna dal 20 al 29 Agosto 1970

QUELQUES PROBLEMÈS DE LA THEORIE DES EQUATIONS

NON LINÉAIRES D'ÉVOLUTION

par

J. L. LIONS

(Institut H. Poincaré - Paris)

CHAP. 1. Généralités sur quelques méthodes d'analyse non lineaire.

Illustrations sur de exemples simples.

Introduction.

1. Procedés d'approximation.

    1.1 Formulation varationnelle.

    1.2 Approximation de Faedo–Galerkin. Eléments finis.

    1.3 Approximation par regularisation elliptique.

    1.4 Approximation par descretisation en t .

2. Passage à la limite par le méthode de compacité.

    2.1. Estimations a priori (I).

    2.2. Estimations a priori (II).

    2.3. Théorème.

3. Méthode de monotonie.

4. Indications sur d'autres méthodes.

    4.1. Approximation par troncature.

    4.2. Approximation par un équation "avec retard".

    4.3. Somme d'opérateurs aux dérivées partielles.

4.4. Méthode de décomposition.

5. Problemès hyperboliques.

    5.1. Exemple 1.

    5.2. Exemple 2.

    5.3. Adaptation des méthodes des N° 1 à 3.

6. Quelques problèmes.

CHAP. 2. Monotonie et changement d'espace pivot.

Introduction

1. Résultat "abstrait" et Exemple.

2. Changement d'espace pivot.

    2.1. Position des problèmes. Opérateur parabolique dégénéré.

    2.2. Résolution au problème avec condition du type "Neumann".

    2.3. Résolution du problème avec condition du type"Dirichlet".

3. Problème.

CHAP. 3. Inequations variationnelles pour des problèmes paraboliques dégénérés.

Introduction.

J. L. Lions

1. Un résultat de compacité.

    1.1. Hypothèses. Enoncé du théorème.

    1.2. Exemple.

    1.3. Les particuliers.

    1.4. Démonstration du Théorème 1.1.

2. Un théorème d'existence.

    2.1. Enoncé du résultat.

    2.2. Démonstration du Théorème 2.1.

3. Conjecture sur le positivité des solutions

4. Inéquations variationnelles.

    4.1. Position du problème. Notations.

    4.2. Nouvelle formulation du problème.

    4.3. Le cas "j = ∞".

5. Problemès.

CHAP. 4. Inéquations variationnelles et écoulement de fluides non newtoniens.

Introduction.

1. Position des problèmes. Résultats.

    1.1. Notations. Problème.

J. L. Lions

1.2. Résultats.

2. Démonstration de l'unicité dans les Théorèmes 1.1 et 1.2.

    2.1. Une inégalité.

    2.2. Démonstration de l'unicité.

3. Démonstration de l'existence dans le Théorème 1.1.

    3.1. Méthode de régularisation.

    3.2. Existence d'une solution de (3.2) , (3.4).

    3.3. Passage à la limite en ε .

4. Les équations de Navies Stokes comme cas limites des inéquations.

5. Problèmes.

CHAP. 5. Inéquations variationnelles hyperboliques. Méthode de pénalisation.

Introduction.

1. Position du problème. Enoncé du résultat principal. Exemples.

    1.1. Notations.

    1.2. Position du problème.

    1.3. Resultat principal.

    1.4. Exemples.

2. <u>Démonstration de l'unicité dans le Théorème 1.1.</u>

3. <u>Démonstration de l'existence dans le Théorème 1.1. par la mé</u>-
   <u>thode de pénalisation.</u>

   3.1. Opérateur de pénalisation.

   3.2. Équation pénalisée.

   3.3. Résolution de l'équation pénalisée.

   3.4. Estimations a priori (I).

   3.5. Estimations a priori (II).

   3.6. Estimation a priori (III).

   3.7. Passage à la limite.

4. <u>Inéquations relatives au système de Maxwell.</u>

   4.1. Équations de Maxwell usuelles.

   4.2. Les inéquations.

5. <u>Problèmes.</u>

CHAP. 6. <u>Perturbations singulières et inéquations variationnel</u>-
   <u>les d'évolution.</u>

<u>Introduction.</u>

1. <u>Perturbations singulières d'inéquations hyperboliques condui</u>-
   <u>sant à des nouvelles inéquations pour des opérateurs elliptiques.</u>

J. L. Lions

J. L. Lions

5. Problèmes.

APPENDICE. **Multiplicateurs de Lagrange.**

BIBLIOGRAPHIE.

J. L. Lions

CHAPITRE 1

GÉNÉRALITÉS SUR QUELQUES MÉTHODES D'ANALYSE NON LINÉAIRE.

ILLUSTRATIONS SUR DES EXEMPLES SIMPLED.

Introduction.

On va pour l'essentiel (N° 1 à 4) étudier le problème parabolique non linéaire suivant: soit Ω un ouvert de $\mathbb{R}^n$ de fron
tiere Γ ; on cherche une fonction $u = u(x,t)$, $x \in \Omega$, $t \in]0,T[$,
telle que

(1) $\dfrac{\partial u}{\partial t} - \Delta u + u^3 = f$, f donnée dans $Q = \Omega \times] 0,T[$,

avec la condition initiable.

(2) $u(x,0) = u_o(x)$, $x \in \Omega$, u_o donnée dans Ω ,

et la condition aux limites

(3) $u = 0$ sur $\sum = \Gamma \times] 0,T[$.

La première idée est"d'approcher" l'équation (1) par des équa
tions "plus simples" - ce qui peut être fait de plusieurs manières,
dont certains sont exposés au N° 1 - puis au N° 4 .

Une fois en "possession" des solutions des equations "simpli
fiées", il faut passer à le limite, les deux methodes "basiques"
étant, pour cela, la méthode de compacité (N°2) et la méthode de

<u>monotonie</u> (N° 3).

Naturellement les méthodes ainsi présentées <u>ne sont pas essen-</u>
<u>tiellement liées</u>à la parabolicité de l'équation (1); on indique <u>ra-</u>
<u>pidement</u> au N° 5 les adaptations des méthodes au <u>problème hyperbo-</u>
<u>lique</u>

$$(4) \qquad \frac{\partial^2 u}{\partial t^2} - \Delta u + \left| \frac{\partial u}{\partial t} \right|^\beta \frac{\partial u}{\partial t} = f$$

avec les <u>conditions initiales</u>

$$(5) \qquad u(x,0) = u_o(x) \ , \ \frac{\partial u}{\partial t}(x,0) = u_1(x) \ , \quad x \in \Omega$$

et (par ex.) la condition aux limites (3).

1. <u>Procédés d'approximation.</u>

   1.1 <u>Formulation variationnelle.</u>

   Toutes les fonctions considerées sont <u>à valeurs réelles.</u>

   Les <u>espace de Sobolev</u>:

$$H^1(\Omega) = \left\{ v \mid v \in L^2(\Omega) \ , \ \frac{\partial v}{\partial x_i} \in L^2(\Omega) \ , \quad i = 1,\ldots,n \right\} \ ;$$

$$H_o^1(\Omega) = \left\{ v \mid v \in H^1(\Omega) \ , \quad v = 0 \quad \text{sur } \Gamma \right\};$$

$H^{-1}(\Omega)=$ dual de $H_o^1(\Omega)$ <u>lorsque</u> $L^2(\Omega)$ <u>est identifié à son dual</u>
(cf. Chap. 2) ;

$(1.1) \quad (f,g) = \displaystyle\int_\Omega f \, g \, dx$ lorsque cette intégrale a un sens, par
ex. pour $f,g \in L^2(\Omega)$ ou pour $f \in H_o^1(\Omega)$, $g \in H^{-1}(\Omega)$.

   Pour $u,v \in H^1(\Omega)$, on posera:

J. L. Lions

$$(1.2) \qquad a(u,v) = \sum_{i=1}^{n} \int_{\Omega} \frac{\partial u}{\partial x_i} \frac{\partial v}{\partial x_i} \, dx \quad .$$

Pour <u>simplifier l'exposé</u> ([1]) on va se placer dans des condi-tions telles que $v \longrightarrow v^3$ applique $H_o^1(\Omega)$ dans $H^{-1}(\Omega)$; or d'a-près <u>l'inégalité de Sobolev</u> on a :

$$v \in H_o^1(\Omega) \Longrightarrow v \in L^q(\Omega) \ , \ \frac{1}{q} = \frac{1}{2} - \frac{1}{n} \ \text{si} \ n > 2 \quad ,$$

et donc

$$(1.3) \qquad |(v^3, \varphi)| \leqslant c \ \|\varphi\|_{H_o^1(\Omega)} \ \|v\|^3_{H_o^1(\Omega)}$$

si (et seulement si) $\frac{4}{q} \leqslant 1$ i.e. si $n \leqslant 4$ (le cas "n = 2" est im-mediat). On supposera donc desormais

$$(1.4) \qquad n \leqslant 4 \ .$$

La "formulation variationnelle" et ... vague ([2]) du problème (1) (2) (3) de l'introduction est alors: désignant par $u(t)$ la fonc-tion $x \longrightarrow u(x,t)$ trouver une fonction $t \longrightarrow u(t)$ de $[0,T] \longrightarrow H_o^1(\Omega)$ telle que

$$(1.5) \ \left(\frac{\partial u(t)}{\partial t} \, , \, v\right) + a(u(t),v) + (u(t)^3, v) = (f(t),v) \ \forall v \in H_o^1(\Omega),$$

avec

$$(1.6) \qquad u(0) = u_o \ .$$

([1]) Si non il suffit de "travailler" dans l'espace $H_o^1(\Omega) \cap L^4(\Omega)$.

([2]) Il faudra en effet préciser les propriétés de la fonction $t \longrightarrow u(t)$.

J. L. Lions

1.2 Approximation de Faedo-Galerkin. Eléments finis.

Pour simplifier l'écriture, posons:

$$(1.7) \qquad H^1_o(\Omega) = V .$$

L'idée est "d'approcher" V **par une famille de sous espaces de dimension finie**. On introduit donc une famille dénombrable V_m (^{1}) de sous espaces de V ayant les proprietés:

$$(1.8) \qquad \forall_m , \quad V_m \text{ est de dimension finie } N(m) ;$$

$$(1.9) \qquad V_m \to V \text{ au sens: } \forall v \in V , \text{ il existe } v_m \in V_m \text{ tels que } v_m \to v \text{ dans } V \text{ lorsque } m \to \infty .$$

Exemple 1.1.

Soit $w_1 \ldots w_m \ldots$ une "base" de V au sens:

$$(1.10) \qquad \forall_m , \text{ les } w_1 \ldots w_m \text{ sont linéairement indipendants}$$

$$(1.11) \qquad \text{les combinaisons } \sum_{\text{finies}} \xi_j w_j , \quad \xi_j \in \mathbb{R} , \text{ sont denses dans } V .$$

Alors on peut prendre

$$V_m = \text{espace engendré par } w_1 , \ldots, w_m .$$

Dans ce cas - mais ce n'est nullement le cas général - on a:

$$(1.12) \qquad V_m \quad V_{m+1} \quad \ldots .$$

(^{1}) qui existe car V est **séparable**.

Exemple 1.2. Éléments finis ([1]).

Soit Ω un ouvert de $\mathbb{R}^2$ et une triangulation de Ω comme indique schématiquement sur la fig. 1.

Soit M un sommet "inté-
rieur" de la triangulation; on
lui associe le fonction w_M qui
est linéaire par morceaux, vaut
1 en M, et est nulle en ces
points P "voisins" de M et

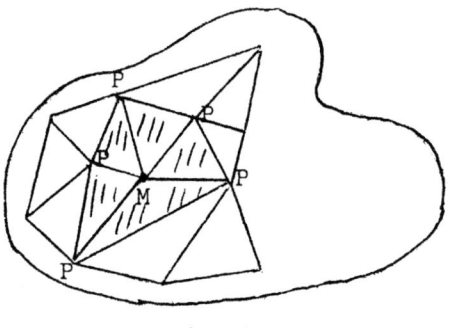

Fig. 1

sommets de la triangulation (cf. Fig. 1) et w_M etant $\equiv 0$

hors du domaine hachuré Fig. 1.

Si m est un indice reférant la triangulation ([2]) alors

$$(1.13) \quad V_m = \text{espace engendré par les } w_M \text{ , } M \in \text{sommets internes}$$
de la triangulation.

Dans ce cas (1.12) n'est pas vrai mais on a encore (1.8) (1.9) .

L' approximation de Faédo-Galerkin consiste alors en le problème suivant:

trouver $t \to u_m(t)$ telle que

([1]) Methode fondamentale en Analyse Numerique. Cf. le cours CIME 1971.

([2]) Par ex. $m = \frac{1}{h}$, h = longueur index de cotés de la triangulation, avec l'hypothèse que l'angle minimum est $> \alpha_o > 0$.

J. L. Lions

(1.14) $u_m(t) \in V_m$,

(1.15) $\left(\dfrac{\partial u_m(t)}{\partial t} , v \right) + a(u_m(t),v) + (u_m(t)^3,v) = (f(t),v)$ $\forall v \in V_m$,

(1.16) $u_m(0) = u_{om}$, $u_{om} \in V_m$, $u_{om} \longrightarrow u_o$ dans $L^2(\Omega)$ $(^1)$.

Il s'agit là d'un système de $N(m)$ <u>équations différentielles</u> <u>ordinaires non lineaires</u>, définissant de facons unique $u_m(t)$ dans un intervalle $[0,t_m]$, $t_m > 0$.

<u>Les estimations a priori effectuées plus loin</u> (N°2) montreront que

(1.17) $\qquad t_m = T$.

Remarque 1.1.

La même méthode donnerait l'existence de $u_m(t)$ dans $[0,t_m]$, $t_m > 0$, pour de formes $a(u,v)$ <u>non elliptiques</u>, par ex. pour

(1.18) $a(u,v) = \displaystyle\sum_{i=1}^{i_o} \int_\Omega \dfrac{\partial u}{\partial x_i} \dfrac{\partial v}{\partial x_i}\, dx - \sum_{i_o+1}^{n} \int_\Omega \dfrac{\partial u}{\partial x_i} \dfrac{\partial v}{\partial x_i}\, dx$.

<u>Mais alors on n'a pas</u> (1.17) (en général).

Avant de passer <u>au problème fondamental: le procédé "d'appro-ximation"</u> (1.14) (1.15) (1.16) <u>fornit-il effectivement uné appro-</u>

$(^1)$ En suffisant $u_o \in L^2(\Omega)$.

J. L. Lions

ximation de la (ou d'une)solution u du probleme initial, nous

allons donner quelques autres procédes d'approximation.

1.3. Approximation par régularisation elliptique.([1])

On admit ici que l'on est en possession d'une "bonne" théo-

rie des problèmes elliptiques non linéaires.

L'idée est alors "d'approcher" le problème (1)(2)(3) de l'in

troduction par un problème elliptique.

Ainsi on"remplace"(1) Introduction, par

$$(1.19) \quad -\varepsilon \frac{\partial^2 u}{\partial t^2} + \frac{\partial u_\varepsilon}{\partial t} - \triangle u_\varepsilon + u_\varepsilon^3 = f \quad \text{dans} \quad Q \ , \ \varepsilon > 0$$

avec

(1.20) $u_\varepsilon(x,0) = u_o(x)$ (identique à (2), Introduction) ,

(1.21) $u_\varepsilon = 0$ sur $\sum$ (identique à (3), Introduction)

et l'on rejointe une condition aux limites pour $t = T$, par ex.

$$(1.22) \qquad \frac{\partial u_\varepsilon}{\partial t} (x,T) = 0 \ , \quad x \in \Omega \ .$$

Le plan (que l'on ne détaillera pas ici) est alors:

(i) on résout (1.19)...(1.22) qui est bien un problème non

linéaire de nature elliptique (on dit que c'est le régularisé

elliptique du problème initial);

(ii) on passe à la limite $(\varepsilon \to 0)$ - ce qui peut être fair

par des méthodes voisines de celles exposées d'après aux N°2 et 3.

([1]) On se borne à des indications générales. Pour des résultat
   précis, cf. LIONS [1] .

Remarque 1.2.

L'idée fondamentale pour résoudre (1.19.) - (1.22) (et d'ail
leurs sous des formes diverses, tous les problèmes elliptiques)
est de se ramener à une approximation a dimension finie dont l'exi
stence est demontrée par un théorème de point fixe ([1]); on passe
ensuite à la limite.

1.4. Approximation par discretisation en t .

Un outre moyen d'"approcher" le problème d'evolution (1.5)
(1.6) par des problèmes elliptiques est de déscrétiser en t .

Si l'on pose:

(1.23) $\begin{cases} u^n = \text{"approximation" de } u \text{ à l'instant } n\,\Delta t \ , \\ \Delta t > 0 \text{ choisi} , \end{cases}$

on definit u^n à partir de u^{n-1} (on part de $u^o = u_o$) par:

(1.24) $(\dfrac{u^n - u^{n-1}}{\Delta t},v) + a(u^n,v) + ((u^n)^3,v) = (f^n,v) \quad \forall v \in V \quad ([2])$

qui est un problème elliptique (non lineaire) $\forall n$.

Remarque 1.3.

Il y a bien des variantes - dont la mise en oeuvre est possi
ble, si non facile - :

---

[1] En schematisant beaucoup: les équations non linéaires elliptiques sont résolues à partir du théorème de Brouwer, les
équations non linéaires d'evolution à partir du théorème
d'Euler sur l'existence de solutions (locales) aux équations
différentielles.

[2] $f^n = $ (par ex) $f(n\,\Delta t)$.

1°) on peut definir u^n par

(1.25) $\quad (\dfrac{u^n - u^{n-1}}{\Delta t} , v) + a(u^{n-1}, v) + ((u^{n-1})^3, v) = (f^n, v) \quad \forall v \in V$

- méthode dite "$\underline{\text{explicite}}$" par opposition à la méthode (1.24) di

te "$\underline{\text{implicite}}$" - ;

   2°) on peut definir u^n par (méthode $\underline{\text{implicite linéarisée}}$):

(1.26) $\quad (\dfrac{u^n - u^{n-1}}{\Delta t} , v) + a(u^n, v) + ((u^{n-1})^2 u^n, v) = (f^n, v) \quad \forall v \in V$.

2. Passage à la limite par la méthode de compacité.

On va maintenant analyser la convergence de la méthode de

Faedo-Galerkin donnée au 1.2.

2.1. Estimations a priori (I).

Il est loisible de prendre dans (1.15) $v = u_m(t)$; on obtient

(en posant:

(2.1) $\quad |v|^2 = (v,v)$, $|v| =$ norme dans $L^2(\Omega)$, $L^2(\Omega) = H$,

(2.2) $\quad a(v) = a(v,v)$,

(2.3) $\quad a(v) + |v|^2 = \|v\|^2$) :

(2.4) $\quad \dfrac{1}{2} \dfrac{d}{dt} |u_m(t)|^2 + a(u_m(t)) + \displaystyle\int_\Omega u_m(t)^4 dx = (f(t), u_m(t))$.

On note (inégalité de Poincaré élémentaire):

(2.5) $\qquad a(v) \geqslant \alpha \|v\|^2$, $\alpha > 0$, $\forall v \in H_o^1(\Omega)$.

On suppose que

(2.6) $\qquad f \in L^2(0, T , H^{-1}(\Omega))$.

J. L. Lions

On designe par $\|f\|_*$ la norme dans $H^{-1}(\Omega)$ duale de $\|\ \|$,

donc:

$$(2.7) \qquad \|f\|_* = \sup. \ \frac{|(f,\varphi)|}{\|\varphi\|} \ , \qquad \varphi \in H_o^1(\Omega) \ .$$

Alors (2.4) donne

$$\frac{1}{2}\frac{d}{dt}|(u_m(t)|^2 + \alpha\|u_m(t)\|^2 + \int_\Omega u_m(t)^4 dx \leqslant \|f(t)\|_* \|u_m(t)\| \leqslant$$

$$\leqslant \frac{\alpha}{2}\|u_m(t)\|^2 + \frac{1}{2\alpha}\|f(t)\|_*^2$$

d'où

$$(2.8) \quad \frac{d}{dt}\left|u_m(t)\right|^2 + \alpha\|u_m(t)\|^2 + 2\int_\Omega u_m(t)^4 dx \leqslant \frac{1}{\alpha}\|f(t)\|_*^2 \ ,$$

d'où

$$(2.9) \quad |u_m(t)|^2 + \alpha\int_0^t \|u_m(\sigma)\|^2 \, d\sigma + 2\int_0^t \int_\Omega u_m(\sigma)^4 dx \, d\sigma \leqslant$$

$$\leqslant |u_{om}|^2 + \frac{1}{\alpha}\int_0^t \|f(\sigma)\|_*^2 \, d\sigma \leqslant \text{constante} \ .$$

On en déduit (1.17) et que

$$(2.10) \quad \begin{cases} u_m \text{ demeure dans un borné de } L^\infty(0,T \ ; \ H) \cap L^2(0,T;V) \\ \text{lorsque } m \to \infty \end{cases}$$

et

$$(2.11) \quad u_m \text{ demeure dans un borné de } L^4(\Omega).$$

<u>Insuffisance de</u> (2.10) (2.11) dans les problèmes non lineai-

res.

D'après (2.10) (et(2.11)) on peut extraire une suite u_μ

de u_m telle que

$$(2.12) \quad \begin{cases} u_\mu \to u \quad \text{dans} \quad L^2(0,T;V) \text{ faible et dans } L^\infty(0,T;H) \\ \text{faible etoile} \end{cases}$$

et

$$(2.13) \quad u_\mu \to u \quad \text{dans} \quad L^4(Q) \quad \text{faible.}$$

Mais cela <u>n'entraine pas</u> en général que $u_\mu^3 \to u^3$ (dans n'importe quelle topologie) de sorte que l'on ne peut passer à la limite.

<u>Remarque</u> 2.1..

Par contre dans les problèmes <u>linéaires</u> (2.12) suffit.

<u>Orientation.</u>

On va montrer qu'en fait <u>on peut</u> ici passer à la limite, <u>en utilisant</u> la structure de (1.15). Dans le présent N° on va obténir <u>une nouvelle estimation</u> pour $\dfrac{\partial u_m}{\partial t}$ qui permettra de passer à la limite; dans le N° suivant on utilisera la monotonie.

2.2. <u>Estimation a priori</u> (II).

<u>But poursuivi:</u>

On veut obténir une nouvelle estimation du type:

J. L. Lions

(2.14) $\dfrac{\partial u_m}{\partial t}$ est borné dans une norme convenable ([1])

de sorte que (2.11) <u>et</u> (2.14) <u>entrainent par ex.</u> que

(2.15) $u_m \in$ <u>rélativement compact</u> de $L^2(0,T;H) = L^2(\mathbf{Q})$.

On peut alors extraire une sous suite u_μ telle que l'on

ait (2.12) (2.13) <u>et</u>

(2.16) $u_\mu \to u$ p.p. dans $\mathbf{Q}$.

Mais (Lemme de la théorie de la mesure; cf. par ex. LIONS

[1], Lemme 1.3, p. 12;13) (2.13) (2.16) entrainent que

(2.17) $u_\mu^3 \to u^3$ dans $L^{4/3}(\mathbf{Q})$ faible.

On verra (au N° 2.3) que cela suffit alors pour passer à la

limite.

<u>Estimation du type</u> (2.14).

On va obtenir une estimation sur une derivée <u>fractionnaire</u>

en t de u_m - et cela suffit.

On introduit:

(2.18) $\begin{cases} \tilde{u}_m, \tilde{f} \dots = \text{prolongement de } u_m, f, \dots \text{ à } \mathbb{R}_t \text{ par 0 hors de} \\ ]0, T[\end{cases}$

([1]) Cela sera précise plus loin!

J. L. Lions

(2.19) $\hat{u}_m$, $\hat{f}$,... = transformées de Fourier en t de $\tilde{u}_m$,$\tilde{f}$,... ([1]).

On déduit de (1.15) que, $\forall v \in V_m$:

(2.20) $\quad \dfrac{d}{dt}(\tilde{u}_m,v) + a(\tilde{u}_m,v) + ((\tilde{u}_m)^3, v) = (\tilde{f},v) + (u_{om},v)\,\delta(t-0) - $

$$ - (u_m(T),v)\,\delta(t-T) . $$

On note que (cf.(1.3))

$$(\tilde{u}_m^3,v) = (g_m,v),\ g_m \in V',\ \|g_m(t)\|_* \leqslant c\ \|u_m(t)\|^3_{L^4(\Omega)}$$

et donc en particulier:

(2.21) $\quad \displaystyle\int_{-\infty}^{+\infty} \|g_m(t)\|_*\ dt \leqslant c \int_0^T \|u_m(t)\|^3_{L^4(\Omega)}\,dt \leqslant$ constante .

Par transformation de Fourier en t de (2.20), il vient

(2.22) $\quad i\tau(\hat{u}_m(\tau),v) + a(\hat{u}_m(\tau),v) + (\hat{g}_m(\tau),v) = (\hat{f}(\tau),v) +$

$$ + (u_{om},v) - e^{-2\pi i\tau T}\,(u_m(T),v) . $$

Faisant dans (2.22) $v = \hat{u}_m(\tau)$ et prenant la partie imaginai-
re et majorant, il vient:

(2.23) $\quad |\tau|\,|\hat{u}_m(\tau)|^2 \leqslant (\|\hat{f}(\tau)\|_* + \|\hat{g}_m(\tau)\|_*)\,|\hat{u}_m(\tau)\| + c\,\|\hat{u}_m(\tau)\|$

(les c designent des constantes diverses).

([1]) Donc $\hat{u}_m(\tau) = \displaystyle\int_{-\infty}^{+\infty} \exp(-2\pi i t\tau)\,u_m(t)dt$

D'après (2.21) on a:

(2.24) $\qquad \|\hat{g}_m(\tau)\|_* \leqslant c$

et donc (2.23) s'écrit:

(2.25) $\qquad |\tau| |\hat{u}_m(\tau)|^2 \leqslant \|\hat{f}(\tau)\|_* \|\hat{u}_m(\tau)\| + c \|\hat{u}_m(\tau)\|$.

Soit alors β tel que $\beta > \dfrac{1}{2}$. On a:

$$\int_{-\infty}^{+\infty} \frac{|\tau|}{1 + |\tau|^\beta} |\hat{u}_m(\tau)|^2 \, d\tau \leqslant \int_{-\infty}^{+\infty} \frac{\|\hat{f}(\tau)\|_* \|\hat{u}_m(\tau)\|}{1 + |\tau|^\beta} \, d\tau +$$

$$+ c \int_{-\infty}^{+\infty} \frac{\|u_m(\tau)\|}{1 + |\tau|^\beta} \, d\tau <$$

$$\leqslant c + c \left(\int_{-\infty}^{+\infty} \|\hat{u}_m(\tau)\|^2 \, d\tau \right)^{1/2} \left(\int_{-\infty}^{+\infty} \frac{d\tau}{(1 + |\tau|^\beta)} \right)^{1/2}$$

$\leqslant c$ (car $\beta > \dfrac{1}{2}$ et comme u_m demeure dans un borné de $L^2(0,T;V)$)

Donc pour β quelconque $> \dfrac{1}{2}$

(2.26) $\qquad \int_{-\infty}^{+\infty} \frac{|\tau|}{1 + |\tau|^\beta} |\hat{u}_m(\tau)|^2 \, d\tau \leqslant c$,

on envoi ([1]) .

(2.27) $\forall \varepsilon > 0$, $D_t^{1/4 - \varepsilon} u_m$ demeure dans un borné de $L^2(0,T;H)$.

L'estimation (2.27) suffit pour arriver à la conclusion (2.15).

On a en fait le résultat général suivant (Lemme de compacité):

([1]) Les dérivées "fractionnaires" sont définies par transformation de Fourier.

J. L. Lions

Théorème 2.1. Soient B_o, B, B_1 trois espaces de Hilbert, avec

(2.28) $\qquad B_o \subset B \subset B_1$,

(2.29) $\qquad$ l'injection $B_o \rightarrow B$ est compacte.

Soit E un ensemble de fonctions $z \rightarrow v(t)$ de $[0,T] \rightarrow B_o$ ayant les propriétés suivantes:

(2.30) $\qquad E$ est borné dans $L^2(0,T;B_o)$

(2.31) $\left\{ \begin{array}{l} \text{il existe } \gamma > 0 \text{ tel que} \\[2mm] \displaystyle\int_{-\infty}^{+\infty} |\tau|^2 \, |\hat{v}(\tau)|^2_{B_1} \; d\tau \leqslant \text{constante } \forall v \in E \; . \end{array} \right.$

Alors E est relativement compact dans $L^2(0,T;B)$

Démonstration: cf. Lions [1], Th. 5.2, p. 61 et 59 .

Application à la situation présente:

$$B_o = H_o^1(\Omega) \; , \quad B = L^2(\Omega) \; , \quad B_1 = L^2(\Omega) \; .$$

L'injection de $B_o \rightarrow B$ est compacte d'après le théorème de Rellich-Kondrachoff ([1]). On obtient alors (2.15) (en prenant pour E l'ensemble des fonctions u_m).

Remarque 2.2.

On peut obtenir des estimations directes sur $\dfrac{\partial u_m}{\partial t}$ en uti-

---

([1]) Cf. J. NECAS [1] ou LIONS-E. MAGENES [1]. On suppose Ω borné. Si Ω n'est pas borné, le raisonnement marche encore: il suffit de restreindre les fonctions u_m à un borné quelconque de Ω .

lisant des "<u>bases speciales</u>" pour construire V_m (selon l'Exemple 1.1), Cf. LIONS [1] .

2.3 Théorème.

On est maintenant en mesure de montrer le

<u>Théorème</u> 2.2. <u>On suppose que</u> f <u>est donné avec</u> (2.6) <u>et que</u> u_o <u>est donné dans</u> $L^2(\Omega)$ (=H). <u>Il existe une fonction</u> u <u>et une seule telle que</u>

(2.32) $u \in L^2(0,T;H_0^1(\Omega)) \cap L^\infty(0,T;L^2(\Omega)), \quad u \in L^4(\mathbb{Q})$.

(2.33) $\dfrac{\partial u}{\partial t} - \Delta u + u^3 = f \quad$ dans a ,

(2.34) $u(0) = u_o$.

<u>Remarque</u> 2.3.

Il resulte de (2.33) que

(2.35) $\dfrac{\partial u}{\partial t} = f + \Delta u - u^3 \in L^2(0,T;H^{-1}(\Omega)) + L^{4/3}(\mathbb{Q})$

de sorte que $u(0)$ <u>a un sens</u>.

<u>Demonstration de l'existence dans le théorème 2.2.</u>

On part de u_m. On extrait u_μ avec (2.12)(2.13)(2.17) et aussi

$$u_\mu(T) \longrightarrow \chi \quad \text{dans} \quad H \quad \text{faible.}$$

Soit $v \in V$ quelconque. On choisit v_μ avec

(2.36) $v_\mu \in V_\mu , \quad v_\mu \to v \quad$ dans V .

J. L. Lions

On utilise (2.20) avec $m = \mu$, $v = v_\mu$. On peut passer à la limite; il vient:

$$(2.37) \quad \frac{d}{dt}(\tilde{u},v) + a(\tilde{u},v) + (\tilde{u}^3,v) = (\tilde{f},v) + (u_o,v)\delta(t-0) - (x,v)\delta(t-T).$$

On en deduit (2.33)(2.34) (et ainsi que $\chi = u(T)$).

Démonstration de l'unicité.

Soient u_1 et u_2 deux solutions eventuelles. Alors posant

$$w(t) = u_1(t) - u_2(t) ,$$

on a:

$$(2.38) \quad (\frac{\partial w(t)}{\partial t},v) + a(w(t),v) + (u_1(t)^3 - u_2(t)^3,v) = 0 , \forall v \in V .$$

Prenant $v = w(t)$ et notant la propriété de monotonie

$$(2.39) \qquad (u_1^3 - u_2^3 , u_1 - u_2) \geqslant 0$$

on en deduit que~

$$(2.40) \qquad (\frac{\partial w(t)}{\partial t} , w(t)) + a(w(t)) \leqslant 0 .$$

Utilisant (2.35) on montre que l'on peut intégrer (2.40):

$$|w(t)|^2 + \int_0^t a(w(\sigma))d\sigma \leqslant 0 \qquad (\text{cas } w(0) = 0)$$

d'où $w = 0$.

J. L. Lions

3. Méthode de monotonie.

On reprend la situation du N° 2.1 et l'on va montrer dire-ctement (i.e. sous les estimations du N° 2.2) que u_μ converge vers une (la) solution.

On extrait donc de la suite u_m une suite u_μ telle que l'on ait (2.12)(2.13) et telle que $u_\mu(T) \to \chi$ dans H faible et que

(3.1) $\qquad u_\mu^3 \to \xi$ dans $L^{4/3}(\Omega)$ faible.

Tout revient à montrer que $\xi = u^3$.

Utilisant (2.20) pour $m = \mu$ et $v = v_\mu$ comme en (2.36), on trouve:

(3.2) $\quad \dfrac{d}{dt}(\tilde{u},v) + a(\tilde{u},v) + (\xi,v) = (\tilde{f},v) + (u_o,v)\delta(t-0) - (\chi,v)\delta(t-T)$.

Il en resulte que

$$\frac{\partial u}{\partial t} - \Delta u + \xi = f \ .$$

donc que

$$\frac{\partial u}{\partial t} \in L^2(0,T;H^{-1}(\Omega)) + L^{4/3}(\Omega) \ ,$$

et que

$$u(T) = \chi \ .$$

Si maintenant v_μ est une fonction quelconque de

J. L. Lions

$L^2(0,T;V)$ $L^4(\Omega)$ à valeurs dans V_μ , on introduit:

$$(3.3) \quad X_\mu = \int_0^T a(u_\mu - v_\mu)dt + \int_0^T (u_\mu^3 - v_\mu^3, u_\mu - v_\mu)dt \ .$$

<u>D'après la monotonie de</u> $\lambda \rightarrow \lambda^3$, on a:

$$(3.4) \qquad\qquad X_\mu \geqslant 0 \ .$$

Mais d'après (2.20)

$$\int_0^T [a(u_\mu, u_\mu) + (u_\mu^3, u_\mu)]dt = \int_0^T (f, u_\mu)dt + \frac{1}{2}|u_{o\mu}|^2 - \frac{1}{2}|u_\mu(T)|^2$$

donc

$$(3.5) \quad X_\mu = \int_0^T (f, u_\mu)dt + \frac{1}{2}\lceil u_{o\mu}|^2 - \frac{1}{2}|u_\mu(T)|^2 -$$

$$- \int_0^T \left[a(u_\mu, v_\mu) + (u_\mu^3, v_\mu)\right]dt -$$

$$- \int_0^T \left[a(v_\mu, u_\mu - v_\mu) + (v_\mu^3, u_\mu - v_\mu)\right]dt$$

On peut supposer que $v_\mu \rightarrow v$ dans $L^2(0,T;V) \cap L^4(Q)$;

comme $u_\mu(T) \rightarrow u(T)$ dans H faible, on déduit de (3.5) que

$$(3.6) \begin{cases} \text{lim} - \text{sup.} \ X_\mu \leqslant \int_0^T (f, u)dt + \frac{1}{2}|u_o|^2 - \frac{1}{2}|u(T)|^2 - \\ \\ - \int_0^T [a(u,v) + (\xi, v)]dt - \int_0^T [a(v, u-v) + (v^3, u-v)]dt \ . \end{cases}$$

Mais d'après (3.2)

J. L. Lions

$$\int_0^T (f,u)dt + \frac{1}{2}|u_o|^2 - \frac{1}{2}|u(T)|^2 = \int_0^T \left[a(u) + (\xi,u)\right]dt ,$$

ce qui, joint à (3.6) donne

(3.7)   lim sup. $X_\mu \leq \int_0^T \left[a(u-v) + (\xi-v^3, u-v)\right]dt .$

Grâce à (3.4) on a donc:

(3.8) $\int_0^T \left[a(u-v) + (\xi-v^3, u-v)\right]dt \geqslant 0 .$

On choisit alors, avec MINTY [1] ,

(3.9) $v = u - \lambda w , \quad \lambda > 0 , \quad w \in L^2(0,T;V) \cap L^4(Q) .$

Il vient, après division par λ :

(3.10) $\lambda \int_0^T a(w)dt + \int_0^T (\xi-(u-\lambda w)^3, w)dt \geqslant 0 .$

Faisant $\lambda \to 0$ on en déduit

$$\int_0^T (\xi-u^3, w)dt \geqslant 0 \quad \forall w$$

d'où $\xi = u^3 .$

Remarque 3.1.

Nous avons démontré (de deux manières assez différentes)
que les u_m réalisent une approximation de u (sans extraction

de sous suite, à cause de l'unicité). On a donc obtenu en outre des methodes constructives d'approximation.

4. Indications sur d'autres méthodes.

4.1. Approximation par troncature.

On introduit:

$$(4.1) \quad \beta_M(\lambda) = \begin{cases} \lambda^3 & \text{si } |\lambda| \leqslant M , \\ \pm M^3 & \text{si } \lambda \geqslant M \text{ ou } \lambda \leqslant -M \end{cases}$$

On résout ensuite l'équation approchée (par ex. par appro
ximations successives).

$$(4.2) \quad \frac{\partial u_M}{\partial t} - \Delta u_M + \beta_M(u_M) = f$$

avec

$$(4.3) \quad u_M(x,0) = u_{0M}(x) \quad \text{(troncature de } u_0)$$

et

$$(4.4) \quad u_M = 0 \quad \text{sur } \Sigma .$$

puis on passe à la limite en M (par le même genre de méthode
N° 2 ou 3).

4.2. Approximation par une équation "avec retard".

Pour $\tau > 0$ (destiné à tendre vers 0) on introduit u_τ
solution de

J. L. Lions

$$(4.5) \begin{cases} \dfrac{\partial u_{\tau}(t)}{\partial t} - \Delta u_{\tau}(t) + (u_{\tau}(t-\tau))^2 \, u_{\tau}(t) = f \ , \\[2ex] u_{\tau}(t) = u_0 \quad \text{dans} \quad [-\tau, 0] \end{cases}$$

avec toujours la condition aux limites

$$(4.6) \qquad u_{\tau} = 0 \quad \text{sur} \quad \textstyle\sum .$$

On résout (4.5) de proche en proche dans $(0,\tau), (\tau, 2\tau),\ldots$

puis l'on passe à la limite en τ .

Naturellement il s'agit là d'une variante <u>des méthodes de</u>

<u>déscrétisation en</u> t (N° 1.4).

4.3. <u>Sommes d'opérateur aux dérivées partielles.</u>

4.3.1. <u>Cas linéaire.</u>

L'équation linéaire de la chaleur

$$(4.7) \qquad \frac{\partial u}{\partial t} - \Delta u = f$$

entre dans le cadre général suivant: A et B etant deux opé-

rateur linéaires non borné dans un Banach E , de domaines re-

spectifs D(A) et D(B) , on cherche une solution u dans

D(A) $\cap$ D(B) $(^1)$ de

$$(4.8) \qquad Au + Bu = f \ .$$

Dans le cas (4.7) on prendra:

$(^1)$ Condition généralement trop restrictive qu'il faut "rela-
xer". Cf. GRISVARD [1] .

J. L. Lions

$$A = \partial/\partial t \; , \quad E = L^2(\Omega) \; ,$$

$$D(A) = \left\{ v \mid v, \frac{\partial v}{\partial t} \in L^2(\Omega) \; , \quad v(x,0) = 0 \quad (^1) \right\} \; ,$$

$$B = - \triangle \; ,$$

$$D(B) = \left\{ v \mid v, \frac{\partial v}{\partial x_i} \; , \frac{\partial^2 v}{\partial x_i \partial x_j} \in L^2(Q) \; , \quad v = 0 \quad \text{sur} \quad \textstyle\sum \right\} \; .$$

On peut alors, selon en particulier P. GRISVARD [1], donner des conditions suffisantes portant sur le spectre de A et
de B pour que (4.8) ait une solution dans divers espaces - in-
troduits à l'aide de la théorie de l'interpolation des espaces
de Banach.

4.3.2 Cas non linéaire (I) .

G. DA PRATO [1] à etudié de facon systematique les proprie
tes des sommes d'applications non linéaires. Pour un opérateur
non lineaire A on introduit - lorsque cela à un sens - l'approximation A_n par .

(4.9) $A_n(v) = n(n - A)^{-1} (nv) - nv \quad (^2) \; .$

Sous des hypothèses convenable on résout directement

(^{1}) On resoudra alors le problème avec donnée initiale nulle.
(^{2}) Dérodage: si A est générateur infinitesimal d'un semi
  group, il est classique $\left(\text{YOSIDA [1]} \right)$ d'approcher A par
  $A_n = nA(n-A)^{-1} = n(n-A)^{-1}(A-n+n) = n^2(n-A)^{-1} - n I \; .$

J. L. Lions

$$A_n(u_n) + B(u_n) = f$$

puis l'on passe à la limite.

Voir details et applications dans les travaux de DA PRATO.

4.3.3. Cas non linéaire (II).

Un autre aspect sous lequel on peut aborder - entre autres - l'equation (1), Introduction est celui des semi groupes non li-néaires. Cf. le cours de PAZY.

4.4. Méthode de décomposition.

Revenons à (1), Introduction.

Supposons u calculé à l'instant $n\Delta t$. On considère alors dans l'intervalle $n\Delta t$, $(n+1)\Delta t$:

1°) l'équation linéaire:

$$(4.10) \quad \begin{cases} \dfrac{\partial \varphi}{\partial t} - \Delta \varphi = f \ , \\[2mm] \varphi(n\Delta t) = u(n\Delta t) \\[2mm] \varphi = 0 \quad \text{sur } \Gamma \times \,]\,n\Delta t, \ (n+1)\Delta t\,[\end{cases}$$

ce qui définit $\varphi(n+1)\Delta t)$;

2°) l'equation non linéaire (mais triviale!)

$$(4.11) \quad \begin{cases} \dfrac{\partial \psi}{\partial t} + \psi^3 = 0 \ , \\[2mm] \psi(n\Delta t) = \varphi((n+1)\Delta t) \ ; \end{cases}$$

J. L. Lions

on pose alors

$$u^{n+1/2} = \varphi((n+1) \Delta t) \ ,$$

$$u^{n+1} = \psi((n+1) \Delta t) \ .$$

On montre (cf. TEMAM [1]) que l'on obtient ainsi des appro ximations de la solution u cherchée.

Cette méthode est dite de décomposition, ou de splitting up. Elle joue un rôle important en Analyse Numerique (cf. MARCHOUK [1], YANENKO [1] et TEMAM, loc. cit).

Remarque 4.1.

Dans les problème linéaires, l'idée précédente rejoint la formule de Trotter dans la théorie des semi-groupes. Pour les extensions récentes aux semi-groupes non linéaires, cf. le cours de PAZY [1] .

5. Problèmes hyperboliques.

Nous donnons seulement deux exemples simples. On reviendra plus loin sur des problèmes de ce type.

5.1. Exemple 1.

On cherche u solution de

$$(5.1) \quad \frac{\partial^2 u}{\partial t^2} - \Delta u + \left| \frac{\partial u}{\partial t} \right|^{\rho} \frac{\partial u}{\partial t} = f \qquad (\rho > 0 \quad \text{donné})$$

avec les conditions initiales.

(5.2) $u(x,0) = u_o(x)$, $\dfrac{\partial u}{\partial t}(x,0) = u_1(x)$, $x \in \Omega$.

et la condition aux limites (pour ex.)

(5.3) $u = 0$ sur $\sum$.

5.2. Exemple 2.

On cherche u solution de

(5.4) $\dfrac{\partial^2 u}{\partial t^2} - \Delta u + |u| f_u = f$

avec les conditions (5.2)(5.3).

5.3. Adaptation des méthodes des N° 1 à 3.

Les details sont laissés en exercices - cf. d'ailleurs
LIONS [1] , debut au chap. 1.

L'adaptation de la méthode de Faedo-Galerkin est evidente.

On peut passer à la limite par la méthode de compacite (N° 2)
dans les deux problèmes (5.1) et (5.6).

On ne peut utiliser la méthode de monotonie que dans le cas
(5.1).

On arrive à montrer:

l'existence et l'unicité (d'ailleurs dans plusieurs classes fon
ctionnelles differentes) dans le cas (5.1);

l'existence d'une solution faible dans le cas (5.4), pour laquelle l'unicité n'est pas toujours connue. (On connait l'exi-

stence et l'unicité si $\rho \leq \dfrac{2}{n-2}$, ce resultat pouvant être ame-
lioré selon JORGENS [1] lorsque n=3 et pour le problème de
Cauchy jusqu'à $\rho = 4$).

6. Quelques problèmes.

Problème 6.1.

Peut-on construire une classe fonctionnelle où le problème (5.4)(5.2)(5.3) admette une solution unique (et cela sous la
restriction $\rho \leq \dfrac{2}{n-2}$)?

Problème 6.2.

Considerons l'équation

$$(6.1) \qquad \frac{\partial^2 u}{\partial t^2} - \Delta u + \left|\frac{\partial u}{\partial t}\right|^\rho \frac{\partial u}{\partial t} = f$$

avec les conditions initiales (5.2) et une condition aux limites du type "derivée oblique":

$$(6.2) \quad \begin{cases} \dfrac{\partial u}{\partial n} + Bu = 0 \quad \text{sur} \quad \sum , \\[2mm] B = \text{opérateur differentiel tangentiel du } 1^{er} \text{ ordre sur } \sum . \end{cases}$$

Ce problème est-il bien posé?

Question anologue pour l'équation (5.4).

Dans les cas linéaires, avec la condition (6.2), des résultats positifs sont dûs à CHAZARAIN [1] - utilisant, entre autres,

J. L. Lions

des fonctionnelles de Gevrey (de telle sorte pour l'adaptation aux problèmes non linéaires n'est pas simple...).

J. L. Lions

CHAPITRE 2

MONOTONIE ET CHANGEMENT D'ESPACE PIVOT

Introduction.

Le but de ce Chap. est de montrer comment, en prenant un "espace pivot" non standard (cf. N°3), on peut resoudre simple ment des problèmes non linéaires plus compliqués dans un cadre fonctionnel différent.

On illustrera ainsi sur des exemples simples les deux re~ marques générales suivantes:

(i) la notion de monotonie dépend de l'espace fonctionnel choisi;

(ii) le choix du cadre fonctionnel joue dans les problè- mes non linéaires un rôle absolument essentiel(et, en particulier, beaucoup plus important que dans les pro blèmes linéaires).

1. Resultat "abstrait".

La méthode donnée aux N° 1 et 3 du Chap. 1 permet de mon- trer, exactement de la même manière, le Th. 1.1 ci après ([1]).

_____

([1]) Dont on trouvera également une démonstration complete par ex. au debut du Chap. 2 de LIONS [1] .

J. L. Lions

On donne: <u>un espace de Hilbert</u> H dit "espace pivot"; cet espace est <u>identifié à son dual</u> ([1]); le produit scalaire dans H est noté (f,g) et l'on pose

(1.1) $|f| = (f,f)^{1/2}$.

On donne ensuite <u>un espace de Banach réflexif</u> V , de norme notée $\| \ \|$. On suppose que

(1.2) $V \subset H$, V dense dans H, l'injection $V \to H$ étant continue.

Soit V' le dual de V que grâce à (1.2) l'on peut identifier à un sur-espace de H :

(1.3) $V \subset H \subset V'$;

On désigne pas $\| \ \|_*$ la norme dans V' , duale de la norme de V , et par (f,v) le produit scalaire entre $f \in V'$ et $v \in V$.

On suppose (cela n'est pas essentiel) V séparable.

Soit maintenant A un opérateur (non linéaire) de $V \to V'$ ayant les propriétés suivantes ([2]):

([1]) <u>Très généralement</u> H est un espace L^2 (ou un produit de tels espaces); on donnera au N° 3 des exemples où il en est autrement.

([2]) Tout cela peut être assez considérablement généralisé. Cf. BREZIS [1], BROWDER [1], LERAY-LIONS [1], LIONS [2] .

J. L. Lions

(1.4) A transform les bornés de V en bornés de V' , ([1])

(1.5) $\forall u,v,w \in V$, la fonction

$\lambda \longrightarrow (A(u+\lambda v),w)$ est continue sur $\mathbb{R}$, ([2])

(1.6) (<u>monotonie</u>) $\forall u,v \in V$, $(A(u)-A(v),u-v) \geqslant 0$,

(1.7) (coercivité) $(A(v),v) \geqslant \alpha \|v\|^p, \alpha > 0$, $p > 1, \forall v \in V$.

<u>L'exemple "modèle"</u>.

Pour $p > 1$, fini, on introduit les <u>espaces de Sobolev</u>:

(1.8) $W^{1,p}(\Omega) = \left\{ v \mid v, \dfrac{\partial v}{\partial x_i} \in L^p(\Omega),\ i = 1,\ldots,n \right\}$,

(1.9) $W_0^{1,p}(\Omega) = \left\{ v \mid v \in W^{1,p}(\Omega),\ v = 0 \ \text{sur}\ \Gamma \right\}$;

$W^{1,p}(\Omega)$ est un espace de Banach separable et reflexif(il est

même uniformément convexe) pour la norme

$$\|v\| = \left(\int_\Omega (|v|^p + \sum_{i=1}^n \left| \dfrac{\partial v}{\partial x_i} \right|^p) dx \right)^{1/p} \ .$$

On prend alors

$V = W_0^{1,p}(\Omega)$, $H = L^2(\Omega)$ et l'on a bien $V \subset H$ si

l'on suppose

([1]) On dit que A est <u>borné</u>

([2]) On dit que A est <u>biensi continu</u>.

J. L. Lions

(1.10) $\qquad$ $p > 2$ $(^1)$.

Alors $(^2)$

(1.11) $\qquad$ $V^1 = W^{-1,P'}(\Omega)$, $\dfrac{1}{p} + \dfrac{1}{p'} = 1$.

On definit alors A par

(1.12) $A(v) = -\displaystyle\sum_{i=1}^{n} \dfrac{\partial}{\partial x_i} \left(\left| \dfrac{\partial v}{\partial x_i} \right|^{P-2} \dfrac{\partial v}{\partial x_i} \right)$ (on dit parfois que A

est le "pseudo-Laplacien").

On verifie que (1.4)...(1.7) ont lieu.

Cela posé, on a le

Théorème 1.1. On donne f et u_o avec

(1.13) $f \in L^{P'}(0,T;V')$,

(1.14) $u_o \in H$.

Il existe alors une fonction u et une seule telle que

(1.15) $\qquad$ $u \in L^P(0,T;V)$,

(1.16) $\qquad$ $u' = \dfrac{\partial u}{\partial t} \in L^{P'}(0,T;V')$,

---

$(^1)$ Les cas $p = 2$ correspond aux problèmes linéaires. Le théo reme de plongement de Sobolev montre que $V \subset H$ si (Ω étant borné) $p \geqslant \dfrac{2n}{n+2}$. Nous n'insistons pas sur ce point car de toutes facons on peut s'affranchir de l'hypothèse "$V \subset H$".

$(^2)$ C'est en fait la définition de $W^{-1,P'}(\Omega)$!

J. L. Lions

(1.17) $\qquad \dfrac{\partial u}{\partial t} + A(u) = f \quad$ dans $\quad Q$,

(1.18) $\qquad u(0) = u_o$.

Appliqué à l'Exemple "modèle", on obtient l'existence et l'unicité de u solution de

$$(1.19)\begin{cases} \dfrac{\partial u}{\partial t} - \displaystyle\sum_{i=1}^{n} \dfrac{\partial}{\partial x_i} \left(\left| \dfrac{\partial u}{\partial x_i} \right|^{p-2} \dfrac{\partial u}{\partial x_i} \right) = f \\[4mm] u = 0 \quad \text{sur} \quad \textstyle\sum \; , \\[3mm] u(x,0) = u_o(x) \; , \; x \in \Omega \; . \end{cases}$$

Orientation.

On va maintenat appliquer le Théorème 1.1. à des situations plus "sophisticées"...

2. Changement d'espace pivot.

2.1. Position des problèmes. Opérateur parabolique dégénéré.

Soit p donné avec

(2.1) $\qquad p \geqslant 2$.

On cherche une fonction $u = u(x,t)$ solution de (^{1})

_____

(^{1}) Lorsqu'on obtiendra des solutions très faibles(dans $L^p(Q)$) de (2.2) il faudra comprendre (2.2) au sens:
$$\dfrac{\partial u}{\partial t} - \dfrac{1}{(p-1)} \Delta \left(|u|^{p-2}u \right) + |u|^{p-2}u = f_o \; .$$

J. L. Lions

(2.2) $\quad \dfrac{\partial u}{\partial t} - \sum\limits_{i=1}^{n} \dfrac{\partial}{\partial x_i} \left(|u|^{P-2} \dfrac{\partial u}{\partial x_i} \right) + |u|^{P-2} u = f_0 \quad$ dans $Q = \Omega \times]0,T[$,

avec la condition initiale

(2.3) $\qquad u(x,0) = u_0(x)$

et l'une des conditions aux limites suivantes:

(2.4) Condition du type "Dirichlet": $|u|^{P-2} u = g$ donné sur Σ $(^1)$,

ou

(2.5) Condition du type "Neumann" $\dfrac{\partial}{\partial n} (|u|^{P-2} u) = g$ sur Σ.

L'opérateur intervenant dans le 1er membre de (2.2) est parabolique, dégénéré (lorsque $u = 0$).

2.2. Résolution du problème avec condition du type "Neumann".

On applique le Théorème 1.1. dans les conditions suivantes.

Espace $H = (H^1(\Omega))!$

Pour $u, v \in H^1(\Omega)$, on pose

(2.6) $\quad b(u,v) = \dfrac{1}{(P-1)} \displaystyle\int_\Omega \sum\limits_{i=1}^{n} \dfrac{\partial u}{\partial x_i} \dfrac{\partial v}{\partial x_i} dx + \int_\Omega u\, v\, dx ,$

forme bilineaire continue coercive sur $H^1(\Omega)$. Donc

(2.7) $\qquad b(u,v) = \langle Bu , v \rangle \quad , \quad Bu \in (H^1(\Omega))'$

$(^1)$ Ce qui évidemment équivaut à se donner u sur Σ.

où $< , >$ désigne le produit scalaire entre $H^1(\Omega)$ et $(H^1(\Omega))!$, B est <u>un isomorphisme de</u> $H^1(\Omega)$ <u>sur</u> $(H^1(\Omega))!$

Donc <u>en particulier</u> si $v \in L^p(\Omega)$, il existe u unique dans $H^1(\Omega)$ tel que

$$(2.8) \qquad b(u,v) = \int_\Omega w \, v \, dx \qquad \forall v \in H^1(\Omega) \ .$$

On interprète (2.8) comme suit: d'abord prenant $v \in D(\Omega)$ (fonctions C^∞ à support compact dans Ω) on trouve que

$$(2.9) \qquad - \frac{1}{(p-1)} \Delta u + u = w \quad \text{dans } \Omega \ ;$$

puis (^{1}) en utilisant la formule de Green (2.9) donne

$$- \frac{1}{(p-1)} \int_\Gamma \frac{\partial u}{\partial n} v \, d\Gamma + b(u,v) = \int_\Omega w \, v \, dx$$

d'où

$$(2.10) \qquad \frac{\partial u}{\partial n} = 0 \ .$$

Pour $f, g \in (H^1(\Omega))'$ on posera

$$(2.11) \quad \begin{cases} (f,g) = (\text{produit scalaire de } f \text{ et } g \text{ dans } H) = \\ \qquad = \langle B^{-1} f, g \rangle = \langle f, B^{-1} g \rangle \ . \end{cases}$$

<u>Espace</u> V .

On prend

(^{1}) Justification dans LIONS-MAGENES [1], Vol. 1, Chap. 2 .

J. L. Lions

(2.12) $\qquad V = L^P(\Omega)$.

On a l'identification

(2.13) $\qquad V \subset H$

(si $v \in V$ alors il définit $\tilde{v} \in H$ par

$$< \tilde{v}, \varphi > = \int_\Omega v \varphi \, dx \qquad \forall \varphi \subset H^1(\Omega)$$

et l'application linéaire $v \to v$ étant biunivoque, on identifie $\tilde{v}$ à v).

Opérateur A.

Pour $u, v \in V$ on pose

(2.14) $\qquad a(u,v) = \int_\Omega |u|^{P-2} u \, v \, dx$.

La forme $v \to a(u,v)$ est linéaire contenu sur V donc
s'écrit

(2.15) $\qquad a(u,v) = (A(u),v)$, $A(u) \in V'$.

L'opérateur A ainsi défini de $V \to V'$ vérifie (1.4)...
(1.7).

Fonction f avec (1.13).

On fait les hypothèses suivantes ([1]):

([1]) qui peuvent être notablement étendues.

J. L. Lions

$$(2.16) \qquad f_o \in L^{p'}(\Omega) \; ,$$

$$(2.17) \qquad g \in L^{R'}(\textstyle\sum) \; .$$

On <u>définit</u> $f(t) \in V'$ par

$$(2.18) \quad \begin{cases} (f(t),v) = \displaystyle\int_\Omega f_o(x,t)B^{-1}v \, dx \; + \; \dfrac{1}{(p-1)} \int_\Gamma g(x,t)B^{-1}v \, d\Gamma \, , \\[2ex] v \in V \; . \end{cases}$$

On note que B^{-1} (operateur de Green du problème de Neumann)

envoie $L^p(\Omega)$ dans $W^{2,p}(\Omega)$ (cf. AGMON-DOUGLIS-NIRENBERG [1])

donc <u>en particulier</u> $B^{-1}v \in L^p(\Omega)$ et $B^{-1}v|_\Gamma \in L^p(\Gamma)$ et $(^1)$

$$(2.19) \; |(f(t),v)| \leqslant c(\|f_o(t)\|_{L^{p'}(\Omega)} + \|g(t)\|_{L^{p'}(\Gamma)}) \, \|v\|$$

ce qui avec (2.16)(2.17) montre que f satisfait à (1.13).

<u>Fonction</u> u_o .

On suppose que

$$(2.20) \qquad u_o \in L^2(\Omega) \; .$$

Alors u_o s'identifie à un élément de H .

On peut appliquer le Théorème 1.1. <u>On a existence et unici-</u>
<u>té de</u> u <u>avec</u> (1.15)...(1.18).

$(^1) \quad \|v\| = \|v\|_{L^p(\Omega)}$.

J. L. Lions

Interprétations (1.17).

L'équation (1.17) équivaut à

(2.21) $\quad (\frac{\partial u}{\partial t}, v) + (A(u),v) = (f,v) \qquad \forall v \in V$

ou encore en utilisant (2.11),(2.15) et (2.18):

$$(2.22) \begin{cases} \langle \frac{\partial u}{\partial t}, B^{-1}v \rangle + a(u,v) = \int_{\Omega} f_o B^{-1}v \, dx + \int_{\Gamma} \frac{1}{(p-1)} g(B^{-1}v)d\,\Gamma, \\ \forall v \in V . \end{cases}$$

Si l'on pose

(2.23) $\qquad\qquad B^{-1}v = \varphi$

On note que φ parcourt l'espace des fonctions telles que
(cf.(2.10))

(2.24) $\quad \varphi \in W^{2,P}(\Omega) \quad , \quad \frac{\partial \varphi}{\partial n} = 0$;

(2.22) s'ecrit alors:

(2.25) $\langle \frac{\partial u}{\partial t}, \varphi \rangle + \int_{\Omega} |u|^{P-2}u(B\varphi)dx = \int_{\Omega} f_o \varphi dx + \int_{\Gamma} \frac{1}{(p-1)} g \varphi \, d\,\Gamma.$

Prenant en particulier $\varphi \in D(\Omega)$ on trouve que u satisfait à (2.2), équivalent à

(2.26) $\quad \frac{\partial u}{\partial t} - \frac{1}{(p-1)} \Delta (|u|^{P-2}u) + |u|^{P-2}u = f_o$.

Effectuons maintenant une intégration par parties formelle, on déduit de (2.26), pour φ avec (2.24):

J. L. Lions

$$\langle \frac{\partial u}{\partial t}, \varphi \rangle - \frac{1}{(p-1)} \int_\Gamma \frac{\partial}{\partial n} (|u|^{p-2} u) \varphi \, d\Gamma + \int_\Omega |u|^{p-2} u \, B\varphi \, dx =$$

$$= \int_\Omega f_o \varphi \, dx$$

d'où comparant à (2.25):

$$(2.27) \quad \int_\Gamma \frac{\partial}{\partial n} (|u|^{p-2} u) \varphi \, d\Gamma = \int_\Gamma g \, \varphi \, d\Gamma \quad \forall \varphi \text{ avec } (2.24)$$

donc on a (2.5).

En résumé: on a obtenu l'existence et l'unicité d'une fonction u satisfaisant à

$$(2.28) \qquad u \in L^p(Q) ,$$

solution "faible" (au sens (2.25)) de (2.2)(2.3)(2.5).

Remarque 2.1.

On peut donner un sens à $\frac{\partial}{\partial n} (|u|^{p-2} u)$ avec les méthodes
de LIONS-MAGENES [1], Vol. 2, Chap. 4.

Remarque 2.2.

Nous avons utilisé la methode de monotonie. Changeant de
cadre fonctionnel on peut utiliser la methode de compacité.
On cherche alors une fonction u telle que

$$(2.29) \qquad u \in L^\infty(0,T; L^2(\Omega)) ,$$

$$(2.30) \qquad |u|^{\frac{p-2}{2}} \frac{\partial u}{\partial x_i} \in L^2(Q) \quad \forall i \ ,$$

$$(2.31) \begin{cases} \int_\Omega \frac{\partial u}{\partial t} v \, dx + \sum_{i=1}^{n} \int_\Omega |u|^{p-2} \frac{\partial u}{\partial x_i} \frac{\partial v}{\partial x_i} \, dx = \int_\Omega f_o(x,t)v \, dx + \\ \qquad\qquad\qquad\qquad + \frac{1}{(p-1)} \int_\Gamma g v \, d\Gamma \ , \end{cases}$$

(2.31) ayant lieu $\forall v$ tel que $v \in L^2(\Omega)$, $|v|^{\frac{p-2}{2}} \frac{\partial v}{\partial x_i} \in L^2(\Omega)$,

avec la condition initiale (2.3).

On peut montrer l'existence et l'unicité $(^1)$ d'une solu-

tion (cf. LIONS $[1]$, Chap. 1 ; cf. également le Chap. 3 ci après).

Remarque 2.3.

La situation précédente amène à deux remarques importan-

tes:

(i) le problème étant résolu dans deux classes fonctionnelles

F_1 e F_2 (sous de hypothèses differentes sur les données),

peut-on interpoler "entre" ces situations?

C'est le probleme de l'interpolation non linéaire (cf.

BROWDER 2   J. PEETRE 1   , LIONS 3   , L. TARTAR 1   );

$(^1)$ L'unicité est d'ailleurs consequence de ce qu'on vient de
voir, jusqu'on a maintenant une solution dans une classe fon
ctionnelle plus petite.

(ii) comme il est assez naturel dans les problèmes non li
néaires (^{1}), on cherche les solutions dans une classe fonctionnelle non linéaire. L'un des problèmes de base est alors d'o-
btenir des théorèmes de traces non linéaires; cf. L. TARTAR
[2] .

2.3. Resolution du problème avec condition du type "Dirichlet".

Des considérations très voisines des précédentes sont valables pour le problème avec condition du type "Dirichlet".
Donnons les briévement (^{2}). On prend

$$H = H^{-1}(\Omega) \; ;$$

la forme $b(u,v)$ est continue coercive sur $H_o^1(\Omega)$ donc

(2.32) $b(u,v) = \langle\!\langle B_o u,\ v \rangle\!\rangle$, $B_o u \in H^{-1}(\Omega) = (H_o^1(\Omega))'$,

où $\langle\!\langle , \rangle\!\rangle$ designe le produit scalaire entre $H_o^1(\Omega)$ et $H^{-1}(\Omega)$.

On prend encore $V = L^p(\Omega)$, $a(u,v)$ défini par (2.14);
alors

(2.33) $\begin{cases} a(u,v) = (A(u),v) , A(u) \in V' = \text{dual de } V \text{ lorsque} \\ H \text{ est identifié à son dual } (^3) \end{cases}$

(^{1}) Mais les exemples effectif ne semblent pas abonder jusqu'ici.
(^{2}) Cf. LIONS [1], Chap. 2, N° 3.2.
(^{3}) Donc V' et A sont différents de ceux du N° 2.2.

J. L. Lions

On définit (comparer à (2.18)) $f(t)$ par

$$(2.34) \quad (f(t),v) = \int_\Omega f_o(x,t)\, B_o^{-1} v\, dx - \frac{1}{(p-1)} \int_\Gamma g\, \frac{\partial}{\partial n}(B_o^{-1}v)d\Gamma;$$

on verifie encore que si f_o et g satisfait à (2.16)(2.17) on a (1.13).

On peut donc appliquer le Théorème 1.1 ; on arrive aussi à l'existence et l'unicité de $u \in L^p(Q)$ telle que, en posant

$$(2.35) \qquad\qquad B_o^{-1}v = \psi :$$

$$(2.36) \quad \begin{cases} \left\langle\!\!\left\langle \dfrac{\partial u}{\partial t},\psi \right\rangle\!\!\right\rangle + \displaystyle\int_\Omega |u|^{p-2} u(B_o\psi)dx = \int_\Omega f_o\psi\, dx - \int_\Gamma \dfrac{g}{(p-1)}\dfrac{\partial\psi}{\partial n}\,d\Gamma, \\[2mm] \forall \psi \in W^{2,p}(\Omega), \psi = 0 \quad \text{sur} \quad \Gamma, \end{cases}$$

et $u(0) = u_o$.

L'interprétation formelle de (2.36) conduit à (2.2)(2.3) (2.4).

Remarque 2.4.

On a encore des Remarques analogues à 2.1, 2.2, 2.3.

Remarque 2.5.

Prenons $g = 0$ dans chacun des deux problèmes précedents.

Soit u_o(risp. u_1) la solution faible du problème de Dirichlet (risp. Neumann). Donc, en un sens faible:

i.e. $\qquad |u_o|^{p-2}u_o = 0 \quad \text{sur} \quad \sum$

J. L. Lions

(2.37) $u_o = 0$ sur Σ

et

(2.38) $\dfrac{\partial}{\partial n}(|u_1|^{p-2} u_1) = 0$ sur Σ .

<u>Formellement</u> (2.37) implique $|u_o|^{p-2} \dfrac{\partial u_o}{\partial n} = 0$ soit

(2.39) $\dfrac{\partial}{\partial n}(|u_o|^{p-2} u_o) = 0$ sur Σ

<u>mais cela n'est pas justifié en général</u>, comme montre le contre

exemple suivant: partons d'une fonction u régulière dans $\bar{Q}$,

non nulle sur Σ pour $t > 0$ et nulle sur Γ pour $t = 0$, te<u>l</u>

le que:

(2.40) $\dfrac{\partial u}{\partial n} = 0$ sur Σ .

<u>Calculons</u> f_o <u>et</u> u_o par

(2.41) $\dfrac{\partial u}{\partial t} - \displaystyle\sum_{i=1}^{n} \dfrac{\partial}{\partial x_i} \left(|u|^{p-2} \dfrac{\partial u}{\partial x_i}\right) + |u|^{p-2} u = f_o$,

(2.42) $u(x,0) = u_o(x)$.

Resolvons ensuite le problème de Dirichlet (resp. de Neu-

mann) relatif aux données f_o, u_o et avec $g = 0$. La solution

du problème de Neumann est évidentemment u . Celle de proble-

me de Dirichlet, soit u_* , est <u>distincte</u> de u car u n'est pas nul sur $\sum$. Si (2.39) était justifié on aurait

$$\frac{\partial}{\partial n}(|u_*|^{P-2}u_*) = 0 \text{ sur } \sum , \text{ donc } u_* \text{ serait solution du pro}$$

blème de Neumann, ce qui n'est pas.

Notons encore que partant de u arbitrairement régulière

on peut avoir - au moins si p est un entier pair - f_0 et u_0

arbitrairement réguliers et néanmoins la solution du problème de

Dirichlet non c^1 dans $\bar{Q}$ car sinon (2.39) serait justifié ce

qui conduit, comme on vient de le voir, à une contradiction: il

y a généralement dans les problèmes non linéaires un "seuil de

régularité" pour la solution, que l'on ne peut dépasser quelle

que soit la régularité en général des données.

3. Problèmes.

Quelle est la situation pour l'équation

$$(3.1) \qquad \frac{\partial^2 u}{\partial t^2} - \sum_{i=1}^{n} \frac{\partial}{\partial x_i}(|u|^{P-2}\frac{\partial u}{\partial x_i}) = f_0$$

avec les conditions aux limites du type Dirichlet ou Neumann et

les données initiales $u(x,0), \frac{\partial u}{\partial t}(x,0)$?

J. L. Lions

CHAPITRE 3

INEQUATIONS VARIATIONNELLES POUR DES PROBLÈMES

PARABOLIQUES DÉGÉNÉRÉS

Introduction.

Notre objet principal dans ce chapitre est l'étude du pro_

blème suivànt (qui intervient dans l'étude de l'hydrodynamique

des milieux pareux; cf. DUVAUT-LIONS [1]): trouver dans $Q =$

$= \Omega \times]0,T[$ une fonction u telle que

(1) $\quad \dfrac{\partial u}{\partial t} - \displaystyle\sum_{i=1}^{n} \dfrac{\partial}{\partial x_i} \left(u \dfrac{\partial u}{\partial x_i} \right) = f_o \quad$ dans Q ,

(2) $\quad u(x,0) = u_o(x) \quad$ dans Ω

avec la condition aux limites

(3) $\quad u \dfrac{\partial u}{\partial n} + \phi(u) = g \quad$ sur $\sum$ $\quad$ (g donnée sur $\sum$)

où ϕ est une fonction dont le graphe est indiqué Fig. 1

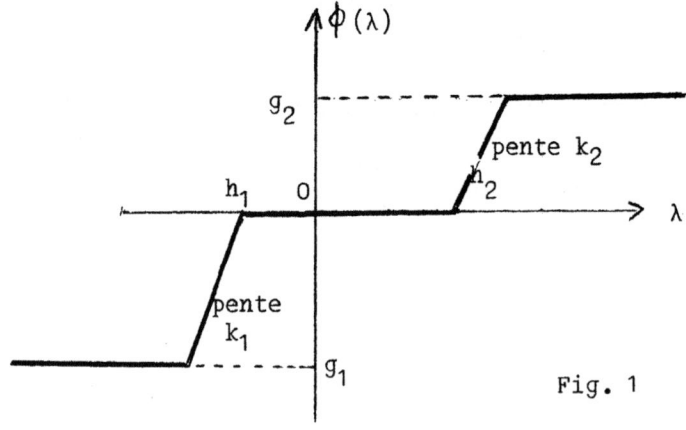

Fig. 1

J. L. Lions

Il se pose également le problème de faire tendre (cf. notations sur la Fig. 1) k_1 et k_2 vers $+ \infty$ - ce qui necessite une nouvelle formulation des problèmes: les inéquations variationnelles.

Dans un premier temps on remplace le premier membre de (1) par

$$\frac{\partial u}{\partial t} - \sum_{i=1}^{n} \frac{\partial}{\partial x_i} \left(|u| \frac{\partial u}{\partial x_i} \right)$$

puis l'on montre que, sous des conditions réalisées dans le pratique, u est $\geqslant 0$, de sorte que l'on retrouve (1).

On utilise des méthodes de compacité nécessitant un nouveau théorème, donné au N° 1.

On suppose dans tout ce Chapitre que Ω est un ouvert borné.

1. Un résultat de comparité.

Le résultat donné ci après étend un résultat de DOUBINSKI [1] extension d'un résultat d'AUBIN [1]. Si le résultat semble nouveau, les méthodes ne le sont pas. (cf. par ex. LIONS [1] Chap. 1 N° 5 et N° 12.2).

J. L. Lions

1.1. Hypothèses. Enoncé du Théorème.

On donne:

(1.1) deux espaces de Banach B_1 et F $(^1)$,

(1.2) un cône $S \subset B_1$ de sommet $\{0\}$ (donc $v \in S \Rightarrow \lambda v \in S$ $\forall \lambda \geqslant 0$),

(1.3) une fonction $v \to M(v)$ de $S \to \mathbb{R}_+$ $(M(v) \geqslant 0)$ telle que

$$M(\lambda v) = \lambda M(v) \quad \forall \lambda \geqslant 0 \; , \quad \forall v \in S \; ,$$

(1.4) un opérateur π de $S \to F$ tel que

$$\pi(\lambda v) = \lambda \pi(v) \quad \forall v \in S \; , \quad \forall \lambda \geqslant 0 \; ,$$

et tel que

$$\|\pi(v)\|_F \leqslant c \, M(v)^\theta \; , \quad 0 < \theta \leqslant 1 \; .$$

Définissons alors

(1.5) $S_1 = \left\{ v \mid v \in S \; , \quad M(v) \leqslant 1 \right\}$.

On suppose que

(1.6) S_1 est relativement compact dans B_1 ,

(1.7) $\pi(S_1)$ est relativement compact dans F ,

(1.8) si u_n , $v_n \in S_1$ avec $u_n - v_n \to 0$ dans B_1 et

$$\pi(u_n) - \pi(v_n) \longrightarrow \varphi \text{ dans } F \text{ alors } \varphi = 0 \; .$$

$(^1)$ De façon générale $\| \; \|_X$ désigne la norme dans X .

J. L. Lions

On a alors le

Théorème 1.1. On suppose que (1.1)...(1.8) ont lieu. On définit $\widetilde{\mathcal{F}}$ par:

(1.9) $\widetilde{\mathcal{F}} = \left\{ v \mid v \text{ localement } L^1 \text{ sur } [0,T] \text{ à valeurs dans } B_1 \text{ telle que} \right.$

$v(t) \in S \quad \text{p.p. avec} \quad \displaystyle\int_0^T M(v(t))^{p_0} dt \leq c_0 \ ,$

$\dfrac{\partial v}{\partial t} \in L^{p_1}(0,T;B_1) \ , \quad \left\| \dfrac{\partial v}{\partial t} \right\|_{L^{p_1}(0,T;B_1)} \leq c_1 \left. \right\} .$

(les p_i sont donnés avec $1 < p_i < \infty$).

Alors l'ensemble $\pi(\widetilde{\mathcal{F}})$ est relativement compact dans $L^{p_0}(0,T;F)$.

Avant la Demonstration, un exemple péut être utile....

1.2. Exemple.

On prendra

(1.10) $\quad B_1 = (H^r(\Omega))'$ ($r > 1$ fixé plus loin),

(1.11) $\quad F = L^3(\Gamma)$,

(1.12) $\quad S = \left\{ v \mid |v|^{1/2} v \in H^1(\Omega) \right\}$,

(1.13) $\quad M(v) = \left\| \, |v|^{1/2} v \right\|_{H^1(\Omega)}^{2/3}$,

(1.14) $\quad \pi v = \text{trace de } v \text{ sur } \Gamma = \gamma v , \quad v \in S .$

Comme $|v|^{1/2} v \in H^1(\Omega)$ on a: $\gamma(|v|^{1/2} v) \in L^2(\Gamma)$ (en par-

J. L. Lions

ticulier) donc $\gamma v \in L^3(\Gamma)$ et $\|\gamma(|v|^{1/2}v)\|_{L^2(\Gamma)} \leqslant c \||v|^{1/2}v\|_{H^1(\Omega)}$

donne

(1.15) $\|\pi v\|_{L^3(\Gamma)} \leqslant c \, M(v)$.

On a maintenant:

(1.16) $S_1 = \left\{ v \mid |v|^{1/2}v \in H^1(\Omega), \; \||v|^{1/2}v\|_{H^1(\Omega)} \leqslant 1 \right\}$.

Mais $|v|^{1/2}v \in H^1(\Omega) \Longrightarrow |v|^{1/2}v \in L^2(\Omega) \Longleftrightarrow v \in L^3(\Omega)$; on

sait (LIONS [1] , Prop. 12.1 p. 143), que

(1.17) S_1 est relativement compact dans $L^3(\Omega)$

donc a fortiori (1.6) a. t. il lieu.

Par ailleurs l'application "trace sur Γ " est <u>compacte</u> de

$H^1(\Omega) \longrightarrow L^2(\Gamma)$; donc si $|v|^{1/2}v \in$ borné de $H^1(\Omega)$ il en ré

sulte que $\gamma(|v|^{1/2}v) \in$ relativement compact de $L^2(\Gamma)$ donc

(1.7) a lieu.

<u>Vérifions</u> (1.8).

Si u_n , $v_n \in S_1$ on peut extraire des sous suites u_μ, v_μ

telle que $|u_\mu|^{1/2}u_\mu \to |u|^{1/2}u$, $|v_\mu|^{1/2}v_\mu \to |v|^{1/2}v$ dans

$H^1(\Omega)$ faible; alors $\gamma(|u_\mu|^{1/2}u_\mu) \longrightarrow \gamma(|u|^{1/2}u)$ dans $L^2(\Gamma)$

fort donc $\gamma u_\mu \to \gamma u$ (et des même $\gamma v_\mu \to \gamma v$) dans $L^3(\Gamma)$ fort.

Par ailleurs $u_\mu \to u$ et $v_\mu \to v$ dans $L^2(\Omega)$ (par ex.)
et comme $u_\mu - v_\mu \to 0$ dans B_1 on a: $u - v = 0$ donc

$\gamma u_\mu - \gamma v_\mu \to \varphi$ (par hypothèse) et $\to 0$, dans $L^3(\Gamma)$ donc

$\varphi = 0$.

<u>Le Théorème</u> 1.1 <u>est donc applicable.</u>

1.3. <u>Cas particuliers.</u>

<u>Cas particulier "linéaire"</u> (AUBIN [1]).

On donne trois <u>espaces</u> de Banach:

(1.18) $\qquad B_o \subset B \subset B_1$,

et l'on prend:

$$S = B_o \quad , \quad M(v) = \|v\|_{B_o} \quad ,$$
$$F = B \quad , \quad \pi = \text{injection}$$

et l'on suppose l'injection $B_o \to B$ compacte.

<u>Cas particulier "non linéaire"</u> (DOUBINSKI [1]).

On donne deux espaces B , B_1 et l'ensemble S avec

(1.19) $\qquad S \subset B \subset B_1$.

On donne M avec (1.3). On prend

$$F = B \quad , \quad \pi = \text{injection} .$$

On suppose que S_1 est relativement compact dans B ce qui equivaut~à (1.7) et, dans ce cas, implique (1.6).

1.4. Démonstration du Théorème 1.1.

On utilise de façon essentielle le

Lemme 1.1. **Hypothèse du Théorème** 1.1. $\forall \eta > 0$ il existe c_η telle que

$$(1.8) \quad \| \pi(u) - \pi(v) \|_F \leqslant \eta (M(u) + M(v)) + c_\eta \, \| u - v \|_{B_1}$$
$$\forall u, \, v \in S .$$

Démonstration.

Par l'absurde. Si ce n'etait pas vrai, il existerait $\eta_0 > 0$ et des suites $u_n, v_n \in S$ telles que

$$(1.9) \quad \| \pi(u_n) - \pi(v_n) \|_F \geqslant \eta_0 (M(u_n) + M(v_n)) + n \, \| u_n - v_n \|_{B_1} .$$

On introduit alors

$$(1.10) \quad \widetilde{u}_n = \frac{u_n}{M(u_n) + M(v_n)} \, , \, \widetilde{v}_n = \frac{v_n}{M(u_n) + M(v_n)} \, ; \, \widetilde{u}_n, \widetilde{v}_n \in S_1 .$$

On a alors:

$$(1.11) \quad \| \pi(\widetilde{u}_n) - \pi(\widetilde{v}_n) \|_F \geqslant \eta_0 + n \, \| \widetilde{u}_n - \widetilde{v}_n \|_{B_1} .$$

D'après (1.7) on peut supposer - par extraction éventuelle de sous-suites - que $\pi(\widetilde{u}_n)$, $\pi(\widetilde{v}_n)$ convergent dans F donc

$$\pi(\tilde{u}_n) - \pi(\tilde{v}_n) \longrightarrow \varphi \text{ dans } F \ .$$

Mais d'après (1.11) $\tilde{u}_n - \tilde{v}_n \to 0$ dans B_1 et donc d'après (1.8), $\varphi = 0$ et donc $\pi(\tilde{u}_n) - \pi(\tilde{v}_n) \to 0$ dans F - ce qui est contradictoire à (1.11).

Passons à la démonstration du Théorème 1.1. Soit u_n une suite de $\tilde{\mathcal{F}}$. D'après le Lemme 1.1 on a:

$$\int_0^T \|\pi(u_{n+m}) - \pi(u_n)\|_F^{P_o} dt \ \leqslant$$

$$\leqslant \eta \int_0^T \left[M(u_{n+m})^{P_o} + M(u_n)^{P_o} \right] dt + c_\eta \int_0^T \|u_{n+m} - u_n\|_{B_1}^{P_o} dt$$

$$\leqslant 2c_o \eta + c_\eta \int_0^T \|u_{n+m} - u_n\|_{B_1}^{P_o} dt \ .$$

On aura donc le résultat si l'on montre que u_η appartient à un ensemble relativement compact de $L_-^{P_o}(0,T;B_1)$; on va en fait démontrer plus, à savoir:

(1.12) $u_\eta \in$ ensemble relativement de $C^o([0,T]; B_1)$.

On observe d'abord que, pour presque tout t (disons pour $t \notin Z$, Z ensemble de mesure nulle) on peut extraire un sous-suite dépendant de t telle que

(1.13) $M(u_k(t)) \leqslant K_t < \infty$;

si non il existerait $E \subset [0,T]$, mesure $E > 0$ avec

$$M(u_n(t))^{P_0} \to + \infty \qquad \forall t \in E ,$$

et alors $\displaystyle\int_0^T M(u_n(t))^{P_0} dt \geqslant \int_E M(u_n(t))^{P_0} dt \to + \infty$, ce qui

est impossible si $u_n \in \widetilde{\mathcal{F}}$.

On peut alors, d'après (1.6), extraire pour $t \notin Z$ une

sous-suite dépendant de t telle que

(1.14) $\qquad u_k(t) \longrightarrow u(t)$ dans B_1 fort .

Soit alors $t_1, t_2 \ldots$ une suite dense dans $[0,T]$, $t_i \notin Z$.

D'après (1.14) et le procédé diagonal, on peut extraire une

suite u_μ telle que

(1.15) $\qquad u_\mu(t_i) \to u(t_i)$ dans B_1 fort, $\forall i$

Mais $\forall t \in [0,T]$ on a :

$$\|u_\mu(t_i) - u_\mu(t)\|_{B_1} = \left\| \int_t^{t_i} u'_\mu(\sigma) \right\|_{B_1} \leqslant$$

$$\leqslant |t_i - t|^{1/P'_1} \left(\int_t^{t_i} \|u'_\mu(\sigma)\|_{B_1}^{P_1} d\sigma \right)^{1/P_1} \leqslant c |t_i - t|^{1/P'_1} ,$$

ce qui joint à (1.15) et à la densité de la suite $\{t_i\}$ montre

que u_μ converge dans B_1 uniformément en $t \in [0,T]$, d'où

(1.12).

J. L. Lions

2. Un théorème d'existence.

2.1. Enoncé du résultat.

Pour u,v,w fonctions données sur Ω , on pose lorsque cela a un sens:

$$(2.1) \qquad a(\overset{o}{u},v,w) = \int_{\Omega} |u| \sum_{i=1}^{n} \frac{\partial v}{\partial x_i} \frac{\partial w}{\partial x_i} \, dx \; .$$

Théorème 2.1. On donne f et g avec

$$(2.2) \qquad f_o \in L^2(\Omega) \; , \; g \in L^2(\textstyle\sum) \; .$$

Il existe une fonction u telle que

$$(2.3) \qquad u \in L^{\infty}(0,T;L^2(\Omega)) \; ,$$

$$(2.4) \qquad |u|^{1/2} u \in L^2(0,T;H^1(\Omega)) \; ,$$

$$(2.5) \quad (\frac{\partial u}{\partial t}(t),v) + a(u(t);u(t),v) + \int_{\Gamma} \phi(u(t))v \, d\Gamma =$$

$$= (f_o(t),v) + \int_{\Gamma} g(t) \, v \, d\Gamma \qquad \forall v \in H^r(\Omega)$$

où r est un nombre réel quelconque tel que

$$(2.6) \qquad r \geqslant 1 + \frac{n}{6} \; ,$$

avec

$$(2.7) \qquad u(0) = 0 \; .$$

(Dans (2.5) la fonction ϕ est definie comme à la Fig. 1, Introduction).

Le problème de l'unicité est ouvert.

Remarque 2.1.

La condition (2.7) a un sens. Il résulte en effet de (2.5) que

(2.8) $\quad \dfrac{\partial u}{\partial t} - \displaystyle\sum_{i=1}^{n} \dfrac{\partial}{\partial x_i} \left(|u| \dfrac{\partial u}{\partial x_i} \right) = f_o \quad$ dans Q .

Mais d'apres (2.4) $\quad |u|^{1/2} u \in L^2(Q)$ donc $u \in L^3(Q)$ et

$|u| \dfrac{\partial u}{\partial x_i} = \dfrac{2}{3} |u|^{1/2} \dfrac{\partial}{\partial x_i} (|u|^{1/2} u) \in L^6(Q) \times L^2(Q) \subset L^{3/2}(Q)$

donc $|u| \dfrac{\partial u}{\partial x_i} \in L^{3/2}(0,T;L^{3/2}(\Omega))$ et

$\dfrac{\partial}{\partial x_i} (|u| \dfrac{\partial u}{\partial x_i}) \in L^{3/2}(0,T;W^{-1,3/2}(\Omega))$ donc

(2.9) $\quad \dfrac{\partial u}{\partial t} \in L^{3/2}(0,T;W^{-1,3/2}(\Omega))$.

Donc u **est en particulier** continue de $[0,T] \to W^{-1,3/2}(\Omega)$ et (2.7) a un sens.

Remarque 2.2.

Utilisant entre autres le Remarque précédente, **on voit que** (2.5) **a lieu** $\forall v$ tel que

(2.10) $\quad \dfrac{\partial v}{\partial x_i} \in L^3(Q) \quad \forall i$.

J. L. Lions

En effet:

1°) $H^r(\Omega) \subset \left\{ v \mid \dfrac{\partial v}{\partial x_i} \in L^3(\Omega) \; \forall i \right\}$ dès que (2.6) a lieu, car

si $v \in H^r(\Omega)$ alors $\dfrac{\partial v}{\partial x_i} \in H^{r-1}(\Omega)$ et d'après les Théorème de

Sobolev "fractionnaire" (J. PEETRE [2]) cela entraine que

$\dfrac{\partial v}{\partial x_i} \in L^q(\Omega)$, $\dfrac{1}{q} = \dfrac{1}{2} - \dfrac{r-1}{n} \leqslant \dfrac{1}{3}$ si r satisfait à (2.6); par ai_l_

leurs $H^r(\Omega)$ est _dense_ dans l'espace des v telles que l'on ait (2.10) ;

2°) les deux membres de (2.5) sont continus en v pour la

norme

$$\sum_{i=1}^{n} \left\| \frac{\partial v}{\partial x_i} \right\|_{L^3(\Omega)} + \left\| v \right\|_{L^3(\Omega)} \; .$$

Interprétation du problème.

On a déja (2.8). Multipliant (2.8) par v et intégrant par

parties, il vient:

$(\dfrac{\partial u(t)}{\partial t} , v) - \displaystyle\int_{\Gamma} |u| \dfrac{\partial u}{\partial n} v \, d\Gamma + a(u;u,v) = (f_o(t),v)$ d'où

en comparant à (2.5):

(2.11) $\qquad\qquad |u| \dfrac{\partial u}{\partial n} + \Phi(u) = g \; .$

2.2. Démonstration du Théorème 2.1.

2.2.1. Approximation de Faedo-Galerkin.

On choisit une "_base speciale_": soient w_i les fonctions propres de l'injection $H^r(\Omega) \rightarrow L^2(\Omega)$, i.e. les éléments de $H^r(\Omega)$ tels que

$$(2.12) \quad \begin{cases} (w_j,v)_{H^r(\Omega)} = \lambda_j(w_j,v) \quad \forall v \in H^r(\Omega) \ , \\ |w_j| = 1 \ , \quad 0 < \lambda_1 \leqslant \lambda_2 \leqslant \ldots \end{cases}$$

On choisit V_m comme dans l'Exemple 1.1 N° 1 Chap. 1.

On définit donc $u_m = u_m(t)$ par

$(2.13) \quad u_m(t) \in V_m = \text{espace } [w_1,\ldots,w_m] \text{ engendré par les } w_i \ ,$

$(2.14) \quad (u_m'(t),v) + a(u_m(t); u_m(t),v) + \int_\Gamma \phi(u_m(t))v \, d\Gamma =$

$$= (f_o(t),v) + \int_\Gamma g(t)v \, d\Gamma \quad \forall v \in V_m \ ,$$

$(2.15) \quad u_m(0) = 0 \ .$

Cela définit u_m dans $[0,t_m]$. On va voir que $t_m = T$.

On pose

$(2.16) \qquad a(v;v,v) = a(v) \ .$

On déduit de (2.14) que

$(2.17) \quad \frac{1}{2}\frac{d}{dt}|u_m(t)|^2 + a(u_m(t)) + \int_\Gamma \phi(u_m(t))u_m(t)d\Gamma = (f_o(t),u_m(t)) +$

$$+ \int_\Gamma g(t)u_m(t)d\Gamma \ .$$

Notons que

J. L. Lions

(2.18) $\|v\|_{L^3(\Gamma)} \leqslant c|v| + c\, a(v)^{1/3} \quad \forall v$ tel que $|v|^{1/2} v \in H^1(\Omega)$;

en effet il existe une constante c telle que

(2.19) $\left\| |v|^{1/2} v \right\|_{L^2(\Gamma)} \leqslant c \left\| |v|^{1/2} v \right\|_{L^{4/3}(\Omega)} + c \left(\sum_{i=1}^{n} \left\| D_i \left(|v|^{1/2} v \right) \right\|_{L^2(\Omega)}^2 \right)^{1/2}$

d'où

$$\left(\int_{\Gamma} |v|^3 d\Gamma \right)^{1/2} \leqslant c \left(\int_{\Omega} |v|^2 dx \right)^{3/4} + c\, a(v)^{1/2}$$

d'où (2.18) en élévant à la puissance $2/3$.

Donc (^{1}):

(2.20) $\left| \int_{\Gamma} g(t) u_m(t) d\Gamma \right| \leqslant c \|g(t)\|_{L^2(\Gamma)} (|u_m(t)| + a(u_m(t))^{1/3}) \leqslant$

$\leqslant c|u_m(t)|^2 + \dfrac{1}{2} a(u_m(t)) + c(\|g(t)\|_{L^2(\Gamma)}^2 + \|g(t)\|_{L^2(\Gamma)}^{3/2})$.

On déduit de (2.20) et (2.17), en utilisant le fait que

$\lambda \tilde{\phi}(\lambda) \geqslant 0$ donc que

$$\int_{\Gamma} \tilde{\phi}(u_m(t))\, u_m(t) d\Gamma \geqslant 0 :$$

(2.21) $\dfrac{d}{dt} |u_m(t)|^2 + a(u_m(t)) \leqslant c|u_m(t)|^2 + c(\|g(t)\|_{L^2(\Gamma)}^2 + \|g(t)\|_{L^2(\Gamma)}^{3/2}) +$

$+ c|f_0(t)|^2$.

(^{1}) On pourrait étendre les hypothèses faites sur g .

On en déduit que

(2.22) u_m demeure dans un borné de $L^\infty(0,T;L^2(\Omega))$,

et que $\dfrac{\partial}{\partial x_i}(|u_m|^{1/2} u_m)$ demeure dans un borné de $L^2(Q)$ ce

qui joint à (2.22) montre que

(2.23) $(u_m)^{1/2} u_m$ demeure dans un borné de $L^2(0,T;H^1(\Omega))$.

2.2.2. <u>Estimations a priori supplémentaires.</u>

On va maintenant estimer $\dfrac{\partial u_m}{\partial t}$. D'après ce qu'on a vu aux

Remarques 2.1 et 2.2 on a:

(2.24) $|a(u;u,v)| \leqslant c \, \||u|^{1/2} u\|_{H^1(\Omega)} \, \|v\|_{H^r(\Omega)}$.

donc

(2.25) $\begin{cases} a(u_m;u_m,v) = (h_m,v) \ , \ h_m \in (H^r(\Omega))' \ , \\[2mm] \|h_m(t)\|_{(H^r(\Omega))'} \leqslant c \, \||u_m(t)|^{1/2} u_m(t)\|_{H^1(\Omega)} \end{cases}$

et donc

(2.26) h_m demeure dans un borné de $L^2(0,T;(H^r(\Omega))')$.

On déduit alors de (2.14) que

(2.27) $\begin{cases} (u_m',v) = (k_m,v) \qquad \forall v \in V_m \ , \\[2mm] k_m \in \text{borné de } L^2(0,T;(H^r(\Omega))') \end{cases}$

donc (grâce au choix de la base spéciale (2.12)):

(2.28) $\dfrac{\partial u_m}{\partial t} \in \text{borné de } L^2(0,T;(H^r(\Omega))')$.

2.2.3. <u>Passage à la limite en</u> m .

D'après les estimations $(2.22)(2.23)(2.28)$ et ce qu'on a

m au N° 1.2 ou est dans les conditions d'application du Théo-

rème 1.1, avec $p_o = 3$, $p_1 = 2$.

Par conséquent on peut extraire une sous suite u_μ telle

que

(2.28) $\quad u_\mu \rightharpoonup u$ $\quad$ dans $\quad L^\infty(0,T;L^2(\Omega))$ $\quad$ faible étoile,

(2.29) $\quad u_\mu \rightarrow u$ $\quad$ dans $\quad L^3(Q)$ $\quad$ fort $\left(^1\right)$,

(2.30) $\quad |u_\mu|^{1/2} u_\mu \rightharpoonup \chi$ dans $\quad L^2(0,T;H^1(\Omega))$ $\quad$ faible,

(2.31) $\quad u_\mu \rightarrow u$ $\quad$ dans $\quad L^3(\textstyle\sum)$ $\quad$ fort.

De (2.29) et (2.30) on déduit que

(2.32) $\qquad\qquad \chi = |u|^{1/2} u$.

On peut supposer enfin que

(2.33) $\quad u_\mu' \rightharpoonup u'$ dans $\quad L^2(0,T; (H^r(\Omega))')$ $\quad$ faible.

On utilise alors (2.14) avec $m = \mu$ et avec v fixé

dans ℓ_{μ_o}, $\mu_o \leqslant \mu$.

$\left(^1\right)$ On applique le Théorème 1.1 dans le cadre du cas particulier "non linéaire" du N° 1.3.

Grâce aux estimations ci dessus on peut passer à la limi
te; il vient

$$(u'(t),v) + a(u(t);u(t),v) + \int_{\Gamma} \phi(u(t))v d\Gamma = (f_o(t),v) + \int_{\Gamma} g(t)v d\Gamma$$

et cela $\forall v \in V_{\mu_o}$ d'où le résultat.

Remarque 2.3 (Cette remarque est due à L. Tartar).

On peut éviter le recours au Théorème de compacité du N°1
sous forme complète en montrant que, si $|u_m|^{1/2} \dot{u}_m \in$ borné de
$L^2(0,T;H^1(\Omega))$ alors

(2.34) $u_m \in$ borné de $L^3(0,T;W^{\sigma,3}(\Omega))$, σ quelconque $< 2/3$

(et donc s pouvant être choisi $> 1/3$).

Si l'on admet pour l'instant (2.34), on applique le Théorème de compacité du N°1 sous forme linéaire (cf. N° 1.3), avec

$$B_o = W^{\sigma,3}(\Omega) \ , \ B = W^{\sigma,3}(\Omega) \ (\frac{1}{3} < \sigma < \sigma), \ B_1 = (H^r(\Omega))' \ .$$

Alors $u_m \in$ relativement compact de $L^3(0,T;W^{\sigma,3}(\Omega))$ et
on peut donc supposer que

(2.35) $u_\mu \to u$ dans $L^3(0,T;W^{\sigma,3}(\Omega))$ fort .

Comme l'opération "trace sur Γ " est continue de
$W^{\sigma,3}(\Omega) \to L^3(\Gamma)$ en particulier, il résulte de (2.35) que

J. L. Lions

$u_\mu \to u$ dans $L^3(\Sigma)$ fort, d'où le Théorème comme ci dessus.

Pour démontrer (2.34) on utilise la définition de $W^{\sigma,3}(\Omega)$ à l'aide des quotients différentiels $\left(\iint \dfrac{|u(x)-u(y)|^3}{|x-y|^{n+3\sigma}} \, dx \, dy < \infty \right)$.

3. <u>Sur le positivité des solutions.</u>

On a le résultat suivant - dont on va donner un début de démonstration (^{1}) : on se place dans les hypothèses du Théorème 2.1 et on suppose en outre que

(3.1) $f_o \geqslant 0$ p.p. dans Q , $g \geqslant 0$ dans Σ .

<u>Alors toute solution</u> (^{2}) u <u>de</u> (2.3)(2.4)(2.5)(2.7) <u>vérifie</u>

(3.2) $u \geqslant 0$ p.p. dans Q .

Voici un "début" de démonstration, montrant la difficulté du problème.

On prend dans (2.5)

(3.3) $v = u^-(t)$

ce qui est loisible pour toute les expressions intervenant dans (2.5), sauf peut être pour le 1er terme (source de la principa-

(^{1}) Le résultat complet est dû à R. Temann; cf. Remarque 3.2.

(^{2}) Et probablement - conjecturant l'unicité.....-<u>la</u> solution.

J. L. Lions

le difficulté):

$$(3.4) \qquad \xi(t) = (\frac{\partial u(t)}{\partial t}, u^-(t)) .$$

On obtient:

$$(3.5) \quad \xi(t) + a(u(t);u(t),u^-(t)) + \int_\Gamma \phi(u(t))u^-(t)d\Gamma = (f_o(t),u^-(t)) +$$

$$+ \int_\Gamma g(t)u^-(t)d\Gamma .$$

Mais

$$(3.6) \quad a(u(t) ; u(t),u^-(t)) = -\int_\Omega |u(t)| \sum_{i=1}^n (\frac{\partial}{\partial x_i} u^-(t))^2 dx,$$

$$\int_\Gamma \phi(u(t)) u^-(t)d\Gamma = -\int_{\Gamma \atop u(t)\leq 0} \phi(u(t)) u(t) d\Gamma, \quad \text{et comme}$$

$$\lambda\phi(\lambda) \geq 0 \quad \forall \lambda \in \mathbb{R} \quad \text{on a :}$$

$$(3.7) \qquad \int_\Gamma \phi(u(t)) u^-(t)d\Gamma \leq 0 .$$

Alors avec (3.6) et (3.7) , (3.5) donne

$$(3.8) \quad -\xi(t) + \int_\Omega |u(t)| \sum_{i=1}^n \left(\frac{\partial}{\partial x_i} u^-(t)\right)^2 dx + (f_o(t), u^-(t)) +$$

$$+ \int_\Gamma g(t) u^-(t)d\Gamma \leq 0 .$$

<u>**Formellement**</u> (et la justification de ce point semble être la difficulté principale, cf. <u>Remarque</u> 3.1) on a:

J. L. Lions

$$(3.9) \qquad - \int_0^T \xi(t)dt \geqslant 0$$

Comme d'après (3.1) $(f_o(t), u^-(t)) \geqslant 0$ et $\int_\Gamma g(t)u^-(t)d\Gamma \geqslant 0$

on "déduit" alors de (3.8) que

$$\int_0^T \int_\Omega |u(t)| \sum_{i=1}^n \left(\frac{\partial}{\partial x_i} u^-(t)\right)^2 dx \leqslant 0$$

donc

$$(3.10) \qquad |u(t)|^{1/2} \frac{\partial}{\partial x_i} u^-(t) = 0 \quad \forall i \ .$$

Donc on bien $u(t) = 0$ et dans la région où $u(t) \neq 0$ on

a: $u^-(t) =$ constante (localement) et nécessairement alors

$$u^- = 0 \ .$$

ce qui equivaut à (3.2) .

Remarque 3.1.

La difficulté principale semble être que l'on n'obtient

pas des estimations en dualité pour u et pour $\frac{\partial u}{\partial t}$. Il faut

remarquer que les estimations obtenues sur u indiquent l'ap-

partenance de u à une classe fonctionnelle non linéaire de sor

te qu'il faudrait préciser ce qu'on entend par "estimations en

"dualité" !

Remarque 3.2. (dûe à R. Temam).

On peut montrer que $u \geqslant 0$ par la méthode suivante, due à Temam.

On définit $\varphi_M(\lambda) = |\lambda|$ si $|\lambda| \leqslant M$, M si $\lambda \geqslant M$, $-M$ si $\lambda \leqslant -M$ et on définit u_M solution de

$$\left(\frac{\partial u_M}{\partial t}, v\right) + \int_\Omega \left[\varphi_M(u_M) + \varepsilon_M\right] \operatorname{grad} u_M \ \operatorname{grad} v \ dx + \int_\Gamma \phi(u_M)v \ d\Gamma =$$

$$= \int_\Omega f_0(x,t)v \ dxt \ \int_\Gamma gv d\Gamma \ ,$$

où $\varepsilon_M > 0$, $u_M(0) = 0$.

On montre facilement l'existence et l'unicité de u_M. On montre ensuite que $u_M \rightarrow u$ solution du problème lorsque $M \rightarrow + \infty$ et $\varepsilon_M \rightarrow 0$.

On vérifie, par le raisonnement ci dessus - <u>correct pour</u> u_M - que si $f_0 \geqslant 0$, $g \geqslant 0$, on a: $u_M \geqslant 0$ d'où $u \geqslant 0$.

4. <u>Inéquations variationnelles.</u>

4.1. <u>Position du problème. Notations.</u>

On reprend la fonction ϕ donnée à la Fig. 1, Introduction et l'on pose:

$$(4.1) \qquad \phi_j(\lambda) = \text{fonction} \ \phi \ \text{avec} \ k_1 = k_2 = j \ .$$

J. L. Lions

<u>Le problème est d'etudier la situation lorsque</u> $j \rightarrow +\infty$.

On introduit - ce qu'est indispensable pour la formulation des problèmes.

(4.2) $\qquad \psi_j^q(\lambda) = \int_0^\lambda \phi_j(u)d\mu$,

(4.3) $\qquad \psi(\lambda) = \lim_{j \rightarrow +\infty} \psi_j^j(\lambda)$ (cf. graphes sur Fig. 2 et 3).

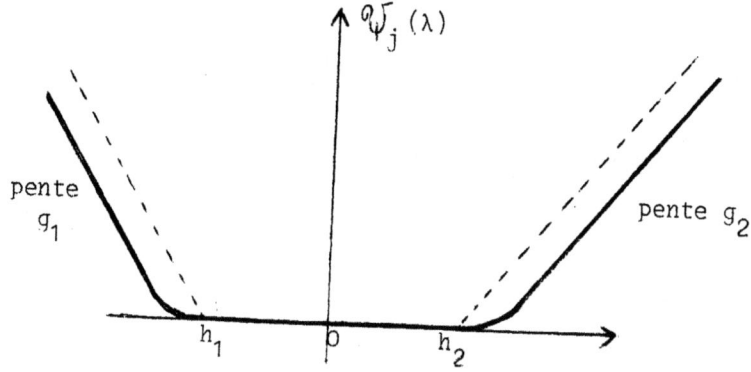

Fig. 2

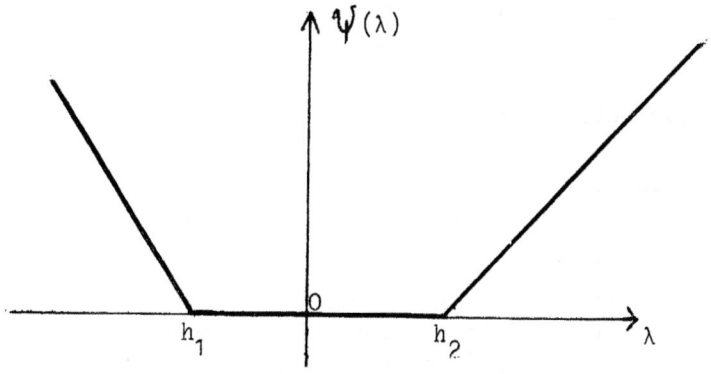

Fig. 3

J. L. Lions

4.2. Nouvelle formulation du problème.

Le Théorème 2.1 montre l'existence d'une fonction u_j tel le que

(4.4) $u_j \in$ borne de $L^{\infty}(0,T;L^2(\Omega))$ (^{1}) lorsque $j \to +\infty$,

(4.5) $|u_j|^{1/2} u_j \in$ borne de $L^2(0,T;H^1(\Omega))$,

(4.6) $u_j' \in$ borné de $L^{3/2}(0,T;W^{-1,3/2}(\Omega))$,

(4.7) $\begin{cases} (u_j'(t),v) + a(u_j(t);u_j(t),v) + \int_\Gamma \Phi_j(u_j(t))v \, d\Gamma = \\ \quad = (f_o(t),v) + \int_\Gamma g(t)v \, d\Gamma \qquad \forall v \in H^r(\Omega) , \end{cases}$

(4.8) $\qquad\qquad u_j(0) = 0$.

On note maintenant (convexité de ψ_j) que

$$\psi_j(\lambda) - \psi_j(\mu) - (\lambda - u) \psi_j'(\mu) \geq 0$$

donc que

(4.9) $\int_\Gamma \left[\psi_j(v) - \psi_j(u_j(t)) \right] d\Gamma \geq \int_\Gamma \Phi_j(u_j(t)) \, (v - u_j(t)) d\Gamma$.

Alors (4.7) "entraine" (^{2})

(^{1}) Toutes les estimations obtenues dans le Théorème 2.1 sont en effet indépendantes de j .

(^{2}) Avec la difficulté deja rencontrée au N° 3 que $(u_j'(t),u_j(t))$ est __formel__. Mais on fait un calcul formel conduisant à une formulation nouvelle qui est ensuite justifiée.

J. L. Lions

(4.10) $\quad (u'_j(t), v-u_j(t)) + a(u_j(t); u_j(t), v-u_j(t)) + \int_\Gamma [\psi'_j(v) - \psi_j(u_j(t))] d\Gamma \geq$

$$\geq (f_0(t), v-u_j(t)) + \int_\Gamma g(t) (v-u_j(t)) d\Gamma$$

et réciproquement (4.10) entraine (4.7) (toujours formollement

prendre $v = u_j(t) + \lambda w \, \lambda > 0$, diviser par λ et faire $\lambda \to 0$).

Introduisons maintenant l'espace suivant:

(4.11) $\quad \mathcal{V} = \left\{ v \mid v, \dfrac{\partial v}{\partial x_i} \in L^4(Q) \; \forall i, \; \dfrac{\partial v}{\partial t} \in L^2(Q), v(x,0) = 0 \right\}.$

On note alors que

(4.12) $\quad \begin{cases} \displaystyle\int_0^T [(v', v-u_j) + a(u_j; u_j, v-u_j)] dt + \int_\Sigma [\psi_j(v) - \psi_j(u_j)] d\Sigma \geq \\[2mm] \geq \displaystyle\int_0^T (f_0, v-u_j) d\Gamma + \int_\Sigma g(v-u_j) d\Sigma \quad \forall v \in \mathcal{V}. \end{cases}$

En effet le 1er membre de (4.12) est égal au 1er membre de

(4.10) intégré sur $[0,T] + \displaystyle\int_0^T (v'-u'_j, v-u_j) dt = \frac{1}{2} |v(T) - u_j(T)|^2 \geq 0.$

L'inéquation (4.12) est le 1er exemple que nous remontrons

ici d'inéquation variationnelle d'évolution. La formulation (4.12)

n'est pas indispensable pour j fini - mais permet par contre

de considérer le cas " $j = \infty$ ".

4.3. Le cas " $j = \infty$ ".

On va montrer le

Théorème 4.1. On se place dans les conditions du Théorème 2.1.

Il existe une fonction u (^{1}) telle que

(4.13) $u \in L^\infty(0,T;L^2(\Omega))$,

(4.14) $|u|^{1/2}u \in L^2(0,T;H^1(\Omega))$,

(4.15) $\dfrac{\partial u}{\partial t} \in L^2(0,T;(H^r(\Omega))')$,

(4.16) $\displaystyle\int_0^T \left[(v',v-u)+a(u;u,v-u)\right]dt + \int_\Sigma \left[\psi(v)-\psi(u)\right]d\Sigma -$

$\displaystyle - \int_0^T (f_o,v-u)dt - \int_\Sigma g(v-u)d\Sigma \geqslant 0 \quad \forall v \in \mathcal{U}$

(où $\mathcal{U}$ est défini en (4.11) et ψ en (4.3)).

Remarque 4.1.

L'interprétation (formelle) des conditions aux limites est:

$-g_2 \leqslant |u|\dfrac{\partial u}{\partial n} - g \leqslant -g_1$,

$|u|\dfrac{\partial u}{\partial n} - g = -g_2 \implies u \geqslant h_2$,

$-g_2 < |u|\dfrac{\partial u}{\partial n} - g < 0 \implies u = h_2$,

$|u|\dfrac{\partial u}{\partial n} - g = 0 \implies h_1 < u < h_2$,

$0 \quad < |u|\dfrac{\partial u}{\partial n} - g < -g_1 \implies u = h_1$

$|u|\dfrac{\partial u}{\partial n} - g = -g_1 \implies u \leqslant h_1$.

(^{1}) Le problème de l'unicité est ouvert.

Démonstration du Théorème 4.1.

Soit u_{jm} la solution de (2.13)(2.14)(2.15) avec ϕ_j au lieu de ϕ , donc en particulier:

$$(4.17) \quad (u'_{jm}(t),v) + a(u_{jm}(t);v_{jm}(t),v) + \int_\Gamma \phi_j(u_{jm}(t))v\,d\Gamma =$$

$$= (f_o(t),v) + \int_\Gamma g(t)v\,d\Gamma \qquad \forall v \in V_m \quad .$$

Les estimations faites au N° 2 sont indépendantes de j donc (cf. (2.22)(2.23) et (2.28)):

(4.18) u_{jm} demeure dans un borné de $L^\infty(0,T;L^2(\Omega))$ (lorsque $j\to\infty$, $m\to\infty$),

(4.19) $|u_{jm}|^{1/2}\,u_{jm}$ demeure dans un borné de $L^2(0,T;H^1(\Omega))$,

(4.20) $\dfrac{\partial u_{jm}}{\partial t}$ demeure dans un borné de $L^2(0,T;(H^r(\Omega))')$.

Par ailleurs soit

$$(4.21) \qquad \mathcal{V}_m = \left\{ v \mid v \in \mathcal{V}, \ v(t) \in V_m \right\} .$$

Alors le calcul analogue à celui conduisant à (4.12) étant maintenant justifié, on a:

$$(4.22) \begin{cases} \displaystyle\int_0^T \left[(v',v-u_{jm})+a|u_{jm},u_{jm},v-u_{jm})\right]dt + \int_\Sigma \left[\psi_j(v)-\psi_j(u_{jm})\right]d\Sigma \geq \\[2mm] \displaystyle\geq \int_0^T (f_o,v-u_{jm})dt + \int_\Sigma g(v-u_{jm})d\Sigma \qquad \forall v \in \mathcal{V}_m \quad . \end{cases}$$

D'après (4.18)(4.19)(4.20) on peut extraire une sous suite,

encore designée par u_{jm} , telle que (en utilisant aussi les résultats du N° 1).

$$u_{jm} \rightarrow u \quad \text{dans} \quad L^{\infty}(0,T;L^2(\Omega)) \quad \text{faible étoile,}$$

$$|u_{jm}|^{1/2} u_{jm} \rightarrow |u|^{1/2} u \quad \text{dans} \quad L^2(0,T;H^1(\Omega)) \quad \text{faible,}$$

(4.23)

$$\frac{\partial u_{jm}}{\partial t} \rightarrow \frac{\partial u}{\partial t} \quad \text{dans} \quad L^2(0,T;(H^r(\Omega))') \quad \text{faible,}$$

$$u_{jm}\big|_{\Sigma} \rightarrow u\big|_{\Sigma} \quad \text{dans} \quad L^3(\Sigma) \quad \text{fort .}$$

On vérifie sans peine que

$$(4.24) \quad \psi_j(u_{jm}) \longrightarrow \psi(u) \quad \text{dans, par ex.,} \quad L^2(\Sigma) .$$

Pour $v \in \mathcal{V}$ on prend une suite (qui existe) $v_m \in \mathcal{V}_m$ telle que $v_m \rightarrow v$ dans $L^4(Q)$, $\dfrac{\partial v_m}{\partial x_i} \rightarrow \dfrac{\partial v}{\partial x_i}$ dans $L^4(Q)$ $\forall i$, $\dfrac{\partial v_m}{\partial t} \rightarrow \dfrac{\partial v}{\partial t}$ dans $L^2(Q)$.

On prend $v = v_m$ dans (4.22), que l'on écrit:

$$(4.25) \quad \int_0^T \Big[(v'_m, v_m - u_{jm}) + a(u_{jm}; u_{jm}, v_m) \Big] dt + \int_{\Sigma} \Big[\psi_j(v_m) - \psi_j(u_{jm}) \Big] d\Sigma$$

$$- \int_0^T (f_o, v_m - u_{jm}) dt - \int_{\Sigma} g(v_m - u_{jm}) d\Sigma \geqslant \int_0^T a(u_{jm}; u_{jm}, u_{jm}) dt .$$

On vérifiera plus loin que

J. L. Lions

$$(4.26) \quad \int_o^T a(u_{jm}; u_{jm}, v_m) \, dt \longrightarrow \int_o^T a(u;u,v) dt \ ,$$

$$(4.27) \quad \lim . \inf . \int_o^T a(u_{jm}; u_{jm}, u_{jm}) dt \geqslant \int_o^T a(u;u,u) dt \ .$$

Alors (4.25) donne grâce à (4.24)

$$\int_o^T \Big[(v',v-u) + a(u;u,v)\Big] dt + \int_\Sigma \Big[\psi(v) - \psi(u)\Big] \, d\Sigma \ -$$

$$-\int_o^T (f_o,v-u) dt - \int_\Sigma g(v-u) d\Sigma \ \geqslant \ \int_o^T a(u;u,u) \, dt$$

d'où (4.16) .

Vérification de (4.26).

On a à considérer

$$(4.28) \quad \int_Q |u_{jm}| \, \frac{\partial u_{jm}}{\partial x_i} \frac{\partial v_m}{\partial x_i} \, dx \, dt = \int_Q |u_{jm}|^{1/2} \Big(|u_{jm}|^{1/2} \frac{\partial u_{jm}}{\partial x_i}\Big) \frac{\partial v_m}{\partial x_i} \, dx \, dt \ ,$$

or $\quad u_{jm} \longrightarrow u \quad$ dans $\ L^2(Q) \ \underline{\text{fort}} \$ donc

$|v_{jm}|^{1/2} \longrightarrow |u|^{1/2} \quad$ dans $\ L^4(Q) \ \underline{\text{fort}} \ ,$

$\dfrac{\partial v_m}{\partial x_i} \longrightarrow \dfrac{\partial v}{\partial x_i} \quad$ dans $\ L^4(Q) \ \underline{\text{fort}} \ ,$

$|u_{jm}|^{1/2} \dfrac{\partial v_{jm}}{\partial x_i} \longrightarrow |u|^{1/2} \dfrac{\partial u}{\partial x_i} \quad$ dans $\ L^2(Q) \ \text{faible}$

et donc l'expression (4.28) tend vers $\int_Q |u|^{1/2}\left(|u|^{1/2}\frac{\partial u}{\partial x_i}\right)\frac{\partial v}{\partial x_i}$ dx dt

d'où (4.26).

Vérification de (4.27).

On a:

$$\int_0^T a(u_{jm};u_{jm},u_{jm})dt = \frac{4}{9} \int_Q \sum_{i=1}^n \left(\frac{\partial}{\partial x_i}\left(|u_{jm}|^{1/2} u_{jm}\right)\right)^2 dx\ dt$$

d'où le résultat.

5. Problèmes.

Problème 5.1.

Y-a-t-il unicité de la solution dans le Théorème 2.1 ?

Question analogue dans le Théorème 5.1.

Problème 5.2

Peut-on justifier et compléter le raisonnement du N° 3 et démontrer que $u \geq .0$?

Problème 5.3

Peut-on dans le Théorème 5.1 faire par ex. $g_2 \to +\infty$?
La difficulté est que si les estimations sur u_{jm} sont indepen-
dentes également de g_2 , il n'on va pas de même pour $\frac{\partial u_{jm}}{\partial t}$ de
telle sorte que l'on ne peut plus utiliser la méthode de compaci

J. L. Lions

té - et qu'il ne semble pas possible d'utiliser la méthode de
monotonie. (Le problème présent rejoint le problème 11.22, p.
304, de LIONS [1]).

J. L. Lions

CHAPITRE 4

INEQUATIONS VARIATIONNELLES ET ÉCOULEMENT DE FLUIDES

NON NEWTONIENS.

Introduction.

Certains problèmes de mécanique, pour lesquels nous renvoyons à DUVAUT-LIONS [1] , conduisent à des problèmes contenant
comme cas particuliers les équations de Navier-Stokes.On touche
donc là à l'un des "grands" problèmes non linéaires, ayant donné lieu aux travaux classiques de J. LERAY [1] [2] [3] - On peut
dire, grosso-modo, qu'il s'agit "d'inéquations variationnelles
attachées aux opérateurs de Navier-Stokes".

On présente ici quelques résultats, relatifs à la dimension
d'espace 2.

Pour les problèmes relatifs à la dimension d'espace 3 nous
renvoyons à DUVAUT-LIONS, loc. cit , Chap. 6 - ainsi que son
l'étude des problèmes stationnaires.

Il __semble__ que l'on puisse étendre toutes les propriétés con
nues relatives aux équations de Navier-Stokes ([1]) à la situation

([1]) Sauf, très certainement, les propriétes de régularité.

J. L. Lions

présente. Pour les propriétés asyntotiques, cf. J.C.SIMON [1].

1. Position des problèmes. Résultats.

    1.1. Notations - Problème.

    Soit Ω un ouvert borné (^{1}) de $\mathbb{R}^2$ (^{2}) .

    On introduit:

(1.1) $\mathcal{V}=$ espace des vecteurs $\varphi = \{\varphi_1, \varphi_2\}$, avec

       $\varphi_i \in \mathcal{D}(\Omega)$, Div $\varphi = 0$ dans Ω ,

(1.2) $H =$ adhérence de $\mathcal{V}$ dans $(L^2(\Omega))^2$, $(f,g) =$ produit sca-

    laire dans H (induit par celui de $(L^2(\Omega))^2$) ,

(1.3) $V =$ adhérence de $\mathcal{V}$ dans $(H^1(\Omega))^2$.

    Pour $u,v \in V$ on pose

(1.4) $\qquad a(u,v) = \displaystyle\sum_{i,j=1}^{2} \int_{\Omega} \frac{\partial u_i}{\partial x_j} \frac{\partial v_i}{\partial x_j} \, dx$

et pour $u,v,w \in V$ on pose (^{3})

(1.5) $\qquad b(u,v,w) = \displaystyle\sum_{i,j=1}^{2} \int_{\Omega} u_i \frac{\partial v_j}{\partial x_i} w_j \, dx$.

(^{1}) Ce qui n'est pas essentiel.

(^{2}) Le fait que la dimension d'éspace soit 2 est essentiel pour la démonstration de l'unicité. Cf. Problème 5.1.

(^{3}) On vérifie à l'aide du théorème de Sobolev que cela a un sens.

J. L. Lions

Enfin pour $v \in V$ on pose

$$(1.6) \qquad j(v) = \int_\Omega \sum_{i,j} \left(\frac{\partial v_i}{\partial x_j}\right)^2\Big)^{1/2} dx \quad (^1) \; .$$

Le problème est le suivant: trouver une fonction $t \rightarrow u(t)$

telle que

$$(1.7) \qquad u(t) \in V \; ,$$

$$(1.8) \quad (\frac{\partial u(t)}{\partial t}, v-u(t)) + a(u(t), v-u(t)) + b(u(t), u(t), v-u(t)) +$$

$$+ gj(v) - gj(u(t)) \geqslant (f(t), v-u(t)) \quad \forall v \in V$$

où g est un scalaire > 0 ,

avec

$$(1.9) \qquad u(0) = u_o \quad donné.$$

Remarque 1.1.

On vérifie que

$$b(u,u,u) = 0$$

de sorte que l'expression $b(u(t), u(t), u(t))$ appaieussent dans

(1.8) n'est explicitée que pour la symétrie de l'énoncé.

Remarque 1.2.

La fonction $v \rightarrow j(v)$ n'est pas différentiable; c'est ce

---

(^{1}) Il s'agit d'une "modélisation" du problème réel - cf. DUVAUT-LIONS, loc. cit.

J. L. Lions

qui nécessite l'énoncé sous la forme "inéquation variationnelle"

(1.8) (^{1}).

Si la fonction j etait différentielle, de derivée j', alors

on pourrait remplacer (1.8) par l'équation

(1.10) $(\frac{\partial u(t)}{\partial t}, v) + a(u(t),v) + b(u(t),u(t),v) + g(j'(u(t)),v) =$

$$= (f(t),v) \quad \forall v \in V \quad .$$

Remarque 1.3.

Si g = 0 , alors (1.8) se reduit à l'équation classique

de Navier Stokes, sous la forme "faible" introduite par LERAY,

loc. cit. (solutions turbolentes).

Remarque 1.4.

Revenons à le "non differentiabilité" de j. On peut intro-

duire une "différentielle généralisée" ou "sous différentielle"(^{2})

qui est un operateur multivoque, j'(u(t)) étant non plus un

élément de V' mais un sous ensemble convexe non nécéssairement

réduit à un point de V' (cf. PAZY [1]).

Alors (1.10) doit être remplacé par

(^{1}) On a déja rencontré un exemple analogue, de ce point de vue
au Chap. 3, N° 4 .

(^{2}) Il s'agit d'un cas particulier – pour les fonctions convexes–
de la theorie du contingent.

J. L. Lions

$$(1.11) \quad \begin{cases} (\dfrac{\partial u(t)}{\partial t}, v) + a(u(t),v) + b(u(t),u(t),v) + g(\chi,v) = (f(t),v) \\ \\ \chi \in j'(u(t)) , \end{cases}$$

ou encore si l'on pose:

$$(1.12) \quad a(u,v) = (Au, v) , \quad Au \in V' \quad (^1)$$

$$(1.13) \quad b(u,u,v) = (B(u,u),v) , \quad B(u,u) \in V' :$$

$$(1.14) \quad -\left[\dfrac{\partial u(t)}{\partial t} + Au(t) + B(u(t),u(t)) - f(t)\right] \in g \, j'(u(t)) ;$$

il s'agit d'une équation d'évolution pour opérateurs multivoques

(cf. H. BREZIS [1][2], F. BROWDER [1] et le cours de PAZY).

1.2. Résultats.

Théorème 1.1.. On suppose que f et u_0 sont donnés avec

$$(1.15) \quad f \in L^2(0,T;V') ,$$

$$(1.16) \quad u_0 \in H .$$

Il existe une fonction u et une seule telle que

$$(1.17) \quad u \in L^2(0,T;V) ,$$

$$(1.18) \quad \dfrac{\partial u}{\partial t} \in L^2(0,T;V')$$

et vérifiant (1.8) (1.9).

$(^1)$ V' = dual de V lorsque H est identifié à son dual.

Remarque 1.5

Il résulte de (1.17)(1.18) que (après modification even-
tuelle sur un ensemble de mesure nulle) t → u(t) est continue de [0,T] → H . Donc (1.9) a un sens.

Pour rappeler la dépendence de la solution u en le paramètre
g > 0 on note u_g cette solution dans le résultat suivant:

Théorème 1.2. Dans les hypothèses du Théorème 1.1, on a:

(1.19) u_g → u dans $L^2(0,T;V)$ faible,

(1.20) $\dfrac{\partial u_g}{\partial t}$ → $\dfrac{\partial u}{\partial t}$ dans $L^2(0,T;V')$ faible

où u désigne la solution des équations classiques de Navier
Stokes:

(1.21) $\left(\dfrac{\partial u(t)}{\partial t},v\right)$ + a(u(t),v) + b(u(t),u(t),v) = (f(t),v) $\forall v \in V$

avec (1. 9).

2. Démonstration de l'unicité dans les Théorème 1.1 et 1.2.

Il suffit de donner la démonstration de l'unicité dans le
Théorème 1.1; elle est d'ailleurs analogue à celle donnée dans
LIONS-PRODI [1] pour le cas des équations.

2.1. Une inégalité.

On rappelle d'abord l'inégalité suivante, du type "interpolation": Ω étant un ouvert de $\mathbb{R}^2$, il existe une constante

J. L. Lions

c telle que

$$(2.1) \quad \|v\|_{L^4(\Omega)} \leqslant c \, \|v\|_{H^1(\Omega)}^{1/2} \, \|v\|_{L^2(\Omega)}^{1/2} \qquad \forall v \in H_0^1(\Omega) \ .$$

Une démonstration élémentaire (O.A. LADYZENSKAYA [1]) :

pour $v \in D(\Omega)$, prolongée par 0 hors de Ω , on note que

$$v^2(x) = 2 \int_{-\infty}^{x_1} v \, \frac{\partial v}{\partial x_1} \, dx_1 \leqslant 2 \, v_1(x_2) \ , \ v_1(x_2) = \int_{-\infty}^{\infty} |v| \, \left|\frac{\partial v}{\partial x_1}\right| dx_1 \ .$$

De même, changeant les indices, $v^2(x) \leqslant 2v_2(x_1)$ d'où

$$\int_{\mathbb{R}^2} v^4 dx \leqslant 4 \left(\int_{\mathbb{R}} v_1(x_2) dx_2 \right) \left(\int_{\mathbb{R}} v_2(x_1) dx_1 \right)$$

$$\leqslant 4 \, \left\|\frac{\partial v}{\partial x_1}\right\|_{L^2} \, \|v\|_{L^2} \, \left\|\frac{\partial v}{\partial x_2}\right\|_{L^2}$$

d'où en particulier (2.1).

<u>Application:</u> on a :

$$(2.2) \quad |b(u,v,w)| \leqslant c \, \|u\|^{1/2} \, |u|^{1/2} \|v\| \, \|w\|^{1/2} \, |w|^{1/2} \quad (^1) \ .$$

En effet on doit estimer $\left| \int_{\Omega} u_i \, \frac{\partial v_i}{\partial x_i} \, w_j \, dx \right| \leqslant$

$$\leqslant \|u_i\|_{L^4(\Omega)} \, \|w_j\|_{L^4(\Omega)} \, \left\|\frac{\partial v_j}{\partial x_i}\right\|_{L^2}$$

$$\leqslant \text{(en utilisant (2.1)} c \|u_i\|_{H^1(\Omega)}^{1/2} \, \|u_i\|_{L^2(\Omega)}^{1/2} \, \|w_j\|_{H^1(\Omega)}^{1/2} \, \|w_j\|_{L^2(\Omega)}^{1/2} \, \|v_j\|_{H^1(\Omega)}$$

(^{1}) On rappelle que $|\ |$ (resp. $\|\ \|$) désigne la norme dans H (resp. V).

d'où (2.2).

2.2. Démonstration de l'unicité.

Soient u et u_* deux solutions eventuelles du problème.
On pose:

$$U = u - u_* .$$

On prend, ce qui est loisible, $v = u'_*(t)$ (resp. $v = u(t)$)
dans l'inéquation (1.8) (resp. l'inéquation analogue relative à
u_*); additionant, il vient:

$$(2.3) \quad -\left(\frac{\partial U(t)}{\partial t} , U(t)\right) - a(U(t),U(t)) - b(u(t),u(t), U(t)) +$$
$$+ b(u_*(t), u_*(t) , U(t)) \geq 0 .$$

Les termes en j ont disparu et le raisonnement est analo‌gue désormais à celui relatif aux équations; on note que

$$b(u_*,u_*,U) - b(u,u,U) = b(u-U,u-U,U) - b(u,u,U) = -b(U,u,U)$$

de suite que (2.3) s'écrit:

$$(2.4) \quad \left(\frac{\partial U(t)}{\partial t} ,U(t)\right) + a(U(t),U(t)) \leq - b(U(t),u(t),U(t)).$$

Mais

$$(2.5) \quad a(v,v) = a(v) \geq \alpha \|v\|^2 , \alpha > 0 , \forall v \in V$$

et d'après (2.2):

$$(2.6) \quad |b(U(t),u(t),U(t))| \leq c \|U(t)\| |U(t)| \|u(t)\| \leq$$
$$< \alpha \|U(t)\|^2 + c' \|u(t)\|^2 |U(t)|^2$$

J. L. Lions

de sorte que avec (2.5)(2.6), (2.4) donne

$$(2.7) \quad \begin{cases} \dfrac{1}{2} \dfrac{d}{dt} |U(t)|^2 \leq c' \, m(t) \, |U(t)|^2 \,, \quad m(t) = \|u(t)\|^2 \,, \\[2mm] U(0) = 0 \,. \end{cases}$$

De (2.7) et de l'inégalité de Gronwall on déduit que

$$U = 0 \,.$$

3. Démonstration de l'existence dans le Théorème 1.1.

3.1. Méthode de régularisation.

On va "approcher" j par des fonctionnelles j_ε différentiables, l'inéquation correspondante se réduisant alors à une équation que l'on résout - puis l'on passe à la limite en ε.

La "régularisation" de j peut se faire d'une infinité de manières ([1]); par ex.

$$(3.1) \quad j_\varepsilon(v) = \frac{1}{1+\varepsilon} \int_\Omega (D(v))^{\frac{1+\varepsilon}{2}} \, dx \,, \quad D(v) = \sum_{i,j=1}^{2} \left(\frac{\partial v_i}{\partial x_j} \right)^2 \,.$$

L'équation correspondante est (cf. Remarque 1.2)

$$(3.2) \quad (u'_\varepsilon, v) + a(u_\varepsilon, v) + b(u_\varepsilon, u_\varepsilon, v) + g(j'_\varepsilon(u_\varepsilon), v) = (f, v) \quad \forall v \in V$$

où

([1]) Le choix de la régularisation peut être important pour l'Analyse Numérique du problème.

$$(3.3) \quad (j_\varepsilon'(u),v) = \int_\Omega D(u)^{\frac{\varepsilon-1}{2}} \sum_{i,j=1}^{2} \frac{\partial u_i}{\partial x_j} \frac{\partial v_i}{\partial x_j} \, dx \quad,$$

avec

$$(3.4) \qquad u_\varepsilon(0) = u_0 \; .$$

3.2. <u>Existence d'une solution de</u> (3.2),(3.4) .

Il s'agit d'un problème du type "Navier Stokes classique"

avec <u>en outre</u> la présence de l'opérateur <u>monotone</u> $u \longrightarrow j_\varepsilon'(u)$.

On construit les solutions approchées $u_{\varepsilon m}$ par la méthode

de Faedo-Galerkin (cf. Chap. 1):

$$(3.5) \qquad u_{\varepsilon m}(t) \in V_m \; ,$$

$$(3.6) \quad \begin{cases} (u_{\varepsilon m}',v) + a(u_{\varepsilon m},v) + b(u_{\varepsilon m},u_{\varepsilon m},v) + g(j_\varepsilon'(u_{\varepsilon m}),v) = (f,v) \\[2mm] \forall v \in V_m \end{cases}$$

$$(3.7) \quad u_{\varepsilon m} = u_{om} \; , \; u_{om} \in V_m \; , \; u_{om} \to u_0 \quad \text{dans} \quad H \quad \text{lorsque} \quad m \to \infty.$$

On prend (mais cela <u>n'est pas indispensable</u>):

$$(3.8) \qquad V_m = \left[w_1,\ldots,w_m \right] \; ,$$

les w_i étant les fonctions propres $(^1)$

$(^1)$ On suppose Ω borne mais cela non plus n'est pas essentiel
(<u>par ex.</u> on peut approcher Ω par des ouverts bornés).

$$(3.9) \begin{cases} a(w_i,v) = \lambda_i(w_i,v) & \forall v \in V \\ w_i \in V, \ |w_i| = 1, \quad 0 < \lambda_1 \leqslant \lambda_2 \leqslant \dots . \end{cases}$$

Estimations a priori.

On peut prendre $v = u_{\varepsilon m}$ dans (3.6). Notant que $b(v,v,v) = 0$ et que $(j'_\varepsilon(v),v) \geqslant 0$, on a:

$$\frac{1}{2} \frac{d}{dt} |u_{\varepsilon m}(t)|^2 + a(u_{\varepsilon m}(t)) \leqslant (f(t), u_{\varepsilon m}(t))$$

d'où l'on déduit:

(3.10) $u_{\varepsilon m}$ demeure dans un borné ([1]) de $L^\infty(0,T;H) \cap L^2(0,T;V)$.

Verifions maintenant que

(3.11) $u'_{\varepsilon m}$ demeure dans un borné de $L^2(0,T;V')$.

On note d'abord que

$$|b(u_{\varepsilon m}, u_{\varepsilon m}, v)| = |-b(u_{\varepsilon m}, v, u_{\varepsilon m})| \leqslant c \|u_{\varepsilon m}\| \ |u_{\varepsilon m}| \ \|v\| \leqslant$$

$$\leqslant \text{(d'après (3.10))} \ c \|u_{\varepsilon m}\| \ \|v\|$$

donc

$$(3.12) \begin{cases} b(u_{\varepsilon m}, u_{\varepsilon m}, v) = (h_{\varepsilon m}, v), \\ h_{\varepsilon m} \in \text{borné de } L^2(0,T;V') \end{cases}$$

Utilisant (1.12), (3.6) équivaut alors à:

―――――――――

([1]) Borné indépendant de ε et de m.

J. L. Lions

(3.13) $(u'_{\varepsilon m} + Au_{\varepsilon m} + h_{\varepsilon m} + g\, j'_\varepsilon(u_{\varepsilon m}) - f, v) = 0 \qquad \forall\, v \in V_m$.

Mais d'après (3.3) on a:

$$|(j'_\varepsilon(u), v)| \leqslant c \left(\int_\Omega D(u)^\varepsilon \, dx \right)^{1/2} \|v\|$$

de sorte que en particulier

(3.14) $\qquad g\, j'_\varepsilon(u_{\varepsilon m}) \in$ borné de $L^2(0,T;V')$.

Si P_m désigne l'opérateur de projection défini par

(3.15) $\qquad P_m f = \displaystyle\sum_{i=1}^{m} (f, w_i) w_i$,

on a alors d'après (3.13)

$$u'_{\varepsilon m} = P_m(f - Au_{\varepsilon m} - h_{\varepsilon m} - g\, j'_\varepsilon(u_{\varepsilon m}))$$

$$= P_m\, k_{\varepsilon m} ,$$

$k_{\varepsilon m}$ borné de $L^2(0,T;V')$

d'où l'on déduit (3.11) grâce au choix de la base $\{w_i\}$.

Passage à la limite en m .

D'après (3.10), (3.11) , et le Théorème de compacité du N° 1, Chap. 3, dans le 1er cas particulier du N° 1.3, on peut extraire une suite, encore notée $u_{\varepsilon m}$, telle que

J. L. Lions

$$(3.16) \quad \begin{cases} u_{\varepsilon m} \longrightarrow u_\varepsilon \quad \text{dans} \quad L^2(0,T;V) \quad \text{faible,} \\ u'_{\varepsilon m} \longrightarrow u_\varepsilon \quad \text{dans} \quad L^2(0,T;V') \quad \text{faible,} \end{cases}$$

$$(3.17) \quad u_{\varepsilon m} \longrightarrow u_\varepsilon \quad \text{dans} \quad (L^2(Q))^2 \quad \underline{\text{fort}} \ ,$$

$$(3.18) \quad j'_\varepsilon(u_{\varepsilon m}) \longrightarrow \quad \text{dans} \quad L^2(0,T;V') \quad \text{faible.}$$

Grâce à (3.16) (3.17) on a:

$$b(u_{\varepsilon m}, u_{\varepsilon m}, v) \longrightarrow b(u_\varepsilon, u_\varepsilon, v) \qquad \forall v \in V$$

<u>par ex.</u> dans $L'(0,T)$ faible et par conséquent on déduit de (3.6)(3.7) que:

$$(3.19) \quad (u'_\varepsilon, v) + a(u_\varepsilon, v) + b(u_\varepsilon, u_\varepsilon, v) + g(\chi_\varepsilon, v) = (f, v) \quad \forall v \in V \ ,$$

$$(3.20) \quad u_\varepsilon(0) = u_o \ .$$

En outre les estimations a priori obtenues étant independan<u>n</u>tes de ε (et également de g , $g \leqslant G_o < \infty$):

$$(3.21) \quad u_\varepsilon \in \text{borné de} \quad L^2(0,T;V) \ ,$$

$$(3.22) \quad u'_\varepsilon \in \text{borné de} \quad L^2(0,T;V') .$$

On aura donc résolu (3.2)(3.4) si l'on montre que

$$(3.23) \quad \chi_\varepsilon = j'_\varepsilon(u_\varepsilon) \ .$$

On utilise à cet effet un "raisonnement de monotonie". Soit φ donnée dans $L^2(0,T;V)$ avec $\varphi' \in L^2(0,T;V')$. Posons

J. L. Lions

$$(3.24) \qquad X_{\varepsilon m} = g \int_0^T (j'_\varepsilon(u_{\varepsilon m}) - j'_\varepsilon(\varphi), u_{\varepsilon m} - \varphi) dt \; +$$

$$+ \int_0^T a(u_{\varepsilon m} - \varphi) dt + \int_0^T (u'_{\varepsilon m} - \varphi', u_{\varepsilon m} - \varphi) dt \; .$$

On a:

$$(3.25) \qquad X_{\varepsilon m} \geqslant \frac{1}{2} |(u_{om} - \varphi(0)|^2 \; .$$

Par ailleurs, utilisant (3.6), on a:

$$\int_0^T \left[(u'_{\varepsilon m}, u_{\varepsilon m}) + a(u_{\varepsilon m}) + g\, j'_\varepsilon(u_{\varepsilon m}) \right] dt = \int_0^T (f, u_{\varepsilon m}) dt$$

et donc

$$X_{\varepsilon m} = \int_0^T (f, u_{\varepsilon m}) dt - g \int_0^T (j'_\varepsilon(u_{\varepsilon m}), \varphi) dt - g \int_0^T (j'_\varepsilon(\varphi), u_{\varepsilon m} - \varphi) dt$$

$$- \int_0^T \left[a(u_{\varepsilon m}, \varphi) + a(\varphi, u_{\varepsilon m} - \varphi) + (u'_{\varepsilon m}, \varphi) + (\varphi', u_{\varepsilon m} - \varphi) \right] dt$$

d'où pour $m \to \infty$

$$(3.26) \quad X_{\varepsilon m} \to \int_0^T \left[(f, u_\varepsilon) - g(\chi_\varepsilon, \varphi) - g(j'_\varepsilon(\varphi), u_\varepsilon - \varphi) - a(u_\varepsilon, \varphi) \right.$$

$$\left. - a(\varphi, u_\varepsilon - \varphi) - (u'_\varepsilon, \varphi) - (\varphi', u_\varepsilon - \varphi) \right] dt = X_\varepsilon \; .$$

Alors (3.25) donne

$$(3.27) \qquad X_\varepsilon \geqslant -\frac{1}{2} |u_o - \varphi(0)|^2 \; .$$

J. L. Lions

Mais de (3.19) (3.20) on déduit

$$\int_0^T (f,u_\varepsilon)dt = \int_0^T \left[(u_\varepsilon',u_\varepsilon) + a(u_\varepsilon) + g(\chi_\varepsilon,u_\varepsilon)\right] dt$$

de sorte que

$$(3.28) \quad X_\varepsilon = g \int_0^T (\chi_\varepsilon - j_\varepsilon'(\varphi), u_\varepsilon - \varphi)dt + \int_0^T a(u_\varepsilon - \varphi)dt +$$

$$+ \int_0^T (u_\varepsilon' - \varphi', u_\varepsilon - \varphi)dt .$$

Donc (3.27) s'écrit alors:

$$(3.29) \quad g \int_0^T (\chi_\varepsilon - j_\varepsilon'(\varphi),u_\varepsilon - \varphi)dt + \int_0^T a(u_\varepsilon - \varphi)dt + \frac{1}{2}|u_\varepsilon(T) - \varphi(T)|^2 \geqslant 0.$$

Prenant

$$\varphi = u_\varepsilon - \lambda\psi, \ \psi \in L^2(0,T;V), \ \psi' \in L^2(0,T;V'), \ \lambda > 0 ,$$

on en déduit, après division par λ :

$$g \int_0^T (\chi_\varepsilon - j_\varepsilon'(u_\varepsilon - \lambda\psi),\psi)dt + \lambda \int_0^T a(\psi)dt + \frac{\lambda}{2}|\psi(T)|^2 \geqslant 0$$

d'où faisant $\lambda \to 0$:

$$g \int_0^T (\chi_\varepsilon - j_\varepsilon'(u_\varepsilon) , \psi)dt \geqslant 0 \qquad \forall \psi$$

d'où (3.23) .

3.3. Passage à la limite en ε.

Pour $v \in L^2(0,T;V)$, on pose:

$$(3.30) \quad Y_\varepsilon = \int_0^T \left[(u_\varepsilon', v - u_\varepsilon) + a(u_\varepsilon, v-u_\varepsilon) + b(u_\varepsilon, u_\varepsilon, v - u_\varepsilon) + \right.$$
$$\left. + g\, j_\varepsilon(v) - g\, j_\varepsilon(u_\varepsilon) - (f, v - u_\varepsilon)\; dt \right].$$

Utilisant (3.2) on voit que

$$Y_\varepsilon = \int_0^T g\left[j_\varepsilon(v) - j_\varepsilon(u_\varepsilon) - (j_\varepsilon'(u_\varepsilon), v-u)\right] dt$$

et comme $j_\varepsilon(v) - j_\varepsilon(u) - (j_\varepsilon'(u), v-u) \geqslant 0$, on a:

$$(3.31) \quad\quad\quad Y_\varepsilon \geqslant 0 .$$

On déduit de (3.31) et de l'expression (3.30) de Y_ε que

$$(3.32) \quad \int_0^T \left[(u_\varepsilon', v) + a(u_\varepsilon, v) + b(u_\varepsilon, u_\varepsilon, v) + g\, j_\varepsilon(v) - (f, v - u_\varepsilon)\right] dt \geqslant$$

$$\geqslant \frac{1}{2}\,|u_\varepsilon(T)|^2 - \frac{1}{2}\,|u_0|^2 + \int_0^T a(u_\varepsilon)dt + g \int_0^T j_\varepsilon(u_\varepsilon)dt .$$

Grâce à (3.21)(3.22) et le Théorème de compacité du Chap.3, on peut extraire de u_ε une sous suite, encore notée u_ε, tel le que

$$(3.33) \quad \begin{cases} u_\varepsilon \longrightarrow u \quad \text{dans} \quad L^2(0,T;V) \quad \text{faible}, \\ u_\varepsilon' \longrightarrow u' \quad \text{dans} \quad L^2(0,T;V) \quad \text{faible}, \\ u_\varepsilon \longrightarrow u \quad \text{dans} \quad (L^2(Q))^2 \quad \text{fort}. \end{cases}$$

J. L. Lions

Grâce à (3.33) on vérifie que le 1er membre de (3.32) converge vers

$$(3.34) \quad \int_0^T \Big[(u',v) + a(u,v) + b(u,u,v) + g\,j(v) - (f,v-u)\Big]dt \ .$$

On vérifie ensuite que

$$\lim. \inf. \; |u_\varepsilon(T)|^2 \geqslant |u(T)|^2 \ ,$$

$$\lim. \inf. \; \int_0^T a(u_\varepsilon)dt \geqslant \int_0^T a(u)dt \ ,$$

$$\lim. \inf. \; \int_0^T j_\varepsilon(u_\varepsilon)dt \geqslant \int_0^T j(u)dt$$

de suite que avec (3.32) et (3.34) on obtient:

$$(3.35) \quad \begin{cases} \int_0^T \Big[(u',v-u) + a(u,v-u) + b(u,u,v-u) + g\,j(v) - g\,j(u) - (f,v-u)\Big]dt \geqslant 0 \\ \\ \forall v \in L^2(0,T;V) \ . \end{cases}$$

Le dernier point de la démonstration est de vérifier que (3.35) implique (1.8) (la réciproque est évidente). Cela est une conséquence du théorème de Lebesgue selon lequel

$$\frac{1}{2\beta} \int_{t-\beta}^{t+\beta} f(\sigma)d\sigma \longrightarrow f(t) \quad \text{p.p. lorsque } \beta \to 0 \ , \quad f \text{ étant une}$$

fonction mesurable à valeurs dans un espace de Banach.

Pour s fixé quelconque dans $]0,T[$ on introduit en effet

$$\mathcal{O}_j =]s - 1/j , s + 1/j [\subset]0,T[\text{ pour } j \text{ assez grand },$$

et pour w quelconque dans V on definit $v = v_j$ par

$$(3.36) \qquad v(t) = \begin{cases} w \text{ dans } \mathcal{O}_j \\ u(t) \text{ hors de } \mathcal{O}_j \end{cases}$$

Choisissant v par (3.36) dans (3.35), il vient:

$$(3.37) \quad \int_{\mathcal{O}_j} \Big[(u',w-u) + a(u,w-u) + b(u,u,w) + g\,j(w) - g\,j(u) - (f,w-u)\Big]dt$$

$$\geq 0 .$$

Divisant (3.37) par $2/j$ et faisant $j \to \infty$ on en déduit que, p.p. en s :

$$(u'(s),w-u(s)) + a(u(s),w-u(s)) + b(u(s),u(s),w) + g\,j(w) -$$

$$- g\,j(u(s)) - (f(s),w-u(s)) \geq 0 .$$

4. Les équations de Navier Stokes comme cas limites des inéquations.

Comme on a souligné dans la démonstration donnée au N° 3, les estimations ne dépendent pas de g pour $0 < g \leq G_0 < \infty$.

Si l'on designe par u_g la solution donnée par le Théorème 1.1, on a donc, lorsque $g \to 0$,

(4.1) u_g(resp. u_g') demeure dans un borné de $L^2(0,T;V)$ (resp. de $L^2(0,T;V')$).

On peut donc extraire une suite, encore notée u_g , telle que

(4.2) $u_g \to u$ (resp. $u_g' \to u'$) dans $L^2(0,T;V)$(resp. $L^2(0,T;V')$)

faible

et

(4.3) $u_g \to u$ dans $(L^2(Q))^2$ __fort__ .

D'après (3.35) on a:

$$\int_o^T \left[(u_g',v-u_g) + a(u_g,v-u_g) + b(u_g,u_g,v) + g\,j(v) - gj(u_g) - (f,v-u_g)\right]dt$$

$$\geqslant 0 .$$

d'où

$$\int_o^T \left[(u_g',v) + a(u_g,v) + b(u_g,u_g,v) + g\,j(v) - (f,v-u_g)\right]dt \geqslant$$

$$\geqslant \int_o^T \left[(u_g',u_g) + a(u_g) + gj(u_g)\right]dt$$

d'où l'on déduit, comme au point 3.3, que

$$\int_o^T \left[(u',v) + a(u,v) + b(u,u,v) - (f,v-u)\right]dt \geqslant$$

$$\geqslant \lim. \inf. \int_o^T \left[(u_g',u_g) + a(u_g) + g\,j(u_g)\right]dt$$

$$\geqslant \int_o^T \left[(u',u) + a(u)\right] dt$$

donc

$$\int_0^T \left[(u',v-u) + a(u,v-u) + b(u,u,v-u) - (f,v-u) \right] dt \geqslant 0$$

ce qui equivaut (d'après le 3.3) à :

$$(u'(t),v-u(t)) + a(u(t),v-u(t)) + b(u(t),u(t),v-u(t)) - (f(t),v-u(t)) \geqslant 0$$

p.p.

d'où (1.21).

5. Problèmes.

Problème 5.1. Le cas de la dimension d'espace 3.

On peut montrer (cf. DUVAUT-LIONS [1] , Chap. 6) l'existence d'une solution dans le cas de la dimension d'espace 3, mais, exactement comme pour les équations de Navier Stokes, on ne connait pas de classe fonctionnelle où l'on puisse montrer l'existence et l'unicité.

Problème 5.2

Est-il possible d'étendre au cas des inéquations les resultats de FOIAS-PRODI [1] ?

J. L. Lions

CHAPITRE 5

INÉQUATIONS VARIATIONNELLES HYPERBOLIQUES. MÉTHODE DE PENALISATION

Introduction.

Nous allons maintenant étudier les inéquations variationnelles hyperboliques, pour une équation aux derivées partielle
du 2eme ordre en t puis brièvement pour le système de Maxwell
(pour la motivation physique et l'étude complète de ce problème,
cf. DUVAUT-LIONS, Chap. 7).

Nous utiliserons la méthode de pénalisation - méthode également valable pour les "inéquations parabolique" que nous n'a-
bordons pas systématiquement dans ce cours, renvoyant à H. BREZIS
[1] [2] [3] et LIONS [1] .

1. Position du problème. Enoncé du résultat principal. Exemples.

1.1. Notations.

Soient V et H deux espaces de Hilbert (sur $\mathbb{R}$) avec
(1.1) $V \subset H$, $V \to H$ continue, V dense dans H ;
on identifie H à son duel de sorte que
(1.2) $V \subset H \subset V'$.

J. L. Lions

On désigne par $(\ ,\ )$ (resp.$|\ |$, resp. $\|\ \|$) le produit scalaire dans H (resp. la norme dans H , resp. la norme dans V) ; $(\ ,\ )$ désigne aussi le produit scalaire entre V' et V.

Soit $a(u,v)$ une forme bilinéaire continue sur V telle que

(1.3) $\qquad a(u,v) = a(v,u) \qquad \forall u,v \in V$,

(1.4) il existe λ et $\alpha > 0$ tels que

$$a(v,v) + \lambda \, |v|^2 \geqslant \alpha \, \|v\|^2 \qquad \forall v \in V \ .$$

On posera dans la suite

(1.5) $\qquad a(v,v) = a(v)$.

La forme a définit $A \in \mathcal{L}(V;V')$ par

(1.6) $\qquad (Au,v) = a(u,v)$.

On posera:

(1.7) $\qquad D(A) = \left\{ v \mid v \in V \ , \ Av \in H \right\}$.

On donne ensuite un ensemble K avec

(1.8) $\qquad K$ est un convexe fermé de V .

1.2. Position du problème.

On cherche une fonction $t \to u(t)$ de $(0,T) \to V$ telle que

(1.9) $\qquad u'(t) = \dfrac{\partial u(t)}{\partial t} \in K$,

J. L. Lions

(1.10) $\quad (u''(t), v-u'(t)) + a(u(t), v-u'(t)) \geqslant (f(t), v-u'(t)) \quad \forall v \in K$,

(1.11) $\quad u(0) = u_o \quad , \quad u'(0) = u_1$,

f , u_o et u_1 donnés.

Remarque 1.1.

Si $K = V$, (1.9) est sous objet et (1.10) se réduit à l'é-quation

(1.12) $\quad (u''(t), v) + a(u(t), v) = (f(t), v) \qquad \forall v \in V$.

Remarque 1.2.

Le problème (1.9)(1.10)(1.11) est dit " problème d'une équation hyperbolique" parce que (cf. les Exemples au N° 1.4 ci après) si l'on prend pour V l'espace $H^1(\Omega)$ (ou un sous espa-ce de $H^1(\Omega)$) l'opérateur $\frac{\partial^2}{\partial t^2} + A$ est hyperbolique. Mais si l'on prend pour V par ex. $H^2(\Omega)$ l'opérateur $\frac{\partial^2}{\partial t^2} + A$ n'est plus hyperbolique; la terminologie est donc un point de refère seulement.

1.3. Résultat principal.

On démontrera ci après le

Théorème 1.1. On suppose que (1.3)(1.4) ont lieu. On donne u_o, u_1, f avec

(1.13) $u_o \in D(A)$,

(1.14) $u_1 \in K$,

(1.15) f , $f' \in L^2(0,T;H)$.

Il existe alors une fonction u et une seule telle que

(1.16) $u, u' \in L^\infty(0,T;V)$,

(1.17) $u'' \in L^\infty(0,T;H)$

et vérifiant (1.9)(1.10)(1.11) .

1.4. Exemples.

Prenons $V = H^1(\Omega)$, $a(u,v) = \sum_{i=1}^{n} \int_\Omega \frac{\partial u}{\partial x_i} \frac{\partial v}{\partial x_i} \, dx$ et K

défini par

(1.18) $K = \left\{ v \mid v \in H^1(\Omega) , v \geqslant 0 \text{ p.p. sur } \Gamma \right\}$

ou

(1.9) $K = \left\{ v \mid v \in H^1(\Omega) , v \geqslant 0 \text{ p.p. sur } \Omega \right\}$.

On a bien (1.8).

La solution u du problème s'interprète de la façon sui-

vante:

Exemple (1.18).

Alors

(1.20) $\dfrac{\partial^2 u}{\partial t^2} = \Delta u = f$ dans Q ,

(1.21) $\frac{\partial u}{\partial t} \geqslant 0$, $\frac{\partial u}{\partial n} \geqslant 0$ (1) , $\frac{\partial u}{\partial t} \frac{\partial u}{\partial n} = 0$ sur $\sum$

(1.22) $u(x,0) = u_0(x)$, $\frac{\partial u}{\partial t}(x,0) = u_1(x)$, $x \in \Omega$.

Exemple (1.19).

Alors

$$(1.23) \begin{cases} \dfrac{\partial^2 u}{\partial t^2} - \Delta u - f \geqslant 0 \quad \text{dans} \quad Q \quad , \\[2ex] \dfrac{\partial u}{\partial t} \geqslant 0 \quad \text{dans} \quad Q \quad , \\[2ex] \left(\dfrac{\partial^2 u}{\partial t^2} - \Delta u - f \right) \dfrac{\partial u}{\partial t} = 0 \quad \text{dans} \quad Q \end{cases}$$

(1.24) $\frac{\partial u}{\partial n} = 0$ sur $\sum$

et (1.22).

2. Démonstration de l'unicité dans le Théorème 1.1.

Soient u et u_* deux solutions eventuelles et $U = u - u_*$.
Prenant, ce qui est loisible, $v = u_*'(t)$ (resp. $v = u'(t)$) dans
(1.10) (resp. dans l'inéquation analogue relative à u_*) et ad-
ditionnant, il vient:

_____

(1) $\frac{\partial}{\partial n}$ = derivée normale à Γ dirigée vers l'extérieur de Ω .

$$- (U'',U') - a(U,U') \geqslant 0$$

donc

$$\frac{d}{dt} \left(|U'(t)|^2 + a(U(t)) \right) \leqslant 0$$

donc

$$(2.1) \quad |U'(t)|^2 + \alpha \|U(t)\|^2 \leqslant \lambda |U(t)|^2 .$$

Mais $U(t) = \displaystyle\int_0^t U'(\sigma)d\sigma$ donc (2.1) entraine en particu-

lier (comme $t \leqslant T < \infty$):

$$|U'(t)|^2 \leqslant c \int_0^t |U'(\sigma)|^2 d\sigma$$

donc $U' = 0$ et donc $U = 0$.

3. Démonstration de l'existence dans le Théorème 1.1 par la mé-

thode de pénalisation.

3.1. Opérateur de pénalisation.

On introduit un opérateur β - dit "de pénalisation" at-

taché à K ᵥ ayant les propriétés suivantes:

(3.1) β est borné hémicontinue et monotone de $V \to V'$,

(3.2) noyau de β = K .

De tels operateur existent. Par exemple si l'on pose:

J. L. Lions

(3.3) P_K = opérateur de projection de $V \longrightarrow K$,

(3.4) Λ = isomorphisme canonique de $V \to V'$, défini par

$$(\Lambda u, v) = (u,v)_V \quad \forall u,v \in V ,$$

alors on peut prendre

(3.5) $$\beta = \Lambda (I - P_K) .$$

Noter qu'alors on a:

(3.6) β est Lipschitzien de $V \to V'$.

3.2. Équation pénalisée.

Pour $\varepsilon > 0$ on cherche u_ε solution de

(3.7) $$u_\varepsilon'' + A u_\varepsilon + \frac{1}{\varepsilon} \beta (u_\varepsilon') = f ,$$

(3.8) $$u_\varepsilon(0) = u_o , \quad u_\varepsilon'(0) = u_1 .$$

L'équation (3.7) est dite "équation pénalisée" attachée à l'inéquation variationnelle à résoudre.

3.3. Résolution de l'équation pénalisée.

Le problème (3.7)(3.8) est évidemment , non linéaire. On le résout par la méthode de monotonie. On commence par les solutions approchées de Faedo-Galerkin puis l'on passe a la limite par monotonie - cf. LIONS-STRAUSS [1], LIONS [1].

3.4. Estimation a priori (I) ([1]).

Prenant le produit scalaire des deux membres de (3.7) avec

$u'_\varepsilon (t) - k$, k fixé dans K , il vient:

(3.9) $\quad (u''_\varepsilon , u'_\varepsilon - k) + a(u_\varepsilon , u'_\varepsilon - k) + \frac{1}{\varepsilon} (\beta(u'_\varepsilon), u'_\varepsilon - k) = (f, u'_\varepsilon - k)$.

Mais $\beta(k) = 0$, donc $(\beta(u'_\varepsilon), u'_\varepsilon - k) = (\beta(u'_\varepsilon) - \beta(k), u'_\varepsilon - k) \geqslant 0$

et donc (3.9) entraine

(3.10) $\quad (u''_\varepsilon , u'_\varepsilon - k) + a(u_\varepsilon , u'_\varepsilon - k) \leqslant (f; u'_\varepsilon - k)$

d'où

$$\frac{1}{2} \left[|u'_\varepsilon (t)|^2 + a(u_\varepsilon (t)) \right] \leqslant \frac{1}{2} \left[|u_1|^2 + a(u_o) \right] + (u'_\varepsilon (t), k) - (u_1, k) +$$

$$+ \int_0^t a(u_\varepsilon, k) d\sigma + \int_0^t (f, u'_\varepsilon - k) d\sigma \leqslant$$

$$< \frac{1}{4} |u'_\varepsilon (t)|^2 + c + \int_0^t (\|u_\varepsilon\|^2 + |u'_\varepsilon|^2) d\sigma \ .$$

On en déduit que

(3.11) $\quad |u'_\varepsilon (t)|^2 + \|u_\varepsilon (t)\|^2 \leqslant c \, |u_\varepsilon(t)|^2 + c \int_0^t (\|u_\varepsilon\|^2 + |u'_\varepsilon|^2) d\sigma$.

Mais

([1]) En fait ces estimations a priori sont établiés en même
temps que le point 3.3. Nous séparons ces points pour la
clarté (?) de l'exposé.

J. L. Lions

$$|u_\varepsilon(t)|^2 \leq c(1 + \int_0^t |u'_\varepsilon|^2 \, d\sigma)$$

et donc (3.11) entraine:

$$(3.12) \quad |u'_\varepsilon (t)|^2 + \|u_\varepsilon (t)\|^2 \leq c + c \int_0^t (|u'_\varepsilon|^2 + \|u_\varepsilon\|^2) d\sigma$$

d'où l'on déduit que

$$(3.13) \qquad |u'_\varepsilon (t)| + \|u_\varepsilon(t)\| \leq c .$$

3.5. Estimations a priori (II).

Dérivons (3.7) en t (on justifie cela par ex. en prenant des quotients différentiels).

Il vient:

$$(3.14) \qquad u'''_\varepsilon + Au'_\varepsilon + \frac{1}{\varepsilon} (\beta(u'_\varepsilon))' = f' .$$

Faisant $t = 0$ dans (3.7) on en deduit, comme $\beta(u_1) = 0$ car $u_1 \in K$:

$$(3.15) \qquad u''_\varepsilon (0) = f(0) - Au_0 \quad (\in H) .$$

Prenant le produit scalaire des deux membres de (3.14) avec $u''_\varepsilon (t)$, on obtient

$$(3.16) \quad (u'''_\varepsilon, u'') + a(u'_\varepsilon, u''_\varepsilon) + \frac{1}{\varepsilon} ((\beta(u'_\varepsilon))', u''_\varepsilon) = (f', u''_\varepsilon) .$$

Mais

$$(\beta(u'_\varepsilon (t+h)) - \beta (u'_\varepsilon (t)) , u'_\varepsilon (t+h) - u'_\varepsilon(t)) \geq 0$$

d'où devisant par h^2 et passant à la limite:

$$(\beta(u_\varepsilon'(t))' , u_\varepsilon''(t)) \geq 0$$

et (3.16) entraine donc que

$$(3.17) \qquad (u_\varepsilon''', u_\varepsilon'') + a(u_\varepsilon', u_\varepsilon'') \leq (f', u_\varepsilon'') \quad ,$$

d'où l'on déduit (grâce à (3.15) en particulier) comme au point

3.4 précédent que

$$(3.18) \qquad |u_\varepsilon''(t)| + \|u_\varepsilon'(t)\| \leq c \quad .$$

Remarque 3.1.

Même avec des hypothèses supplémentaires sur f, u_o, u_1, on

ne peut obtenir en général d'estimations sur u_ε''' etc.... .

3.6. Estimation a priori (III).

On déduit de (3.7)(3.13)(3.18) que

$$(3.19) \quad \beta(u_\varepsilon') = \varepsilon(f - u_\varepsilon'' - Au_\varepsilon) \longrightarrow 0 \quad \text{dans} \quad L^\infty(0,T;V') \quad .$$

3.7. Passage à la limite.

On déduit de (3.13)(3.18) que l'on peut extraire de u_ε

une suite, encore notée u_ε , telle que

$$(3.20) \quad u_\varepsilon \rightarrow u \; , \; u_\varepsilon' \rightarrow u' \quad \text{dans} \quad L^\alpha(0,T;V) \quad \text{faible étoile,}$$

$$u_\varepsilon'' \rightarrow u'' \quad \text{dans} \quad L^\infty(0,T;H) \quad \text{faible étoile.}$$

Si l'on prend $v = v(t)$ avec

$$(3.21) \qquad v \in L^2(0,T;V) \; , \; v(t) \in K \quad \text{p.p.}$$

On déduit de (3.7) que

$$(u_{\varepsilon}^{''}(t) + Au_{\varepsilon}(t) - f(t), v(t) - u_{\varepsilon}^{'}(t)) = \frac{1}{\varepsilon}(\beta(v(t)) - \beta(u_{\varepsilon}^{'}(t)), v(t) - u_{\varepsilon}^{'}(t)) \geq 0$$

p.p.

et donc

$$(3.22) \qquad \int_{0}^{T} (u_{\varepsilon}^{''} + Au_{\varepsilon} - f, v - u_{\varepsilon}^{'}) dt \geq 0 \quad ,$$

ou encore

$$\int_{0}^{T} \left[(u_{\varepsilon}^{''} + Au_{\varepsilon}, v) - (f, v - u_{\varepsilon}^{'}) \right] dt \geq \int_{0}^{T} \left[(u_{\varepsilon}^{''}, u_{\varepsilon}^{'}) + (Au_{\varepsilon}, u_{\varepsilon}^{'}) \right] dt =$$

$$= \frac{1}{2} \left[|u_{\varepsilon}^{'}(T)|^{2} + a(u(T)) \right] - \frac{1}{2} \left[|u_{1}|^{2} + a(u_{0}) \right]$$

d'où l'on déduit

$$\int_{0}^{T} \left[(u^{''} + Au, v) - (f, v - u^{'}) \right] dt \geq \frac{1}{2} \lim.\inf. \left[|u_{\varepsilon}^{'}(T)|^{2} + a(u_{\varepsilon}(T)) \right] -$$

$$- \frac{1}{2} \left[|u_{1}|^{2} + a(u_{0}) \right] \geq \frac{1}{2} \left[|u^{'}(T)|^{2} + a(u(T)) \right] - \frac{1}{2} \left[|u_{1}|^{2} + a(u_{0}) \right] =$$

$$= \int_{0}^{T} (u^{''} + Au, u^{'}) dt \quad .$$

et par conséquent

$$(3.23) \qquad \int_{0}^{T} (u^{''} + Au - f, v - u^{'}) dt \geq 0 \quad \forall v \quad \text{avec (3.21).}$$

On passe de là à l'inégalité ponctuelle (1.10) comme à la fin du N° 3.3, Chap. 4.

Il ne reste plus qu'à montrer (1.9). Si φ est donnée quelconque dans $L^2(0,T;V)$ on a:

$$\int_0^T (\beta(u'_\varepsilon) - \beta(\varphi) , u'_\varepsilon - \varphi) dt \geqslant 0 .$$

D'après (3.19) on en déduit que

$$- \int_0^T (\beta(\varphi) , u' - \varphi) dt \geqslant 0 ,$$

d'où, par le raisonnement habituel

$$\beta(u') = 0 \quad \text{p.p.} \quad \text{d'où (1.9)} .$$

4. Inéquations relatives au système de Maxwell.

On indique dans ce N° les grandes lignes relatives à des problèmes d'inéquations qui se posent relativement aux opérateurs de Maxwell.

Pour la motivation physique et les détails des démonstrations, cf. DUVAUT-LIONS [1] .

4.1. Équations de Maxwell usuelles.

Dans un domaine Ω de $\mathbb{R}^3$ on cherche les couples de vecteurs D,B tels que

$$(4.1) \quad \frac{\partial D}{\partial t} - \text{Rot}(\hat{\mu} B) + J = f , \quad \text{dans} \quad Q = \Omega \times]0,T[,$$

$$(4.2) \quad \frac{\partial B}{\partial t} + \text{Rot}(\hat{\varepsilon} D) = 0 \quad \text{dans} \quad Q ,$$

où $\hat{\varepsilon} = \frac{1}{\varepsilon}$, $\hat{\mu} = \frac{1}{\mu}$, avec ε et μ fonctions constantes par morceaux égales aux constantes dielectriques et permeabilité magnetiques des milieux composants Ω et où J est le vecteur courant.

Les conditions aux limites sont, généralement,

(4.3) $n \wedge D = 0$ sur $\sum$.

On ajoute naturellement les conditions initiales.

(4.4) $D(x,0) = D_o(x)$, $B(x,0) = B_o(x)$, $x \in \Omega$.

Dans les cas usuels J est donné par la loi d'Ohm

(4.5) $J = \sigma \hat{\varepsilon} D$.

Le système (4.1)(4.2)(4.3)(4.4) est avec (4.5) un système hyperbolique du 1er ordre bien posé.

4.2. Les inéquations.

Pour certains milieux (4.5) n'est plus valable et doit être remplacé par la loi suivante:

(4.6) $\begin{cases} \text{si } |D(x,t)| < \mathcal{D}_o \;(^1) \text{ alors } J = \sigma \hat{\varepsilon} D \; ; \\ \text{si } |D(x,t)| = \mathcal{D}_o \text{ alors il existe une constante } \lambda \geqslant 0 \text{ telle que} \\ \qquad J = (\sigma \hat{\varepsilon} + \lambda)D \; . \end{cases}$

_____

(^{1}) Constante - ou plus généralement fonction constante dans
      chacun des milieux physiques composant Ω .

J. L. Lions

Formulation du problème sous forme d'inéquation variationnelle.

On introduit l'ensemble K :

$$(4.7) \quad K = \left\{ \varphi \mid \varphi \in (L^2(\Omega))^3 , |\varphi(x)| \leqslant \mathfrak{D}_0 \text{ p.p. dans } \Omega \right\} ,$$

qui est <u>convexe fermé dans</u> $(L^2(\Omega))^3$.

On vérifie alors simplement que le problème $(4.1)\ldots(4.4)$ avec la loi (4.6) <u>équivaut à la recherche de</u> D et B <u>tels que</u>

$$(4.8) \qquad D(t) \in K ,$$

$$(4.9) \quad \begin{cases} \left(\dfrac{\partial D(t)}{\partial t}, \varphi - D(t)\right) + (\sigma \hat{\varepsilon} D(t), \varphi - D(t)) - (\hat{\mu} B(t), \text{Rot}(\varphi - D(t))) \\[2mm] \qquad \geqslant (f(t), \varphi - D(t)) \quad \forall \varphi \in K \text{ tel que } \text{Rot } \varphi \in (L^2(\Omega))^3 \\[2mm] \qquad\qquad \text{et } n \wedge \varphi = 0 \text{ sur } \Gamma , \text{ } [1] \end{cases}$$

et les équations $(4.2)(4.3)(4.4)$.

On montre (DUVAUT-LIONS, loc. cit) <u>que ce problème est bien posé.</u>

La méthode suivie consiste à approcher, à l'aide d'opera-teurs de pénalisation attachés à K , le problème d'inéquation par des équations (non linéaires).

---

[1] (,) = produit scalaire dans $(L^2(\Omega))^3$.

J. L. Lions

Remarque 4.1.

Lorsque $\mathcal{D}_o \to +\infty$ on montre que la solution relative à

(4.6) converge (en un sens convenable) vers la solution relati-

ve à la loi d'Ohm usuelle.

5. Problèmes.

Problème 5.1

Dans l'exemple (1.18), N° 1.4, il y a sur $\sum$ deux régions:

$$\sum_o \quad \text{où} \quad \frac{\partial u}{\partial t} = 0 \quad \text{et} \quad \sum_1 \quad \text{où} \quad \frac{\partial u}{\partial n} = 0 \quad .$$

Quelle est la nature de la "frontière" entre $\sum_o$ et $\sum_1$?

Question analogue relative à l'exemple (1.19) où il y a

dans Q deux régions, l'une où $\frac{\partial u}{\partial t} = 0$ l'autre où $\frac{\partial^2 u}{\partial t^2} - \Delta u - f =$

$= 0$.

On notera qu'il s'agit donc de problèmes "à frontière libre"

Problème 5.2.

Étude des problèmes d'inéquations variationnelles pour les

systèmes hyperbolique du 1er ordre, non linéaires. Théorie des

chocs.

J. L. Lions

CHAPITRE 6

PERTURBATIONS SINGULIÈRES ET INÉQUATIONS VARIATIONNELLES

D'ÉVOLUTION

Introduction.

Un très vaste champ de recherches est encore, pour l'essentiel, ouvert: dans quelle mesure, et avec quelles techniques
peut-on étendre à certaines inéquations variationnelles la
théorie des perturbations singulières et des couches limites
relative à certaines équations? (Pour le théorie de la couche
limite, cf. en particulier O.A. OLEINIK [1]) .

Nous etudions ici deux exemples.

1. Perturbations singulières d'inéquations hyperboliques conduisant à de nouvelles inéquations pour des opérateurs elliptiques.

1.1. Position du problème ([1]).

Les notations sont celles du Chap. 5, N°1.1. On suppose

que

([1]) Nous n'étudions pas ici le "motivation mécanique" de ce pro
blème, pour lequelle nous renvoyons à DUVAUT-LIONS [1] .

J. L. Lions

$$(1.1) \quad a(u,v) = a(v,u) \quad \forall u,v \in V \ ,$$

$$(1.2) \quad a(v) \geq \alpha \|v\|^2 \ , \quad \alpha > 0 \ , \quad \forall v \in V \ , \quad a(v) = a(v,v) \ .$$

On donne les fonctions f et g avec

$$(1.3) \quad f, f' \ , \ f'' \in L^2(0,T;V') , \ g, g', g'' \in L^2(0,T;V') \ (^1) \ ,$$

et on suppose que

$$(1.4) \qquad g(0) = 0 \ ;$$

On donne u_1 avec

$$(1.5) \qquad u_1 \in K \ ,$$

$$(1.6) \quad K = \text{convexe fermé de } V \ , \quad 0 \in K \ .$$

Soit $\varepsilon > 0$.

Une simple variante du Théorème 1.1, Chap. 5 (cf. démonstra tion ci après) montre <u>l'existence et l'unicité de</u> u_ε <u>satisfaisant à</u>

$$(1.7) \quad u_\varepsilon \ , \ u'_\varepsilon \in L^\infty(0,T;V) \ , \quad u''_\varepsilon \in L^\infty(0,T;H) \ ,$$

$$(1.8) \quad (\varepsilon^2 u''_\varepsilon(t) + A u_\varepsilon(t) - (\varepsilon f(t) + g(t)), v - u'_\varepsilon(t)) \geq 0 \ \text{p.p.} \ , \ \forall v \in K,$$

$$(1.9) \quad u'_\varepsilon(t) \in K \ ,$$

$$(1.10) \quad u_\varepsilon(0) = 0 \ , \quad u'_\varepsilon(0) = u_1 \ .$$

$(^1)$ Il y a une différence avec (1.15),Chap.5 - cela est utile
pour les applications à la plasticité dynamique.

J. L. Lions

Notre but est de faire $\varepsilon \to 0$.

1.2. Résultat principal.

On va démontrer ci après le

Théorème 1.1. On suppose que (1.1)...(1.6) ont lieu. Lorsque $\varepsilon \to 0$, on a:

(1.11) $u_\varepsilon \to u$, $u'_\varepsilon \to u'$ dans $L^\infty(0,T;V)$ faible étoile,

(1.12) $\varepsilon u''_\varepsilon \to 0$ dans $L^\infty(0,T;H)$

où u est la solution de

(1.13) $u \in L^\infty(0,T;V)$, $u' \in L^\infty(0,T;V)$,

(1.14) $u'(t) \in K$,

(1.15) $a(u(t), v-u'(t)) \geqslant (g(t), v-u'(t))$ $\forall v \in K$,

(1.16) $u(0) = 0$.

Remarque 1.1.

Le Théorème 1.1 montre l'existence d'une solution du problème (1.13)...(1.16) - obtenu par NEDELEC [1] par la méthode
des differences finies . Dans le problème (1.13)...(1.16) l'unicité est immediate: si u et u_* sont deux solutions éventuel
les, on pose $U = u - u_*$ et l'on prend $v = u'_*(t)$ (resp.
$v = u'(t)$) dans (1.15) (resp. dans l'inéquation analogue rela
tive à u_* ; additionant il vient

$$- a\,(U(t)\quad,\quad U'(t))\geqslant 0$$

i.e. $\dfrac{d}{dt}\,a(U(t))\leqslant 0$ donc $a(U(t)) = 0$ et $U(t) = 0$.

1.3. Démonstration du Théorème 1.1.

On reprend rapidement la Démonstration du Théorème 1.1, Chap. 5. On designe par $u_{\varepsilon\eta}$ la situation du "problème pénalisé":

$$(1.17)\qquad \varepsilon^2 u''_{\varepsilon\eta} + Au_{\varepsilon\eta} + \frac{1}{\eta}\,\beta\,(u'_{\varepsilon\eta}\,) = \varepsilon\,f + g\;\;,$$

$$u_{\varepsilon\eta}(0) = 0\;\;,\quad u'_{\varepsilon\eta}(0) = u_1\;\;.$$

On établit les estimations a priori analogues à celles du Th. 1.1 , Chap. 5 mais en tenant compte des termes en ε et des hypothèses (1.3).

On prend le produit scalaire des deux membres de (1.17) avec $u'_{\varepsilon\eta}$; il vient, comme $(\beta(v),v)\geqslant 0$ (car $0\in K$) :

$$(1.18)\quad \frac{\varepsilon^2}{2}\frac{d}{dt}\,|u'_{\varepsilon\eta}(t)|^2 + \frac{1}{2}\frac{d}{dt}\,a(u_{\varepsilon\eta}(t)\,) \leqslant (\,\varepsilon\,f(t) + g(t)\,,u'_{\varepsilon\eta}(t))$$

d'où en intégrant en t :

$$\frac{\varepsilon^2}{2}|u'_{\varepsilon\eta}(t)|^2 + \frac{1}{2}\,a(u_{\varepsilon\eta}(t)) \leqslant \frac{\varepsilon^2}{2}\,(u_1)^2 + \int_0^t (\varepsilon f + g\,,\,u'_{\varepsilon\eta})d\sigma =$$

$$= \frac{\varepsilon^2}{2}\,|u_1|^2 + (\varepsilon f(t) + g(t)\,,\,u_{\varepsilon\eta}(t)) - \int_0^t (\varepsilon f' + g'\,,\,u_{\varepsilon\eta})d\sigma$$

$$\leqslant \frac{\varepsilon^2}{2}\,|u_1|^2 + (\varepsilon\|f(t)\|_* + \|g(t)\|_*\,)\,\|u_{\varepsilon\eta}(t)\| + \int_0^t \|\varepsilon f' + g\|_*\,\|u_{\varepsilon\eta}\|\,d\sigma$$

$$\leqslant \frac{\varepsilon^2}{2} |u_1|^2 + \frac{1}{4} a(u_{\varepsilon\eta}(t)) + (\varepsilon \|f(t)\|_* + \|g(t)\|_*)^2 +$$

$$+ \frac{1}{2} \int_0^t \|u_{\varepsilon\eta}(\sigma)\|^2 d\sigma + \frac{1}{2} \int_0^t \|\varepsilon f' + g'\|_*^2 d\sigma$$

d'où

$$(1.19) \; \varepsilon^2 |u'_{\varepsilon\eta}(t)|^2 + \|u_{\varepsilon\eta}(t)\|^2 < c \; \varepsilon^2 |u_1|^2 + c + c \int_0^t \|u_{\varepsilon\eta}\|^2 d\sigma$$

d'où l'on déduit (en utilisant l'inégalité de Gronwall):

(1.20) $\varepsilon |u'_{\varepsilon\eta}(t)| + \|u_{\varepsilon\eta}(t)\| \leqslant c$ (les c désignent des constantes

indépendantes de ε et de η).

Faisant "t = 0" dans (1.17) on trouve:

$$\varepsilon^2 u''_{\varepsilon\eta}(0) = \varepsilon f(0) + g(0) \qquad (\text{car } u_1 \in K) \;(^1)$$

et grâce à (1.4) on a donc:

$$(1.21) \qquad u''_{\varepsilon\eta}(0) = \frac{1}{\varepsilon} f(0) \; .$$

Dérivant (1.17) en t et prenant le produit scalaire des

deux membres avec $u''_{\varepsilon\eta}$, il vient

$$(1.22) \; \frac{\varepsilon^2}{2} \frac{d}{dt} |u''_{\varepsilon\eta}(t)|^2 + \frac{1}{2} \frac{d}{dt} a(u'_{\varepsilon\eta}(t)) + \frac{1}{\eta} (\beta(u'_{\varepsilon\eta}), u''_{\varepsilon\eta}) =$$

$$= (\varepsilon f' + g' , u''_{\varepsilon\eta}) \; .$$

$(^1)$ On pourrait supposer que u_0 est donné avec $u_0 \in D(A)$,
$Au_0 - g(0) = 0$, et imposer $u_\varepsilon(0) = u_0$.

D'apres ce qu'on a vu du Chap. 5 , N° 3.5, on en déduit:

$$\frac{\varepsilon^2}{2} \frac{d}{dt} |u''_{\varepsilon\eta}(t)|^2 + \frac{1}{2} \frac{d}{dt} a(u'_{\varepsilon\eta}(t)) \leqslant (\varepsilon f'(t) + g'(t) , u''_{\varepsilon\eta}(t))$$

d'où par intégration en t (en utilisant (1.21))

$$\frac{\varepsilon^2}{2}|u''_{\varepsilon\eta}(t)|^2 + \frac{1}{2} a(u'_{\varepsilon\eta}(t)) \leqslant \frac{\varepsilon^2}{2} | \frac{1}{\varepsilon} f(0)|^2 + \frac{1}{2} a(u_1) +$$

$$+ \int_0^t (\varepsilon f' + g' . u''_{\varepsilon\eta}) d\sigma \leqslant$$

$$\leqslant \frac{1}{2}|f(0)|^2 + \frac{1}{2} a(u_1) + (\varepsilon f'(t) + g'(t) , u'_{\varepsilon\eta}(t)) -$$

$$- (\varepsilon f'(0) + g'(0), u_1) - \int_0^t (\varepsilon f'' + g'', u'_{\varepsilon\eta}) d\sigma$$

$$\leqslant c + \|\varepsilon f'(t) + g'(t)\|_* \|u'_{\varepsilon\eta}(t)\| +$$

$$+ \|\varepsilon f'(0) + g'(0)\|_* \|u_1\| + \int_0^t \|\varepsilon f'' + g''\|_* \|u'_{\varepsilon\eta}\| d\sigma$$

d'où l'on déduit que

$$(1.23) \quad \varepsilon^2 |u''_{\varepsilon\eta}(t)|^2 + \|u'_{\varepsilon\eta}(t)\|^2 \leqslant c + c \int_0^t \|u'_{\varepsilon\eta}(\sigma)\|^2 d\sigma$$

d'où

$$(1.24) \quad \varepsilon |u''_{\varepsilon\eta}(t)| + \|u'_{\varepsilon\eta}(t)\| \leqslant c .$$

Passant à la limite en η , on en déduit l'existence de

J. L. Lions

u_ε solution de (1.7)...(1.10) avec les estimations:

(1.25) $\|u_\varepsilon(t)\| + \|u_\varepsilon'(t)\| + \varepsilon|u_\varepsilon''(t)| \le c$.

On peut alors extraire de u_ε une suite, encore notée u_ε , telle que

(1.26) $u_\varepsilon \to u$, $u_\varepsilon' \to u'$ dans $L^\infty(0,T;V)$ faible étoile,

(1.27) $\varepsilon u_\varepsilon'' \to \chi$ dans $L^\infty(0,T;H)$ faible étoile.

Mais d'après (1.26) , $u_\varepsilon'' \to u''$ par ex. au sens des distributions sur $]0,T[$ à valeurs dans V , donc

(1.28) $\chi = 0$.

Il reste à vérifier que u est solution de (1.15) (les conditions (1.13)(1.15)(1.16 étant évidentes).

Soit $v \in L^2(0,T;V)$, $v(t) \in K$ p.p. Prenant dans (1.8) $v = v(t)$ et intégrant en t , il vient

(1.29) $\varepsilon^2 \int_0^T (u_\varepsilon'' , v)dt + \int_0^T \left[(Au_\varepsilon , v) - (\varepsilon f + g, v - u_\varepsilon')\right]dt \ge$

$$\ge \varepsilon^2 \int_0^T (u_\varepsilon'' , u_\varepsilon')dt + \int_0^T (Au_\varepsilon , u_\varepsilon')dt =$$

$$= \frac{\varepsilon^2}{2}|u_\varepsilon'(T)|^2 - \frac{\varepsilon^2}{2}|u_1|^2 + \frac{1}{2}a(u_\varepsilon(T)) .$$

On déduit de (1.29) que

$$\int_0^T \left[(Au,v) - (g , v-u')\right]dt \ge \frac{1}{2}a(u(T)) = \int_0^T (Au,u')dt$$

donc

$$(1.30) \quad \int_0^T (Au - g, v-u')dt \geqslant 0 \quad \forall v \in L^2(0,T;V), \ v(t) \in K \quad p.p.$$

On déduit de (1.30) de (1.30) l'inégalité "ponctuelle"
(1.15) comme à la fin du N° 3.3 , Chap. 4 .

1.4. Résolution directe de l'inéquation d'évolution pour
l'opérateur elliptique.

Si l'on a pour seul but la résolution du problème (1.13)...
(1.16) on peut raisonner autrement. On considère l'équation pe-
nalisée

$$(1.31) \quad a(w_\eta(t),v) + \frac{1}{\eta} (\beta(w_\eta'(t)),v) = (g(t),v) \quad \forall v \in V$$

$$(1.32) \quad w_\eta(0) = 0 .$$

On montre l'existence de w_η et l'on obtient les estimations suivantes:

$$(1.33) \quad \|w_\eta(t)\| + \|w_\eta'(t)\| \leqslant c .$$

On peut ensuite passer à la limite en η .

2. Un résultat du type "couche limite".

2.1. Position du problème.

Les notations sont celles du N° 1. On suppose que

$$(2.1) \quad (Av,v) = a(v) > \alpha \|v\|^2 , \ \alpha > 0 ,$$

J. L. Lions

(2.2)   K est un convexe fermé, $0 \in K$;

f et u_o sont donnés avec

(2.3) $f , f' \in L^2(0,T;H)$,

(2.4) $u_o \in D(A)$, $u_1 \in K$.

D'après le Théorème 1.1, Chap. 5, on sait qu'il existe,

$\forall \varepsilon > 0$, une fonction u_ε et une seule telle que

(2.5) $u_\varepsilon, u_\varepsilon' \in L^\infty(0,T;V)$, $u_\varepsilon'' \in L^\infty(0,T;H)$,

(2.6) $u_\varepsilon'(t) \in K$ p.p ,

(2.7) $(\varepsilon u_\varepsilon''(t) + u_\varepsilon'(t) + Au_\varepsilon(t) - f(t), v - u_\varepsilon'(t)) \geq 0 \;\; \forall v \in K$,

(2.8) $u_\varepsilon(0) = u_o$, $u_\varepsilon'(0) = u_1$.

Notre but est de faire $\varepsilon \rightarrow 0$.

2.2. Résultats principaux.

Théorème 2.1. Sous les hypothèses (2.1)...(2.4) on a:

(2.9) $\| u_\varepsilon \|_{L^\infty(0,T;V)} + \| u_\varepsilon' \|_{L^2(0,T;H)} \leq c$,

(2.10) $\sqrt{\varepsilon} \| u_\varepsilon' \|_{L^\infty(0,T;V)} + \sqrt{\varepsilon} \| u_\varepsilon'' \|_{L^2(0,T;H)} \leq c$

(2.11) $\varepsilon \| u_\varepsilon'' \|_{L^\infty(0,T;H)} \leq c$.

Théorème 2.2. On suppose que (2.1)...(2.4) ont lieu et en outre que

$$(2.12) \qquad f(0) - Au_o \in K \quad .$$

Il existe alors une fonction u **et une seule telle que**

$$(2.13) \qquad u \in L^{\infty}(0,T;V) \quad ,$$

$$(2.14) \qquad u' \in L^2(0,T;V) \cap L^{\infty}(0,T;H) \quad ,$$

$$(2.15) \qquad u'' \in L^2(0,T;H)$$

$$(2.16) \qquad u'(t) \in K \quad p.p.$$

$$(2.17) \qquad (u'(t) + Au(t) - f(t), v-u'(t)) \geqslant 0 \quad \forall v \in K$$

$$(2.18) \qquad u(0) = u_o \quad .$$

On a en outre

$$(2.19) \qquad u'(0) = f(0) - Au_o \quad .$$

Voici maintenant le 1er résultat sur le comportement de $u_\varepsilon - u$:

Théorème 2.3. **Les hypothèses sont celles du Théorème 2.2. Lor-**
sque $\varepsilon \to 0$, **on a** :

$$(2.20) \qquad \| u_\varepsilon(t) - u'(t) \| \leqslant c \ \varepsilon^{1/2} \quad ,$$

$$(2.21) \qquad \left(\int_0^T |u'_\varepsilon(t) - u'(t)|^2 dt \right)^{1/2} \leqslant c \ \varepsilon^{1/2} \quad .$$

Le problème du type "couche limite" est maintenant le suivant: on voit que si

$$(2.22) \qquad f(0) - Au_o \neq u_1$$

alors $u'_\varepsilon(0) = u_1 \neq u'(0) = f(0) - Au_o$ et on ne saurait donc
avoir $u'_\varepsilon \to u'$ Uniformément en t . Mais posons

$$(2.23) \qquad \psi = u_1 - (f(0) - Au_0) \quad ,$$

et

$$(2.24) \qquad \psi_\varepsilon(t) = \psi \int_0^t \exp(-\sigma/\varepsilon) d\sigma \quad .$$

Si l'on introduit:

$$(2.25) \qquad w_\varepsilon = u_\varepsilon - u - \psi_\varepsilon$$

on va obtenir - sous des hypothèses supplémentaires.- la con-
vergence uniforme de w_ε' vers 0 ; comme ψ_ε' est "concentré"
au voisinage de $t = 0$, c'est un résultat du type "couche limi
te".

Les hypothèses supplémentaires sont les suivantes; on sup
posera d'abord que

$$(2.26) \qquad \psi \in D(A) \quad ,$$

puis - et c'est cela l'hypothèse principale! -

$$(2.27) \quad \begin{cases} \forall k \in K \text{ et } \forall \theta \text{ avec } |\theta| \le 1, \ \theta \in \mathbb{R} \ , \text{ on a:} \\ k + \theta \psi \in K \ . \end{cases}$$

(Voir des Exemples plus loin).

Alors

Théorème 2.4. On se place dans les conditions du Théorème 2.2
avec en outre (2.26) et (2.27). Alors

(2.28) $\qquad |w'_\varepsilon(t)| \leqslant c \, \varepsilon^{1/2}$.

2.3. Démonstrations.

Démonstrations du Théorème 2.1.

On "approche" (2.7) par le problème "penalisé"

(2.29) $\qquad \varepsilon u''_{\varepsilon\eta} + u'_{\varepsilon\eta} + A u_{\varepsilon\eta} + \frac{1}{\eta}\beta(u'_{\varepsilon\eta}) = f$,

(2.30) $\qquad u_{\varepsilon\eta}(0) = u_o$, $\quad u'_{\varepsilon\eta}(0) = u_1$.

Prenant le produit scalaire des deux membres de (2.29) avec $u'_{\varepsilon\eta}$ on obtient

(2.31) $\qquad \begin{cases} \|u_{\varepsilon\eta}(t)\| \leqslant c \ , \\[2mm] \displaystyle\int_o^T |u'_{\varepsilon\eta}(t)|^2 dt \leqslant c \ , \\[2mm] \sqrt{\varepsilon}\, |u'_{\varepsilon\eta}(t)| \leqslant c \ , \end{cases}$

Prenant $t = 0$ dans (2.29) on trouve (car $\beta(u_1) = 0$) :

(2.32) $\qquad \varepsilon u''_{\varepsilon\eta}(0) = f(0) - A u_o - u_1$.

Dérivant (2.29) en t , il vient:

$$\varepsilon\, u'''_{\varepsilon\eta} + u''_{\varepsilon\eta} + A u'_{\varepsilon\eta} + \frac{1}{\eta}(\beta(u'_{\varepsilon\eta}))' = f'$$

d'où l'on déduit:

$$\frac{\varepsilon}{2}|u''_{\varepsilon\eta}(t)|^2 + \int_o^t |u''_{\varepsilon\eta}(\sigma)|^2 \, d\sigma + \frac{1}{2} a(u'_{\varepsilon\eta}(t)) \leqslant$$

$$\leqslant \frac{\varepsilon}{2}\left|\frac{1}{\varepsilon}(f(0) - A u_o - u_1)\right|^2 + \frac{1}{2} a(u_1) + \int_o^t (f', u''_{\varepsilon\eta}) \, d\sigma$$

d'où

$$\varepsilon^2 |u''_{\varepsilon\eta}(t)|^2 + 2\varepsilon \int_0^t |u''_{\varepsilon\eta}(\sigma)|^2 d\sigma + \varepsilon a(u'_{\varepsilon\eta}(t)) \leq$$

$$\leq |f(0) - Au_o - u_1|^2 + \varepsilon a(u_1) + \varepsilon \int_0^t |f'| |u''_{\varepsilon\eta}| d\sigma \quad .$$

On en déduit que

$$(2.33) \quad \sqrt{\varepsilon} \|u''_{\varepsilon\eta}\|_{L^2(0,T;H)} + \varepsilon |u''_{\varepsilon\eta}(t)| + \sqrt{\varepsilon} \|u'_{\varepsilon\eta}(t)\| \leq c \quad .$$

On déduit de $(2.31)(2.33)$ les estimations $(2.9)(2.10)(2.11)$
par passage à la limite en η .

Démonstration du Théorème 2.2. (Principe).

On considère l'équation pénalisée

$$(2.34) \qquad u'_\eta + Au_\eta + \frac{1}{\eta}\beta(u'_\eta) = f \quad ,$$

$$(2.35) \qquad u_\eta(0) = u_o \quad .$$

Prenant le produit scalaire des deux membres de (2.34)
avec u'_η on en déduit:

$$(2.36) \qquad \|u_\eta(t)\| \leq c \quad , \qquad \int_0^T |u'_\eta|^2 dt \leq c \quad .$$

Prenant $t = 0$ on a:

$$(2.37) \qquad u'_\eta(0) + \frac{1}{\eta}\beta(u'_\eta(0)) = f(0) - Au_o \quad .$$

Mais d'après (2.12), $f(0) - Au_0$ est une solution de l'équation

$$w + \frac{1}{\eta} \beta(w) = f(0) - Au_0 \; .$$

Comme (monotonie de β) cette équation admet au plus une solutions, on a donc

(2.38) $u_\eta' (0) = f(0) - Au_0 \; .$

Dérivant (2.34) en t puis prenant le produit scalaire des deux.membres avec u_η'' , il vient:

(2.39) $|u_\eta'' (t)|^2 + \frac{1}{2} \frac{d}{dt} a(u_\eta'(t)) \leqslant (f'(t), u_\eta''(t))$

d'où

(2.40) $\displaystyle\int_0^T |u_\eta''(t)|^2 dt \leqslant c \; , \quad \|u_\eta' (t)\| \leqslant c \; .$

Des estimations (2.36)(2.40) et de (2.34) on déduit que

(2.41) $\beta(u_\eta') = \eta(f - u_\eta' - Au_\eta) \longrightarrow 0$ dans $L^\infty(0,T;V')$.

On en déduit l'existence de u avec (2.13)...(2.18), et (2.19) résulte de (2.38).

Pour l'unicité de u , on suit la méthode habituelle.

Démonstration du Théorème 2.3.

Prenons $v = u'(t)$ (resp. $v = u_\xi'(t)$) dans (2.7) (resp.(2.17)),

J. L. Lions

si l'on pose: $U_\varepsilon = u_\varepsilon - u$, il vient:

$$- (\varepsilon u''_\varepsilon + U'_\varepsilon + A U_\varepsilon \, , \, U'_\varepsilon \,) \geq 0$$

d'où

$$|U'_\varepsilon(t)|^2 + \frac{1}{2} \frac{d}{dt} a(U_\varepsilon(t)) \leq - \varepsilon(u''_\varepsilon(t), U'_\varepsilon(t)) \leq$$

$$\leq \frac{1}{2} |U'_\varepsilon(t)|^2 + \frac{\varepsilon^2}{2} |u''(t)|^2$$

d'où

$$|U'_\varepsilon(t)|^2 + \frac{d}{dt} a(U_\varepsilon(t)) \leq \varepsilon^2 |u''_\varepsilon(t)|^2$$

et comme $U_\varepsilon(0) = 0$ on a donc

$$\int_0^t |U'_\varepsilon (\sigma)|^2 d\sigma + a(U_\varepsilon(t)) \leq \varepsilon^2 \int_0^t |u''_\varepsilon (\sigma)|^2 d\sigma \leq$$

$$\leq \text{(par la 2}^{\text{eme}} \text{ éstimation (2.10))} \, c \, \varepsilon$$

d'où (2.20) et (2.21).

<u>Démonstration du Théorème</u> 2.4.

On choisit dans (2.7):

(2.42) $v = u'(t) + \psi'_\varepsilon (t) = u'(t) + e^{-t/\varepsilon} \psi , \in K$ <u>d'après</u> (2.27)

et l'on choisit dans (2.17):

(2.43) $v = u'_\varepsilon(t) - \psi'_\varepsilon(t) = u'_\varepsilon(t) - e^{-t/\varepsilon} \psi , \in K$ <u>d'après</u> (2.27)

Dans (2.7) $v - u'_\varepsilon(t) = -w'_\varepsilon(t)$ (avec la notation (2.25))

et dans (2.17), $v - u'(t) = w'_\varepsilon(t)$; donc additionant, il vient:

J. L. Lions

(2.44) $\quad -(\varepsilon u''_\varepsilon + u'_\varepsilon + A u_\varepsilon - f - (u' + A u - f), w'_\varepsilon) \geqslant 0$.

Remplaçant u_ε par $w_\varepsilon + u + \psi_\varepsilon$ et notant (on a fait ce qu'il fallait pour ça) que $\varepsilon \psi''_\varepsilon + \psi'_\varepsilon = 0$, il vient:

$$-(\varepsilon w''_\varepsilon + w'_\varepsilon + A w_\varepsilon + A \psi_\varepsilon + \varepsilon u'', w'_\varepsilon) \geqslant 0$$

d'où

(2.45) $\quad \dfrac{\varepsilon}{2} \dfrac{d}{dt} (w'_\varepsilon(t))^2 + |w'_\varepsilon(t)|^2 + \dfrac{1}{2} \dfrac{d}{dt} a(w_\varepsilon(t)) \leqslant$

$$\leqslant - (\varepsilon u'' + A \psi_\varepsilon , w'_\varepsilon(t)) .$$

Mais $A \psi_\varepsilon = \varepsilon(1 - e^{-t/\varepsilon}) A \psi$ donc

$$|A \psi_\varepsilon| \leqslant \varepsilon |A \psi|$$

et le 2^{em} membre de (2.45) est majoré par

$$\varepsilon |u''(t)| \; |w'_\varepsilon(t)| + \varepsilon |A \psi| \; |w'_\varepsilon(t)| \leqslant$$

$$\leqslant |w'_\varepsilon(t)|^2 + \dfrac{1}{2} \varepsilon^2 (|u''(t)|^2 + |A \psi|^2) \quad ;$$

portant dans (2.45) on en deduit que

(2.46) $\quad \varepsilon \dfrac{d}{dt} |w'_\varepsilon(t)|^2 + \dfrac{d}{dt} a(w_\varepsilon(t)) \leqslant \varepsilon^2 (|u''(t)|^2 + |A \psi|^2)$.

Mais $w_\varepsilon(0) = 0$ et $w'_\varepsilon(0) = u'_\varepsilon(0) - u'(0) - \psi = 0$ d'après (2.8), (2.19) et (2.23). Donc (2.46) donne

(2.47) $\quad \varepsilon |w'_\varepsilon(t)|^2 + a(w_\varepsilon(t)) \leqslant \varepsilon^2 \left(\displaystyle\int_0^t |u''(\sigma)|^2 \, d\sigma + t |A \psi|^2 \right)$

d'où, en particulier:

J. L. Lions

$$(2.48) \qquad |w_{\varepsilon}'(t)|^2 \leq \varepsilon \left[\int_0^t |u''(\sigma)|^2 d\sigma + t|A\psi|^2 \right] .$$

D'après (2.15) on en déduit (2.28).

2.4. Exemples.

1) Prenons

$$V = H^1(\Omega) .$$

$$a(u,v) = \int_{\Omega} \left(\sum_{t=1}^{n} \frac{\partial u}{\partial x_i} \frac{\partial v}{\partial x_i} + u\,v \right) dx ,$$

$$K = \{ v \mid v \in V , \quad v \geq 0 \quad \text{sur } \Gamma \quad \text{p.p.} \} .$$

Alors (Ω étant borné de frontière assez régulière):

$$D(A) = \{ v \mid v \in H^2(\Omega) , \frac{\partial v}{\partial n} = 0 \} .$$

On donne donc $u_o \in H^2(\Omega)$, $\dfrac{\partial u_o}{\partial n} = 0$ et

$$\psi = u_1 - \Delta u_o + u_o - f(0) .$$

On suppose que $\psi \in H^2(\Omega)$, $\dfrac{\partial \psi}{\partial n} = 0$.

Si l'on suppose en outre que

$$\psi = 0 \quad \text{sur } \Gamma$$

alors $k + \vartheta \psi \in K$ $\forall k \in K$ et $\forall \vartheta \in \mathbb{R}$ d'où, en particulier l'hypothèse (2.27).

On est dans les conditions d'applications du Théorème 2.4.

2) Prenons $V = H_o^1(\Omega)$, a comme ci dessus et

$K = \Big\{ v \Big| \ v \geqslant 0 \ \text{p.p. sur} \ E, \ E \ \text{ensemble de mesure} > 0 \ \text{contenu} \\ \text{dans} \ \Omega \Big\}.$

Si l'on suppose que

$$\Psi = 0 \quad \text{sur} \quad E$$

on est encore dans les conditions d'applications du Théorème 2.4.

3. Problèmes.

Comme on a dit dans l'Introduction la théorie des couches limites pour les inéquations variationnelles est un vaste champ à defricher! On signale seulement ici quelques problèmes directement liés à ce qui précède.

Problème 3.1.

Etude du problème du N° 1 lorsque $g(0) \neq 0$

Problème 3.2.

Peut-on élargir les hypothèses (2.27) ?

Il semble plausible que le résultat du Théorème 2.4 n'est pas vrai sans l'hypothèse du type (2.27). Contre exemple?

Problème 3.3.

Peut-on améliorer l'estimation (2.28) (sous hypothèses supplementaires). Si $K = V$ (cas des équations), c'est possible , cf. ZLAMAL [1] [2].

J. L. Lions

CHAPITRE 7

PROBLÈMES VARIÉS

Introduction.

On présente, de manière très succinte, un certain nombre de directions.de recherches.

1. Tableau des problèmes d'inéquations pour les opérateurs paraboliques.

Relativement à un operateur parabolique "abstrait"

$$(1.1) \qquad \frac{\partial}{\partial t} + A$$

A étant un opérateur linéaire ou non, on rencontre les problèmes suivants:

(i). recherche de $u = u(t)$ avec

$$(1.2) \qquad u(t) \in K$$

$$(1.3) \quad (u'(t) + Au(t) - f(t) , v-u(t)) \geqslant 0 \;\; \forall v \in K \;\; ;$$

introduisant la fonction indicatrice de K :

$$(1.4) \qquad \Psi_K(v) = \begin{cases} 0 \;\; \text{si} \;\; v \in K \\ +\infty \;\; \text{si} \;\; v \notin K \;\; , \end{cases}$$

on peut remplacer $(1.2)(1.3)$ par

$(1.5) \quad (u'(t) + Au(t) - f(t), v-u(t)) + \Psi_K(v) - \Psi_K(u(t)) \geq 0 \; \forall v \in V .$

$\underline{\;}$(On a rencontré un problème de ce type mais avec $\Psi_K(v)$ rempl$\underline{a}$ cé par une fonction $j(v)$ partout finie au Chap. 4). Naturelle$\underline{\;}$ ment à (1.2)(1.3) on ajoute

$(1.6) \qquad\qquad u(0) = u_o \;$ donné (dans K) .

(ii) recherche de $u = u(t)$ avec

$(1.7) \qquad\qquad u'(t) \in K ,$

$(1.8) \quad (u'(t) + Au(t) - f(t) , v-u'(t)) \geq 0 \; \forall v \in K$

avec (1.6) .

On a rencontré ce problème au Th. 2.2 , Chap. 6.

Introduisant les puissances frationnaires A^θ de $A (\theta \in]0,1[)$ on peut considérer les deux classes suivantes:

(j) recherche de $u = u(t)$ avec

$(1.9) \qquad\qquad A^\theta u(t) \in K ,$

$(1.10) \quad (u'(t) + Au(t) - f(t) , v-A^\theta u(t)) \geq 0 \quad \forall v \in K ,$

et (1.6) ;

(jj) recherche de $u = u(t)$ avec

$(1.11) \qquad\qquad A^\theta u'(t) \in K ,$

$(1.12) \quad (u'(t) + Au(t) - f(t) , v-A^\theta u'(t)) \geq 0 \quad \forall v \in K ,$

et (1.6).

J. L. Lions

Exemple d'application.

On arrive ainsi à montrer l'existence et l'unicité d'une
fonction u telle que

$$u \in L^{\infty}(0,T ; H^3(\Omega)) \ ,$$

$$\frac{\partial u}{\partial t} \in L^2(0,T;H^3(\Omega)) \cap L^{\infty}(0,T;H^1(\Omega)) \ ,$$

$$\frac{\partial u}{\partial t} + \Delta^2 u - 2\Delta u + u = f \quad \text{dans} \quad Q$$

$$\frac{\partial u}{\partial n} = 0 \quad \text{sur} \quad \Sigma \ ,$$

$$-\Delta u + u \geq 0 \ , \quad \frac{\partial \Delta u}{\partial n} \leq 0 \ , \quad \frac{\partial \Delta u}{\partial n}(-\Delta u + u) = 0 \quad \text{sur} \quad \Sigma$$

$$u(x,0) = 0 \ ,$$

pour f donnée avec $f, \frac{\partial f}{\partial t} \in L^2(Q)$, $f(x,0) = 0$.

2. Interpolation non linéaire.

Dans beaucoup des problèmes qui precèdent (cf. aussi les
travaux de DA PRATO [1]) on peut résoudre les équations (ou iné
quations) dans deux cadres fonctionnels. On a explicité cela aux
Chap. 2 et 3 sur un exemple. On a alors la situation suivante:
l'application (non linéaire)

$$\left.\begin{array}{l} \text{données} \quad (f,u_o) \\ \\ \text{ou} \qquad (f,u_o,u_1) \quad \text{etc} \end{array}\right\} \longrightarrow \text{solution}$$

soit $\qquad \xi \longrightarrow u = G(\xi)$

applique $A_o \longrightarrow B_o$ et $A_1 \longrightarrow B_1$ où A_o , A_1

B_o , B_1 sont des espaces de Banach (ou des classes non linéaires).

Lorsque l'on a affaire à des espaces de Banach, avec $A_o \subset A_1$, $B_o \subset B_1$, alors si G est un operateur localement Lipschitzien de $A_1 \longrightarrow B_1$ et continu localement borné de $A_o \longrightarrow B_o$ alors (LIONS [4] , TARTAR [1]) G envoie les "espaces de traces" entre A_o et $A_1 \longrightarrow$ "les espaces de traces" correspondants entre B_o et B_1 .

Pour les operateurs Lipschitziens de $A_1 \longrightarrow B_1$, cf. PEETRE [1] , LIONS [3], et pour les opérateurs Lipschitziens dans les deux couples A_1, B_1 et A_o, B_o, cf. BROWDER [2] ([1])

Applications.

(i) Résultats de régularité "intermediaire". Cf. LIONS [3], [4], L.TARTAR [1] ;

(ii) intérpolation des hypothèses sur les coefficients ; cf. L. TARTAR [1] [2] .

([1]) Qui a, plus généralement, demontré que dans ce cas G envoie tout espace d'interpolation $\phi(A_o, A_1)$ dans l'espace correspondant $\phi(B_o, B_1)$.

3. <u>Un résultat de "convexité"</u>.

Motivation: <u>contrôle optimal</u> de systèmes gouvernés par des équations aux derivées partielles non linéaires et, en particulier, des <u>inéquations variationnelles</u>. Le résultat qui suit est, à cet égard, un "résultat préliminaire".

Pour g donné dans $L^2(\Sigma)$ on désigne par $u = u(g)$ <u>la solution</u> (^{1}) de

(3.1) $\qquad \dfrac{\partial u}{\partial t} - \Delta u = 0$ dans Q ,

(3.2) $\qquad u \geqslant 0$, $\dfrac{\partial u}{\partial n} \geqslant g$, $u\left(\dfrac{\partial u}{\partial n} - g\right) = 0$ sur Σ ,

(3.3) $\qquad u(x,0) = 0$.

On va démontrer le

<u>Théorème 3.1</u>. <u>Pour</u> $g_1, g_2 \in L^2(\Sigma)$ <u>et</u> $\theta \in [0,1]$, <u>on a</u>

(3.4) $u((1-\theta)g_1 + \theta g_2) \leqslant (1-\theta)u(g_1) + \theta u(g_2)$ p.p. dans Q .

<u>Démonstration</u>.

1) Supposons que

(3.5) $\qquad g \in L^2(\Sigma)$, $\dfrac{\partial g}{\partial t} \in L^2(\Sigma)$, $g(x,0) = 0$, $x \in \Gamma$.

_____

(^{1}) Dans un sens qui est précisé plus loin.

Il existe alors une fonction $u = u(g)$ et une seule telle que

$$(3.6) \qquad u, \frac{\partial u}{\partial t} \in L^2(0,T;H^1(\Omega)) \ ,$$

$$(3.7) \quad (u'(t),v-u(t))+a(u(t), v-u(t)) \geqslant \int_{\Gamma} g(t)(v-u(t))dt \ \forall v \in K$$

$$(3.8) \qquad u(t) \in K \ , \quad u(0) = 0 \ ,$$

où

$$(3.9) \qquad a(u,v) = \int_{\Omega} \sum_{i=1}^{n} \frac{\partial u}{\partial x_i} \frac{\partial v}{\partial x_i} \ dx \ ,$$

$$(3.10) \quad K = \left\{ v \mid v \in H^1(\Omega) \ , \quad v \geqslant 0 \text{ p.p. sur } \Gamma \right\} .$$

<u>Démonstration</u>: par ex. par pénalisation.

2) Si g est dans $L^2(\sum)$, on approche g dans $L^2(\sum)$ par une suite de fonctions g_α satisfaisant à (3.5) (de telles suites existent!) . On montre qu'alors u_α solution de $(3.6)(3.7)$ (3.8) pour $g = g_\alpha$ converge dans $L^2(0,T;H^1(\Omega))$ vers <u>la</u> solution u de (forme "faible" de (3.7)) :

$$(3.11) \qquad \int_0^T \left[(v',v-u) + a(u,v-u) - (f,v-u) \right] dt \geqslant 0$$

$\forall v \in L^2(0,T;H^1(\Omega)), v(t) \in K , v' \in L^2(0,T;L^2(\Omega)) = L^2(Q) ,$

$v(0) = 0 .$

<u>Démonstration</u>: BREZIS [2], LIONS [1].

J. L. Lions

3) D'après le 2), tout revient donc à montrer (3.4) pour

g_1 et g_2 satisfaisant à (3.5). Posons: $u(g_i) = u_i$, $u(g) = u$,

$g = (1-\theta)g_1 + \theta g_2$; alors:

$$(3.12) \begin{cases} (u_i'(t), v_i - u_i(t)) + a(u_i(t), v - u_i(t)) \geqslant \int_\Gamma g(t)(v_i - u_i(t))dt \; \forall v_i \in K, \\[2mm] i = 1, 2, \qquad u_i(0) = 0 , \end{cases}$$

et u satisfait à (3.7).

On introduit

$$(3.13) \qquad U = u - (1-\theta)u_1 - \theta u_2 .$$

On note que sur Γ

$$u(t)|_\Gamma = (1-\theta)u_1(t) + \theta u_2(t) + U(t)|_\Gamma \geqslant U(t)|_\Gamma \quad (\text{cas } u_i(t) \in K)$$

et donc

$$u^+(t)|_\Gamma = u(t)|_\Gamma \geqslant U^+(t)|_\Gamma .$$

Donc

$$(3.14) \qquad v = u(t) - U^+(t) \in K .$$

Il est par ailleurs immediat que

$$(3.15) \qquad v_1 = u_1(t) + (1-\theta) U^+(t) \in K ,$$
$$v_2 = u_2(t) + \theta U^+(t) \in K .$$

Prenant v par (3.14) dans (3.7) et v_i par (3.15) dans

(3.12), il vient, en notant que $v - u(t) = - U^+(t)$, $v_1 - u_1(t) =$

$= (1-\theta) U^+(t)$, $v_2 - u_2(t) = \theta U^+(t)$:

$$- \left| (U'(t), U^+(t)) + a(U(t), U^+(t)) \right] \geqslant 0$$

i.e.

(3.16) $\qquad \frac{1}{2} \frac{d}{dt} \left| U^+(t) \right|^2 + a(U^+(t)) \leqslant 0$.

Mais $U^+(0) = 0$ donc (3.16) entraine $U^+ = 0$ donc

$$U \leqslant 0 \quad \text{p.p. dans} \quad Q$$

d'où (3.4).

4. Solutions presque periodiques des inéquations d'évolution.

Une question naturelle, faisant l'objet des travaux de
M. BIROLI [1] [2] , est de voir dans quelle mesure on peut éten
dre les résultat d'AMERIO-PROUSE [1] relatifs aux solutions presque periodiques des équations d'évolution aux inéquations.

5. Problèmes.

On donne ici quelques problèmes généraux.

Problème 5.1.

Que peut-on dire des inéquations paraboliques d'évolution
dans les espaces de Banach non réflexifs? Cas particulier important: espace

$$W^{',1}(\Omega) = \left\{ v \mid v \in L^1(\Omega) , \frac{\partial v}{\partial x_i} \in L^1(\Omega) \quad \forall i \right\} \quad .$$

Pour des résultat dans ce sens, cf. GRANDALL-LIGGETT [1],

R. TEMAM [2] .

Problème 5.2.

Structure de l'ensemble des valeurs à l'instant T des solutions d'une inéquation variationnelle parabolique lorsque

u(0) varie .

Problème 5.3

Inéquations variationnelles pour les systèmes hyperboliques
du 1er ordre, symetriques ou non.

J. L. Lions

APPENDICE

MULTIPLICATEURS DE LAGRANGE

On va donner un résultat relatif à la situation du Chap.4 .
Mais la méthode est susceptible d'applications dans des situations différentes, c'est pourquoi nous l'avons séparée du Chap.
4.

D'autres résultats de ce genre sont donnés, par des méthodes légèrement differentes, dans un travail à paraitre de R.
Temam ([1]) .

Théorème. Soit u la solution de (1.8)(1.9) donnée au Chap. 4,
Th. 1.1. Il existe alors des fonctions $m = \{m_{ij}\}$ et une distribution p sur Q telles que

(1) $m_{ij} \in L^{\infty}(Q) \qquad \forall i,j$,

(2) $\sum_{i,j} m_{i,j}^2 \leqslant 1$ p.p. dans Q ,

(3) $\sum m_{ij} \dfrac{\partial u_i}{\partial x_j} = D(u)^{1/2}$ p.p. dans Q ([2])

_____

[1] C'est d'ailleurs après discussion avec R. Temam que cet Appendice a été rédigé dans la forme présente.

[2] On rappelle que $D(v) = \left(\sum_{i,j} \left(\dfrac{\partial v_i}{\partial x_j}\right)^2 \right)^{1/2}$.

(4) $\quad \dfrac{\partial u_i}{\partial t} - \Delta u_i + \sum u_j \dfrac{\partial u_i}{\partial x_i} - g \sum \dfrac{\partial}{\partial x_j} m_{ij} = f_i - \dfrac{\partial p}{\partial x_i}$,

$\underline{\text{Réciproquement}}$ $\underline{\text{soient}}$ u , m_{ij},p $\underline{\text{donnés}}$ $\underline{\text{avec}}$

(5) $\qquad u \in L^2(0,T;V) \quad , \quad u' \in L^2(0,T;V')$

(6) $\qquad u(0) = u_o$

$\underline{\text{et les relations}}$ (1)...(4).$\underline{\text{Alors}}$ u $\underline{\text{est la solution}}$ de (1.8) (1.9)

$\underline{\text{donnée au Théorème}}$ 1.1 , Chap. 4.

$\underline{\text{Démonstration.}}$

Posant b(u,v,w) = (B(u,v),w), B(u,v)$\in V'$, on introduit

(7) $\qquad F = u' + Au + B(u,u) - f$.

On a (la dimension d'espace étant égale à 2):

$$F \in L^2(0,T;V') \ .$$

Alors l'inequation variationnelle s'écrit

(8) $\quad (F,v) + g\, j(v) - \left[(F,u) + g\, j(u) \right] \geqslant 0 \qquad \forall v \in V$

et donc, $\forall v \in L^2(0,T;V)$:

(9) $\quad \displaystyle\int_o^T \left[(F,v) + g\, j(v) \right] dt - \int_o^T \left[(F,u) + g\, j(u) \right] dt \geqslant 0$.

Remplaçant v par $\pm \lambda v$ cela équivaut encore à

$$\lambda \left[\pm \int_o^T (F,v)dt + g \int_o^T j(v)dt \right] - \int_o^T \left[(F,u) + g\, j(u) \right] dt \geqslant 0$$

ce qui équivaut encore à

J. L. Lions

(10) $\qquad \left| \int_0^T (F,v)dt \right| \leqslant g \int_0^T j(v)dt .$

(11) $\qquad \int_0^T \left[(F,u) + g \, j(u) \right] dt = 0$

On introduit maintenant

(12) $\qquad \begin{cases} \Phi = (L^1(Q))^4 , \quad \varphi = \left\{ \varphi_{ij} \right\} \epsilon \Phi, \; i,j=1,2 \\[2mm] \|\varphi\|_\Phi = \int_Q (\sum \varphi_{ij}^2)^{1/2} \, dx \, dt \end{cases}$

et l'application

$$ v \longrightarrow \pi \, v = \left\{ \frac{\partial v_i}{\partial x_j} \right\} , $$

de $L^2(0,T;V) \longrightarrow \Phi$.

Alors (10) équivaut à

(13) $\qquad \left| \int_0^T (F,v)dt \right| \leqslant g \; \|\pi v\|_\Phi .$

D'après le Théorème de Hahn-Banach, il existe donc

$m \in (L^\infty(Q))^4, \; \|m\| \leqslant 1 \quad$ où $\quad m = \left\{ m_{ij} \right\} ,$

$$ \|m\| = \text{sup. ess.} \left(\sum_{i,j=1}^2 m_{ij}^2 \right)^{1/2} , $$

de façon que

J. L. Lions

$$(14) \qquad \int_0^T (F,v)dt = - g \sum_{i,j} \int_Q m_{ij} \frac{\partial v_i}{\partial x_j} \, dx \, dt$$

On en déduit (1)(2)(4). Utilisant (14) avec $v = u$ et comparant à (11) on a (3) .

La reciproque est immédiate.

BIBLIOGRAPHIE

S. AGMON, A. DOUGLIS, L. NIREMBERG [1] Estimates near the boun
dary for solutions of elliptic partial differential
equations satisfying general boundary conditions I ,
Comm. Pure Applied Math. 12 (1959), 623-727; II , id.
17 (1964), 35-92.

L. AMERIO-G. PROUSE [1] Abstract almost periodic functions and
functional Analysis. Van Nostrand - New York - 1970.

J.P. AUBIN [1] Un théorème de compacité. C.R. Acad. Sc. t. 256
(1963), 5042-5044.

M. BIROLI [1] Divers travaux en cours de publication.

H. BREZIS [1] Equations et inéquations non linéaires dans les
espaces vectoriels en dualité. Annales Institut Fourier
18 (1968), 115-175.

[2] Inéquations variationnelles associées à des opéra
teurs d'évolution. Nato Summer School - Venise, Juin
1968.

[3] Thèse , Paris - 1970 .

F. BROWDER [1] Non linear operators and non linear equations of
evolution in Banach spaces. Proc. Symp. on non linear
functional analysis. Chicago 1968 , Amer. Math. Soc.
Pub. 1970.

[2] Remarks on non linear interpolation in Banach
spaces. J. Funct. Analysis - 4 (1969), 390-403 .

J. CHAZARAIN [1] Thèse, Paris 1970; à paraitre au J. of Functional
Analysis.

M.G. CRANDALL-T.M. LIGGETT [1] Generation of semi groups of non
linear transformations on general Banach spaces. A
paraitre.

G. DA PRATO [1] Somme d'applications non linéaires dans des cô-
nes et équations d'évolution dans des espaces d'opérateurs . A paraitre

J.A. DOUBINSKI [1] Certaines inégalités integrales et resolution
de systèmes d'équations elliptiques quasi linéaires de
générées. Mat. Sbornik 64 (1964) , 458-480 .

G. DUVAUT et J.L. LIONS [1] Sur les inéquations en Mécanique et
en Physique en préparation.

C. FOIAS et G. PRODI [1] Sur le comportement global des solutions
non stationnaires des equations de Navier Stokes en di
mension 2. Rend. Sem. Mat. Padova XXXIX (1967), 1-34,
et Cours de C. FOIAS , ce Volume.

P. GRISVARD [1] A paraitre aux Annales E.N.S. Paris, 1970.

K. JÖRGENS [1] Das Aufangswertproblem in Grossen für eine klas-
se nichlinearer welle ng leichungen; Math. Zeitsch. 77
(1961) , 295-308.

O.A. LADYZENSKAYA [1] La théorie mathématique des fluides visqueux incompressibles. Moscou, 1961 - Trad. Anglaise,
Gordon Breach, 1963.

J. LERAY [1] Etude de diverses équations, intégrales non linéai
res et de quelques problèmes que posent l'hydrodynamique.
J. Math. Pures et Appl. XII (1933) , 1-82 .

[2] Essai sur le mouvement plan d'un liquide visqueux
que limitent des parois. J. Math. Pures et Appl. XIII
(1934), 331-418.

[3] Sur le mouvement d'un liquide visqueux enplissant
l'éspace. Acta Math. 63 (1934), 193-248 .

J. LERAY et J.L. LIONS [1] Quelques résultats de Visik sur les
problèmes elliptiques non linéaires par les méthodes de
Minty-Browder. Bull. Soc. Math. France 93 (1965),
97-107 .

J.L. LIONS [1] Quelques méthodes de résolution des problèmes aux
limites non linéaires. Dunod - Gauthier Villar, Paris,
1969 .

[2] Sur certaines équations parabolique non linéaires. Bull. Soc. Math. France, 93 (1965), 155-175 .

[3] Some remarks on variational inequalities. Int.
Conference on Functional Analysis and Related topics,
Tokyo, 1969, 269-282.

[4] Sur les inéquations variationnelles d'évolution
pour les opérateurs du 2^{em} ordre en t . Symp. Math.
Publ. dell'Istituto Naz. di Alta Mat. Roma - 1970 .

J.L. LIONS et MAGENES [1] Problèmes aux limites non homogènes et
applications. Dunod, Paris Vol. 1 et 2,1968,Vol.3,1970.

J.L. LIONS et G. PRODI [1] Un théoreme d'existence et unicité
    dans les équations de Navier Stokes en dimension 2.
    C.R. Acad. Sc. Paris, 248 (1959), 3519-3521.

J.L.LIONS et W. STRAUSS [1] Some non linear evolution equations.
    Bull. Soc. Math. France, 93 (1965), 43-96.

G.I. MARCHOUK [1] Conference Congrès International des Math.
    Nice, 1970.

G.J. MINTY [1] Monotone (non linear) operators in Hilbert space.
    Duke Math. J. 29 (1962), 341-346 .

J. NECAS [1] Les méthodes directes dans la théorie des équations
    elliptiques. Ed. Acad. Tchecoslovaque des Sciences.
    Prague, 1967.

J.C. NEDELEC [1] A paraitre.

O.A. OLEINIK [1] Problèmes mathematiques de la théorie des cou-
    ches limites. Ouspechi Math. Nauk. XXIII (1968), 3-65.

A. PAZY [1] Le volume.

J. PEETRE [1] Conférence Congrès International des Math., Nice,
    1970.

    [2] E espaces d'interpolation et théorème de Sobolev.
    Annales Institut Fourier, 16 (1966), 279-317.

L. SCHWARTZ [1] Relévements de fonctions continue on mesurables
    à valeurs dans un quotient d'un Banach (non publié).

J.C. SIMON [1] A paraitre.

J. L. Lions

L. TARTAR [1] C.R. Acad. Sc. Paris, Juillet 1970.

[2] Thèse, Paris, 1971 .

R. TEMAM [1] Sur la stabilité et la convergence de la méthode
des pas fractionnaires. Annali di Mat. Pura ed Appl.
LXXIX (1968), 191-380.

[2] A paraitre

N.N. YANENKO [1] Méthode à pas fractionnaires. A. Colin, 1968.

K. YOSIDA [1] Functional Analysis. Grundleheren - B. 123,
Springer 1965.

M. ZLAMAL [1] Sur le problèmes mixtes pour une équation hyperbo
lique avec petit paramètre. Journal Tchecoslovaque de
Math. 9 (84), 1959, 218-242.

[2] id. 10 (85), 1960, 83-122.

CENTRO INTERNAZIONALE MATEMATICO ESTIVO

(C.I.M.E.)

A. PAZY

SEMI-GROUPS OF NONLINEAR CONTRACTIONS IN HILBERT SPACE

Corso tenuto a Varenna dal 20 al 29 Agosto 1970

SEMI-GROUPS OF NONLINEAR CONTRACTIONS IN HILBERT SPACE

by

A. PAZY

(Hebrew University, Jerusalem)

Lecture 1. - General Introduction.

Let H be a real Hilbert space and consider the abstract Cauchy problem

$$(1.1) \qquad \begin{cases} \dfrac{du}{dt} + Au = 0 \quad t > 0 \\[2mm] u(0) = x \end{cases}$$

where $u(t)$ is a H valued function and A is an operator from H into H . The equation in (1.1) may hold in one of many different senses but for the moment this makes no difference.

Suppose that the initial value problem (1.1) has a unique solution $u(t;x)$ for all initial values $x \in D \subset H$. Denoting by $S(t)$ the operator which maps x into the value of $u(t;x)$ at time t i.e. $u(t;x) = S(t)x$ we clearly have the following properties:

$$(1.2) \qquad S(t) : D \longrightarrow D$$

$$(1.3.) \qquad S(0) = I \qquad (\text{the identity})$$

$$(1.4) \qquad S(t_1+t_2)x = S(t_1) \ S(t_2)x \qquad t_1, \ t_2 \geq 0 \ .$$

These three properties are direct consequences of the assumption that (1.1) has a unique solution for every $x \in D$. Assuming that the equation (1.1) holds in the strong sense we have

$$(1.5) \qquad \lim_{t \to o} \frac{x - S(t)x}{t} = Ax \ .$$

The family of operators $S(t)$ is called a semi-group of operators on D and A is called the generator of $S(t)$. In the theory of semi-groups of operators we study the relations between the semi-group $S(t)$ and its generator A. One of the first and main problems of the theory is to characterize all operators A, $A : H \longrightarrow H$ which generate a semi-group on some subset $D \subset H$. In this generality the problem is not solved and we shall have to restrict ourselves a little but first let us review briefly the "history" of the subject.

At the beginning the main interest was in linear operators. Stone, then Hille, Yosida and others treated the initial value problem (1.1) with a linear operator A. In this case, it is clear that the operators $S(t)$ will also be linear, since

A. Pazy

$$\frac{d}{dt}(\lambda\ S(t)x + \mu\ S(t)y) + A(\lambda S(t)x+\mu S(t)y)=0 \qquad \text{for every } \lambda,\mu$$

by the linearity of A . From the uniqueness of the solution
of (1.1) with x replaced by $\lambda x + \mu y$ it then follows that

$$\lambda\ S(t)x + \mu\ S(t)y = S(t)(\lambda x + \mu y)\ .$$

Since A is linear its domain is naturally a linear
manifold. Assuming that $S(t)$ is a continuous linear operator
for every $t > 0$ we can extend $S(t)$ by continuity to the
closure of the domain of A which is a linear subspace of H
and without loss of generality we can assume that $D = H$. We
thus make the further assumption that $S(t)$ is a bounded
operator on H into H for every $t > 0$. Finally, assuming
that (1.1) holds in some strong enough sense we must have
some kind of continuity of $u(t;x)$. A natural assumption is
that $S(t)x$ is continuous in t or

(1.6)
$$\lim_{t \to o}\ S(t)x = x$$

which is sufficient for $S(t)x$ to be continuous in t for
all $t \geq 0$ by the semi-group property (1.4) .

A. Pazy

A semigroup of bounded operators on H satisfying the condition (1.4) is called a strongly continuous semi-group (or a C_0 semi-group). It is not difficult to see that for such a semi-group $\|S(t)x\| \leq Me^{\omega t}\|x\|$ where $\omega = \lim_{t \to \infty} t^{-1} \log\|S(t)\|$. Considering the semi-group $S_1(t) = e^{-\omega t} S(t)$ we obtain $\|S_1(t)x\| \leq M\|x\|$ i.e. the operators $S_1(t)$ are uniformly bounded by $M \geq 1$. If $M = 1$ $S_1(t)$ is called a semi-group of contractions.

In 1948 Hille and independently Yosida gave a complete characterization of the generators of semi-groups of contractions (on a general Banach space X).

Theorem (Hille-Yosida)

A densely defined linear operator A is a generator of a strongly continuous semi-group of contractions on H if and only if :

i) Every λ with $\text{Re } \lambda > 0$ is in the resolvent set of A.

ii) For $\lambda \in \rho(A)$ $\|R(\lambda;A)\| \leq \dfrac{1}{\text{Re}\lambda}$ $(\text{Re } \lambda > 0)$.

A. Pazy

Shortly after Hille and Yosida, Phillips (1953), Miyadera and Feller generalized the above theorem to strongly continuous uniformly bounded semigroups. In this case a densely defined linear operator A is a generator if and only if (i) holds and $\|R(\lambda:A)^n\| \leq M(\operatorname{Re}\lambda)^{-n}$ $\forall\, n > 0$, $\operatorname{Re} \lambda > 0$.

Note that if $S(t)$ is a uniformly bounded semi-group on X one can always define

$$\|x\|_1 = \operatorname{Sup}_{t \geq 0} \|S(t)x\|$$

$\|\cdot\|_1$ is a norm on X which is clearly equivalent to $\|\cdot\|$ since $\|x\| \leq \|x\|_1 \leq M\|x\|$. In the norm $\|\ \|_1$, $S(t)$ is a semi-group of contractions. On the other hand, it was shown by Feller that the condition: $\|R(\lambda:A)^n\| \leq M(\operatorname{Re}\lambda)^{-n}$ implies that the space X can be renormed so that in the new norm $\|R(\lambda:A)\|_1 \leq (\operatorname{Re}\lambda)^{-1}$. The result of Phillips, Miyadera and Feller can be restated as follows: A is a generator of a strongly continuous uniformly bounded semi-group on X if and only if X can be renormed so that A is a generator of strongly continous semi-group of contractions on X. Thus the generalization to strongly continuous uniformly bounded

A. Pazy

semi-groups amounts to a renormization of X and in a sense

one only characterizes generators of strongly continuous

semi-groups of contractions.

The notion of spectrum and the resolvent of an operator

are strongly connected with the linearity of the operator and

thus it is difficult to generalize the Hille-Yosida theorem

in the form that it was stated above to nonlinear operators.

However, an equivalent form of this theorem due to Phillips

does not contain these notions. Phillips calls a linear

operator   A in a real Hilbert space H monotone if $(Ax,x) \geq 0$

for every $x \in D(A)$, or $(Ax - Ay, x\text{-}y) \geq 0$ for every

$x,y \in D(A)$,   A is called maximal monotone (or rather

1 - maximal monotone) if it has no proper extensions in H

as a linear monotone operator. Then he proves the following

theorem

Theorem (Phillips)

A linear densely defined operator A is a generator of

a strongly continuous semi-group of contractions if and only

if it is maximal monotone.

A. Pazy

Phillips'theorem does not contain any spacial linear
ingredients and will be generalized in these lectures to non
linear operator. The first result in this direction was given
by Y. Kōmura in 1967. The full characterization of the
generators was achieved by Crandall and Pazy (1968) and
independently by Dorroh (1968).

Apart from the characterization of the generator of a
semi-group one is also interested in other problems relating
the semi-group and its generator. We shall mention briefly
some of them.

One is often interested in representation formula i.e.
one wants to obtain the semi-group $S(t)$ from its generator
A and vice versa, via some simple formula. The natural way
to get the semi-group from its generator is by solving the
initial value problem (1.1) . Another way is by the, well
known, exponential formula

$$S(t)x = \lim_{n \to \infty} (I + \frac{t}{n} A)^{-n}x = \lim_{n \to \infty} n \, R(n;A)x \ .$$

The obvious way to get A from $S(t)$ is by (1.5), but
one can also use

A. Pazy

$$R(\lambda;A)x = \int_0^\infty e^{-\lambda t} S(t)x \, dt .$$

There are a lot of other formulas of this type.

Another problem deals with the relations between the convergence of a sequence of generators and the convergence of the corresponding semi-groups. These results, for the linear case, were obtained mainly by Trotter (1958). For the nonlinear case the results are mainly due to Miyadera (1969) and Brezis-Pazy (1969).

Still another problem is concerned with the regularity of $S(t)$. For example one is interested in characterizing those operators A which generate, say, an analytic semi-group of operators. In the linear case this problem was solved by Yosida and Hille. In the nonlinear case there are some results of Kōmura and some very recent results of Brezis.

Finally, one is interested in perturbation theory. The type of question which arises is usually of the form: Let A be a generator of a semi-group having some property, give sufficient conditions on B so that $A+B$ is a generator of a semi-group having the same property.

A. Pazy

In the coming lectures we shall deal with most of these
questions for nonlinear semi-groups of contractions in a
Hilbert space.

We conclude this lecture with the following remark. In
the linear case there was no difficulty in dealing with semi-
groups of type-ω i.e. semi-groups satisfying

$$\| S(t)x \| \leq e^{\omega t} \| x \| .$$

The simple reduction $S_1(t) = e^{-\omega t} S(t)$ reduces such
semi-groups to a semi-groups of contractions with the generator
$(A+\omega I)$. However, it should be noted that in the nonlinear case
this is not always possible since the transformed equation
involves the operator $e^{-\omega t}(A+\omega I)e^{\omega t}$ the domain of which may
depend on t . Nevertheless, we do obtain also in the nonlinear
case all the results for semi-groups of type-ω by a somewhat
more complicated procedure.

A. Pazy

Lecture 2. - Monotone sets.

Let H be a real Hilbert space. We shall be interested
in subsets of H × H and we shall use standard functional
notations for such subsets. In particular, for any subsets A
and B of H × H and any real number α we define:

i) $D(A) = \{x : [x,y] \in A$ for some $y\}$

ii) $R(A) = \{y : [x,y] \in A$ for some $x\}$

iii) $Ax = \{y : [x,y] \in A\}$ if $x \notin D(A)$ then $Ax = \emptyset$

iv) $\alpha A = \{[x, \alpha y] : [x,y] \in A\}$

v) $A^{-1} = \{[y,x] : [x,y] \in A\}$

vi) $A+B = \{[x,y+z] : [x,y] \in A , [x,z] \in B\}$

vii) $AB = \{[x,z] :$ for some $y[x,y] \in B$ and $[y,z] \in A\}$.

Functions are special sets of H × H . The set $I \subset H \times H$
is defined by $I = \{[x,x]: x \in H\}$.

Definition 1.

A subset A of H × H is called monotone if

$$(x_1-x_2, y_1-y_2) \geq 0 \quad \text{for every} \quad [x_i,y_i] \in A \quad i=1,2 \quad .$$

A. Pazy

The set I which we introduce above is clearly monotone.
Another important example is the following; let C be a closed
convex subset of H and define

$$\partial I_c = \left\{ [x,y] : x \in C, (y, x-u) \geq 0 \; \forall u \in C \right\}$$

∂I_c is called the subdifferential of the indicator function
of C and it is easy to show that it is monotone.
Finally if $H = \mathbb{R}^1$ the graph of any monotone increasing
function is a monotone set in $H \times H$.

Theorem 1. (Local boundedness)

Let A be a monotone set. If $x_0 \in D(A)$ is an interior
point of $D(A)$ then there exists a positive number r such
that the set $\bigcup_{|x-x_0| < r} Ax$ is bounded.

Proof:

Suppose the theorem is false then there exists a sequence
of points $\{x_n\}$ such that $x_n \rightarrow x_0$ and $y_n \in Ax_n$ satisfy
$|y_n| = r_n \rightarrow \infty$ as $n \rightarrow \infty$. Define

$$t_n = \max \left\{ \frac{1}{r_n}, \; |x_n - x_0|^{1/2} \right\} .$$

Clearly $t_n \longrightarrow 0$, $r_n t_n \geq 1$ and $t_n \geq t_n^{-1} |x_n - x_0| \longrightarrow 0$

as $n \longrightarrow \infty$. Let $z \in H$ then for n large enough

$u_n = x_0 + t_n z \in D(A)$. Let $v_n \in Au_n$. We shall first prove

that $|v_n| \leq M$, for some $M > 0$. Let $\rho > 0$ be such that

$x_0 + \rho z \in D(A)$ and let $w_0 \in A(x_0 + \rho z)$ then by the

monotonicity of A we have $(v_n - w_0, z)(t_n - \rho) \geq 0$. If

$t_n < \rho$ this implies $(v_n, z) \leq (w_0, z)$ i.e. (v_n, z) is

bounded. Now let $x_0 + su \in D(A)$ and $w \in A(x_0 + su)$, $u \in H$.

By the monotonicity of A we have

$$0 \leq (v_n - w, t_n z - su) = t_n(v_n, z) - s(v_n, u) - (w, t_n z - su)$$

which implies

$$\overline{\lim_{n \to \infty}} (v_n, u) \leq (w, u) \qquad \text{for every } y \in H$$

thus $|v_n| \leq M$ by the uniform boundedness theorem.

Now,

$$(y_n - v_n, x_n - (x_0 + t_n z)) \geq 0$$

implies

$$(y_n, z) \leq \left(y_n, \frac{x_n - x_0}{t_{n\cdot}}\right) - \left(v_n, \frac{x_n - x_0}{t_n} - z\right) \leq |y_n| \left|\frac{x_n - x_0}{t_n}\right| + M\left(\frac{|x_n - x_0|}{t_n} + |z|\right)$$

$$\leq r_n t_n + M(t_n + |z|)$$

and therefore

$$\varlimsup_{r \to \infty} \left(\frac{y_n}{r_n t_n} , z \right) \leq 1 + M|z| < \infty \quad \text{for every } z \in H$$

but this contradicts the uniform boundedness theorem since

$$\left| \frac{y_n}{r_n t_n} \right| = \frac{1}{t_n} \longrightarrow \infty \quad \text{as } n \longrightarrow \infty \text{ and the proof is complete.}$$

Let A be a monotone set. We introduce two sets in $H \times H$ which are related to A namely:

$$J_\lambda = \left\{ [x+\lambda y, x] : [x,y] \in A \right\}$$

and

$$A_\lambda = \left\{ [x+\lambda y, y] : [x,y] \in A \right\} .$$

It is clear that $D(J_\lambda) = D(A_\lambda) = R(I+\lambda A)$ and that for every $x \in R(I+\lambda A)$

$$x = J_\lambda x + \lambda A_\lambda x .$$

Lemma 1.

Let A be a monotone set and $\lambda > 0$ then

a) J_λ is a function and

$$|J_\lambda x - J_\lambda y| \leq |x-y| \quad \text{for every } x,y \in R(I+\lambda A)$$

i.e. J_λ is a contraction

b) A_λ is a lipschitz function with constant $1/\lambda$

c) A_λ is monotone.

Proof:

Let $[x_i, y_i] \in A \quad i = 1, 2$

$$|x_1 + \lambda y_1 - (x_2 + \lambda y_2)|^2 = |x_1 - x_2|^2 + 2\lambda(x_1 - x_2, y_1 - y_2) + \lambda^2 |y_1 - y_2|^2 \geq$$

$$\geq |x_1 - x_2|^2 + \lambda^2 |y_1 - y_2|^2 .$$

Hence J_λ and A_λ are functions, J_λ is a contraction and A_λ is lipschitz with constant $1/\lambda$.

Now,

$$(A_\lambda x_1 - A_\lambda x_2, x_1 - x_2) = \frac{1}{\lambda}(x_1 - J_\lambda x_1 - (x_2 - J_\lambda x_2), x_1 - x_2) \geq$$

$$\geq \frac{1}{\lambda}\left\{|x_1 - x_2|^2 - |J_\lambda x_1 - J_\lambda x_2| \, |x_1 - x_2|\right\} \geq 0 .$$

We conclude this lecture with the following important extension theorem.

A. Pazy

Theorem 2. (Extension)

Let $C \subset H$ be a closed convex subset of H and let A be monotone. If

$$A_{|C} = \left\{ [v,z] : v \in D(A) \cap C, \ z \in Av \right\}$$

then for every $\rho > 0$ and every $f \in H$ there exists an element $u \in C$ such that:

(1) $\qquad (z+\rho u - f, \ v-u) \geq 0 \quad$ for every $[v,z] \in A_{|C}$.

Proof:

Without loss of generality we assume that $f = 0$ (otherwise one considers the set $A_f = \left\{ [x,y-f]:[x,v] \in A \right\}$ which also satisfies the assumption of the theorem).

Let

$$c[v,z] = \left\{ u : u \in C \ (z+\rho u, v-u) \geq 0 \right\}$$

Let $v \in C$. It is easy to check that $C[v,z]$ is bounded, weakly closed and clearly $v \in C[v,z]$. To prove the result we have to show that

$$\bigcap_{[v,z] \in A_{|C}} C[v,z] \neq \emptyset .$$

A. Pazy

Suppose this is false then there exists a finite numbers

of points $[v_i, z_i] \in A_{|C}$ such that

$$\bigcap_{i=1}^{n} C[v_i, z_i] = \emptyset .$$

Let $K = \left\{ \lambda = (\lambda_1, \ldots, \lambda_n) \in \mathbb{R}^n : \sum_{i=1}^{n} \lambda_i = 1 \quad \lambda_i \geq 0 \right\}.$

K is a compact convex set in $\mathbb{R}^n$. Consider the function

$$(2) \qquad f(\mu, \lambda) = \sum_{i=1}^{n} \mu_i(\rho\hat{u} + z_i, \hat{u} - v_i), \quad \hat{u} = \sum_{j=1}^{n} \lambda_i v_j .$$

Clearly $f(\mu, \lambda)$ is continuous in μ and λ and is

concave in μ and convex in λ. By the minimax theorem there

exist μ_0, λ_0 such that

$$(3) \qquad f(\mu, \lambda_0) \leq f(\mu_0, \lambda_0) \leq f(\mu_0, \lambda) \qquad \forall \mu, \lambda \in K .$$

We now prove that $f(\lambda, \lambda) \leq 0$ for every $\lambda \in K$. Indeed

$$f(\lambda, \lambda) = \sum_{i=1}^{n} \lambda_i(\rho\hat{u} + z_i, \hat{u} - v_i) = \sum_{i=1}^{n} \lambda_i(z_i, \hat{u} - v_i) =$$

$$= \sum_{i=1}^{n} \lambda_i(z_i, \sum_{j=1}^{n} \lambda_j v_j - v_i) = \sum_{i=1}^{n} \sum_{j=1}^{n} \lambda_i \lambda_j(z_i, v_j - v_i) \leq 0$$

A. Pazy

Since by the monotonicity of A

$$0 \leq \sum_{i=1}^{n} \sum_{j=1}^{n} \lambda_i \lambda_j (z_i - z_j, v_i - v_j) = 2 \sum_{i=1}^{n} \sum_{j=1}^{n} \lambda_i \lambda_j (z_i, v_i - v_j) \ .$$

Choosing $\lambda = \mu_0$ in (3) we obtain $f(\mu, \lambda_0) \leq 0$ for every $\mu \in K$. Defining $\hat{u}_0 = \sum_{j=1}^{n} \lambda_{oj} v_j$ and taking $\mu = (\delta_{ij})$ in (2) we have

$$(\rho \hat{u}_0 + z_i, v_i - \hat{u}_0) \geq 0 \qquad i = 1, 2, \ldots, n$$

and therefore $\hat{u}_0 \in \bigcap_{i=1}^{n} C[v_i, z_i]$ a contradiction.

Remark:

The main idea of the previous proof is due to Minty.

A. Pazy

Lecture 3. — Maximal Monotone Sets

Definition 1.

A monotone set which is not property contained in any other monotone set is called maximal monotone.

Theorem 1. (Minty)

Let A be monotone. If for some $\lambda > 0$ $R(I+\lambda A) = H$ then A is maximal monotone. If A is maximal monotone then for every $\lambda > 0$ $R(I+\lambda A) = H$.

Proof:

Let $R(I+\lambda_0 A) = H$ and let

(1) $\qquad (v-y, u-x) \geq 0 \qquad \forall\, [x,y] \in A$.

Let ξ be the solution of $u+\lambda_0 v = \xi + \lambda_0 \eta$ with $\eta \in A\,\xi$. Substituting $[\xi, \eta] \in A$ into (1) yield $u = \xi$ and $v = \eta$ i.e. $[u,v] \in A$ and A is maximal.

Let A be maximal monotone, $\lambda > 0$ and $f \in H$. By theorem 2 of the previous lecture there exists an element $v \in H$

such that

$$(y + \frac{1}{\lambda}(u-f), \ x-u) \geq 0 \qquad \forall \ [x,y] \in A \ .$$

By the maximality of A we have $u \in D(A)$ and $f \in u+\lambda Au$ i.e. $R(I+\lambda A) = H$.

Example:

The set ∂I_c which was introduced in the previous lecture is maximal monotone. To prove it let P_c be the projection on the set C then for every $x \in H$ we have

$$(x - P_c x \ , \ P_c x - u) \geq 0 \ \forall u \in C$$

and therefore $[P_c x, \ x-P_c x] \in \partial I_c \ \forall x \in H$ and $x = P_c x+[x-P_c x$ implies $R(I+\partial I_c) = H$.

Theorem 2.

Let A be monotone, C closed and convex. There exists a maximal monotone set $\tilde{A}$ such that $\tilde{A} \supset A_{|c}$ and $D(\tilde{A}) \subset C$.

Proof:

Extend $A_{|c}$ to a maximal monotone set with domain in C and denote the extension by $\tilde{A}$. Given $f \in H$ let $u \in C$

A. Pazy

satisfy

$$(w+u-f, \ v-u) \geq 0 \qquad \forall [v,w] \in A_{|c}$$

(the existence of such a u is implied by theorem 2 of the previous lecture). Since $\tilde{A}$ is maximal on C , $[u,f-u] \in \tilde{A}$ and $f \in R(I+\tilde{A})$. Thus $R(I+\tilde{A}) = H$ and the proof is complete.

Lemma.

Let A be maximal monotone then

a) For every $x \in D(A)$, Ax is closed and convex. The element of minimum norm in Ax is denoted by $A^{\circ}x$,

b) If $[x_n, y_n] \in A$, $x_n \longrightarrow x$ and $|y_n| \leq M$ then $x \in D(A)$. If moreover $y_n \longrightarrow y$ then $[x,y] \in A$,

c) If $[x_n, y_n] \in A$, $x_n \longrightarrow x$ and $\overline{\lim_{n \to \infty}} |y_n| \leq |A^{\circ}x|$ then $y_n \longrightarrow A^{\circ}x$

d) If for some sequence $\lambda_n \longrightarrow 0$, $x_{\lambda_n} \longrightarrow x_0$ and $|A_{\lambda_n} x_{\lambda_n}| \leq M$ then $x_0 \in D(A)$

e) If $x_0 \notin D(A)$ and $[x_n, y_n] \in A$ $x_n \longrightarrow x_0$ then $|A^{\circ}x_n| \longrightarrow \infty$

A. Pazy

Proof:

a) Let $y_i \in Ax$ $i = 1, 2$ $[u, v] \in A$ we have $(y_i - v, x - u) \geq 0$
and therefore also $(ty_1 + (1-t)y_2 - v, x - u) \geq 0$ for
every $[u, v] \in A$. By the maximality of A it follows that
$ty_1 + (1-t)y_2 \in Ax$. The proof that Ax is closed is
similar.

b) Let $y_{n_k} \rightharpoonup y$ and $[u, v] \in A$, then $(y_{n_k} - v, x_{n_k} - u) \geq 0$.
Passing to the limit yields $(y-v, x-u) \geq 0$ for every
$[u, v] \in A$ and therefore $[x, y] \in A$ by the maximality of A.

c) The element y of part (b) satisfies $|y| \leq |A^0x|$ and
since $y \in Ax$ we have $y = A^0x$ and $\lim |y_{n_k}| = |A^0x|$.
Therefore $y_{n_k} \rightarrow A^0x$. From the uniqueness of the limit
we have $y_n \rightarrow A^0x$.

d) Let $x_{\lambda_n} = J_{\lambda_n} x_{\lambda_n} + \lambda_n A_{\lambda_n} x_{\lambda_n}$ therefore $J_{\lambda_n} x_{\lambda_n} \rightarrow x_0$
as $n \rightarrow \infty$, $J_{\lambda_n} x_{\lambda_n} \in D(A)$ and $A_{\lambda_n} x_{\lambda_n} \in A J_{\lambda_n} x_{\lambda_n}$
therefore $x_0 \in D(A)$ by (b).

e) If a subsequence of $|A^0 x_n|$ is bounded it follows by (b)
that $x_0 \in D(A)$ contradicting $x_0 \notin D(A)$.

A. Pazy

If A is maximal monotone then J_λ and A_λ which where defined in the previous lecture are defined on all of H . We collect in our next theorem several properties of maximal monotone sets and of the sets J_λ and A_λ which correspond to them.

Theorem 3.

Let A be maximal monotone, then

a) $\overline{D(A)}$ and $\overline{R(A)}$ are convex $(\overline{D(A)}$ and $\overline{R(A)}$ are the closures of $D(A)$ and $R(A)$ respectively).

b) For every $x \in H$ $\quad \lim_{\lambda \to 0} J_\lambda x = \text{Proj}_{\overline{D(A)}} x$

c) For every $x \in D(A)$, $\quad |A_\lambda x| \leq |A^\circ x|$ and $\lim_{\lambda \to 0} A_\lambda x = A^\circ x$

d) If $x \in H$ and $|A_\lambda x|$ is bounded as $\lambda \to 0$ then $x \in D(A)$.

Proof:

Let $x \in H$, $x = J_\lambda x + \lambda A_\lambda x$ $\quad [J_\lambda x, A_\lambda x] \in A$ and let $[u,v] \in A$ then :

$$|x - J_\lambda x|^2 = (x - J_\lambda x, x - J_\lambda x) = (x - J_\lambda x, x - u) + (x - J_\lambda x, u - J_\lambda x) =$$

$$= (x - J_\lambda x, x - u) + \lambda(A_\lambda x, u - J_\lambda x) \leq$$

A. Pazy

$$\le (x-J_\lambda x, x-u)+\lambda(v,u-J_\lambda x) \le (x-J_\lambda x, x-u) + \lambda(v,u-x)+\lambda(v,x-J_\lambda x)$$

This implies that $|x-J_\lambda x|$ is bounded as $\lambda \longrightarrow 0$ and therefore $|J_\lambda x|$ is bounded. Let $\lambda_n \longrightarrow 0$ such that $J_{\lambda_n} x \longrightarrow \xi$ then

(2) $$\overline{\lim_{\lambda_n \to o}} \; |x-J_{\lambda_n} x|^2 \le (x-\xi, x-u) \quad \forall u \in D(A)$$

It is clear that (2) holds also for every $u \in \overline{conv(D(A))}$. If $x \in \overline{conv(D(A))}$ then substituting $u = x$ in (2) yields $J_{\lambda_n} x \longrightarrow x$ and since $J_{\lambda_n} x_i \in D(A)$ we have $\overline{D(A)} \supset \overline{conv(D(A))}$ i.e. $\overline{D(A)}$ is convex. Since A^{-1} is maximal monotone together with A and $D(A^{-1}) = R(A)$ also $\overline{R(A)}$ is convex.

b) From (2) we have $|x-\xi|^2 \le (x-\xi, x-u)$ for every $u \in D(A)$ and therefore also for every $u \in \overline{D(A)}$ i.e. $(x-\xi, u-\xi) \le 0$ for every $u \in \overline{D(A)}$ which implies that $\xi = \text{Proj}_{\overline{D(A)}} x$ since clearly $\xi \in \overline{D(A)}$. Substituting $u = \xi$ in (2) yields

$$|x-\xi|^2 \le \lim_{\lambda_n \to o} |x-J_{\lambda_n} x|^2 \le \overline{\lim_{\lambda_n \to o}} |x-J_{\lambda_n} x|^2 \le |x-\xi|^2$$

and therefore $J_{\lambda_n} x \longrightarrow \xi$ and by the uniqueness of the

A. Pazy

limit $J_\lambda x \longrightarrow \xi$.

c) Let $x \in D(A)$, $y \in Ax$ and $x = J_\lambda x + \lambda A_\lambda x$ then

$0 \leq (y-A_\lambda x, x-J_\lambda x) = \lambda(y-A_\lambda x, A_\lambda x)$ and therefore $|A_\lambda x| \leq |y|$

Since y was arbitrary $|A_\lambda x| \leq |A^o x|$. By part (b)

$J_\lambda x \longrightarrow x$ as $\lambda \longrightarrow 0$ and therefore $\lim_{\lambda \to o} A_\lambda x = A^o x$ by

the previous lemma part (c) .

d) If $|A_\lambda x|$ is bounded for some x then $x = J_\lambda x + \lambda A_\lambda x$

implies $J_\lambda x \longrightarrow x$ as $\lambda \longrightarrow 0$ and by the previous lemma

(b) $x \in D(A)$.

Definition 2.

Let A be maximal monotone. A subset A_1 of A is

called a principal subset if any maximal monotone extension

$\tilde{A}_1$ of A_1 with $D(\tilde{A}_1) \subset \overline{D(A)}$ coincides with A .

A subset A^i of A is called a section of A if it

is singlevalued i.e. it is a graph. A section A^i which is

a principal subset of A is called a principal section.

Theorem 4.

Let A be maximal monotone and let A' be a section of

A. Pazy

A with the following property: For every $x \in \overline{D(A)}$ there

exists a sequence $x_n \in D(A)$· such that $x_n \to x$ and

(3) $$\overline{\lim_{n \to \infty}} (A'x_n, x-x_n) \geq 0$$

then A' is a principal section.

<u>Proof</u>:

Consider the set

$$M = \left\{ [x,y] : x \in \overline{D(A)}, y \in H, (y-A'z,x-z) \geq 0 \ \forall z \in D(A) \right\}$$

It is clear that $M \supset A$. We shall prove that M is

monotone. From the maximality of A we then have $M = A$ and

A' is principal.

Let $[x_i,y_i] \in M$ $i = 1,2$ and let $x = \frac{1}{2}(x_1+x_2)$. Since

$\overline{D(A)}$ is convex $x \in \overline{D(A)}$ and we have

$$(y_1 - A'z, \frac{x_1-x_2}{2} + x - z) \geq 0, \quad (y_2-A'z, \frac{x_2-x_1}{2}+x - z) \geq 0$$

for every $z \in D(A)$. Adding the two inequalities we obtain:

(4) $$\frac{1}{2}(y_1-y_2,x_1-x_2) \geq (y_1+y_2,z-x)+2(A'z,x-z) \ \forall z \in D(A) .$$

A. Pazy

Let $x_n \in D(A)$, $x_n \to x$ such that $\overline{\lim_{n \to \infty}} (A'x_n, x-x_n) \geq 0$

then substituting x_n for z in (4) and passing to the

limit yields $(y_1 - y_2, x_1 - x_2) \geq 0$ i.e. M is monotone.

Theorem 5.

Let A be maximal monotone then A^0 is a principal

section of A .

Proof:

It is sufficient to show that the conditions of theorem 4

are satisfied. Let $x \in \overline{D(A)}$ and $x_\lambda = J_\lambda x$. Clearly $J_\lambda x \to x$

as $\lambda \to 0$. Since $A_\lambda x \in A J_\lambda x$ we have

$$(A_\lambda x, A^0 J_\lambda x) > |A^0 J_\lambda x|^2$$

and therefore

$$\overline{\lim_{\lambda \to 0}} (A^0 J_\lambda x, x-J_\lambda x) = \overline{\lim_{\lambda \to 0}} \lambda(A^0 J_\lambda x, A_\lambda x) \geq 0 .$$

We give two more corollaries of theorem 4 .

Corollary 1.

Let A be maximal monotone with $D(A) = H$. Then every

A. Pazy

section A with $\overline{D(A')} = H$ is principal.

Proof:

Condition (3) is an immediate consequence of the local boundedness of A.

Corollary 2.

Let A be maximal monotone and let ω be a real number. Then the section $A^i = (A-\omega I)^o + \omega I$ is a principal section of A.

Proof:

We first prove that $|A^i x - A^o x| \le |\omega||x|$. Let $A^i x = z_1$, $A^o x = z_0$ then

$$|z_1 - z_0|^2 = |z_1 - \omega x - (z_0 - \omega x)|^2 = |z_1 - \omega x|^2 - 2(z_1 - \omega x, z_0 - \omega x)$$

$$+ |z_0 - \omega x|^2 \le |z_0 - \omega x|^2 - |z_1 - \omega x|^2$$

since $\quad |z_1 - \omega x|^2 \le (z_1 - \omega x, z_0 - \omega x)$

therefore

$$|z_1 - z_0|^2 \le |z_0|^2 - |z_1|^2 + (z_1 - z_0, \omega x) \le (z_1 - z_0, \omega x) \le |z_1 - z_0||\omega||x|$$

since $|z_0| \leq |z_1|$.

Now,

$$\lim_{\lambda \to 0} \left| (A^0 J_\lambda x - A^i J_\lambda x, \; x - J_\lambda x) \right| \leq \lim_{\lambda \to 0} |\omega| \, |J_\lambda x| \, |x - J_\lambda x| = 0 \quad \forall \; x \in \overline{D(A)}$$

therefore

$$\overline{\lim_{\lambda \to 0}} \; (A^i J_\lambda x, \; x - J_\lambda x) \geq \overline{\lim_{\lambda \to 0}} \; (A^0 J_\lambda x, x - J_\lambda x) - \lim_{\lambda \to 0} (A^0 J_\lambda x - A^i J_\lambda x, x - J_\lambda x)$$

$$= \overline{\lim_{\lambda \to 0}} \; (A^0 J_\lambda x, x - J_\lambda x) \geq 0$$

Thus A^i satisfies the conditions of theorem 4 and is a principal section.

A. Pazy

Lecture 4. - $\mathcal{M}(\omega)$ sets and perturbations.

Definition 1.

A subset A of H×H is called a $\underline{\mathcal{M}(\omega)}$ set if A+ωI

is monotone, or equivalently if $[x_i, y_i] \in A$ i=1,2 implies

$$(x_1 - x_2, \ y_1 - y_2) \geq - \omega \ |x_1 - x_2|^2 \ .$$

A is called a $\underline{\text{maximal } \mathcal{M}(\omega) \text{ set}}$ if it is not a proper subset

of any $\mathcal{M}(\omega)$ set.

It is clear that A is a maximal $\mathcal{M}(\omega)$ set if and only

if A+ωI is a maximal monotone set. It is also clear by the

definition and the properties of monotone and maximal monotone

sets that $\mathcal{M}(\omega)$ sets are locally bounded at interior points

of their domain, that the closure of the domain of maximal $\mathcal{M}(\omega)$

sets is convex and that for every $x \in D(A)$, Ax is closed

and convex if A is a maximal $\mathcal{M}(\omega)$ set. Thus for maximal

$\mathcal{M}(\omega)$ sets $A^o x$ exists.

A. Pazy

Theorem 1.

Let A be a $\mathfrak{M}(\omega)$ set and $C \subset H$ be closed and convex.
If $\overline{D(A)} \subset C$ then there exists a maximal $\mathfrak{M}(\omega)$ set $\tilde{A}$ such
that $\tilde{A} \supset A$ and $D(\tilde{A}) \subset C$.

Proof:

The set $A+\omega I$ is monotone $\overline{D(A+\omega I)} = \overline{D(A)} \subset C$, therefore
there exists a maximal monotone set B such that $B \supset A+\omega I$,
$D(B) \subset C$. Consider the set $B - \omega I = \tilde{A}$. $\tilde{A}$ is a maximal
$\mathfrak{M}(\omega)$ set $\overline{D(\tilde{A})} = \overline{D(B)} \subset C$ and $\tilde{A} \supset A$.

Theorem 2.

Let A be a maximal $\mathfrak{M}(\omega)$ set then A° is a principal
section of A .

Proof:

From corollary 2 of the previous lecture it follows that
$A^{\circ}+\omega I$ is a principal section of $A+\omega I$ (since $A^{\circ}+\omega I =$
$= ((A+\omega I) - \omega I)^{\circ} + \omega I)$. Thus, the only maximal $\mathfrak{M}(\omega)$ extension
of A° with domain in $\overline{D(A)}$ can be A since for any maximal

A. Pazy

extension A_1 of A^0 we have $A_1 + \omega I$ is a maximal monotone

extension of $A^0 + \omega I$ and therefore coincides with $A + \omega I$.

Many of the results for maximal monotone sets hold for

maximal $\mathcal{M}(\omega)$ sets we mention for example.

Lemma 1.

Let A be a maximal $\mathcal{M}(\omega)$ set, then

a) If $[x_n, y_n] \in A$, $x_n \longrightarrow x$ and $|y_n| \leq M$ then $x \in D(A)$.
    If moreover $y_n \longrightarrow y$ then $[x, y] \in A$.

b) If $[x_n, y_n] \in A$, $x_n \longrightarrow x \in D(A)$ and $\overline{\lim} |y_n| \leq |A^0 x|$
    then $y_n \longrightarrow A^0 x$.

c) If $x_0 \notin D(A)$ and $D(A) \ni x_n \longrightarrow x_0$ then $|A^0 x_n| \longrightarrow \infty$

the proof of this lemma is the same as the proof of the

corresponding lemma for maximal monotone sets.

As for monotone sets we define also for $\mathcal{M}(\omega)$ sets

the sets :

$$J_\lambda = \left\{ [x + \lambda y, x] : [x, y] \in A \right\}$$

and

$$A_\lambda = \left\{ [x+\lambda y, \; y] \; : \; [x,y] \in A \right\} .$$

Clearly we have $A_\lambda = \lambda^{-1}(I-J_\lambda)$ and for $[x_i,y_i] \in A$ we have

$$(x_1+\lambda y_1-(x_2+\lambda y_2),x_1-x_2) = |x_1-x_2|^2+\lambda(y_1-y_2,x_1-x_2) \geq$$

$$\geq (1.-\lambda\omega) \; |x_1-x_2|^2$$

and therefore

$$|x_1+\lambda y_1 - (x_2+\lambda y_2)| \geq (1-\lambda\omega)|x_1-x_2|^2$$

which implies that for $\lambda\omega < 1$ J_λ is a function and therefore also A_λ is a function and

(1) $$| J_\lambda x - J_\lambda y| \leq (1-\lambda\omega)^{-1} \; |x-y| .$$

In a similar way we obtain,

$$|(x_1+\lambda y_1)-(x_2+\lambda y_2)|^2 = |x_1-x_2|^2 + 2\lambda(x_1-x_2,y_1-y_2)+\lambda^2|y_1-y_2|^2 \geq$$

$$\geq (1-2\lambda\omega)|x_1-x_2|^2 + \lambda^2|y_1-y_2|^2$$

and therefore for $\lambda\omega < 1/2$ we have

(2) $$|A_\lambda x - A_\lambda y| \leq \frac{1}{\lambda} \; |x - y| .$$

Finally we also have

$$(3) \qquad A_\lambda \in \mathfrak{M}\left(\frac{\omega}{1-\lambda\omega}\right) \qquad \text{for all } \lambda < 1/_\omega .$$

Furthermore, if A is a maximal $\mathfrak{M}(\omega)$ set then $R(I+\lambda A) = H$ for all $0 < \lambda < 1/_\omega$. This follows directly from the fact that $A+\omega I$ is maximal monotone and therefore $R(I+\lambda(A+\omega I)) = H$ for every $\lambda > 0$. Thus $R(I+ \frac{\lambda}{1+\lambda\omega} A) = H$ for every $\lambda > 0$ which implies $R(I+\lambda A) = H$ for every $0 < \lambda < 1/_\omega$.

Finally since $[J_\lambda x, A_\lambda x] \in A$ we have for every $[x,y] \in A$

$$(y - A_\lambda x, \ x - J_\lambda x) \geq - \omega |x - J_\lambda x|^2 = - \omega \lambda^2 |A_\lambda x|^2$$

which implies $|A_\lambda x| \leq (1 - \lambda\omega)^{-1} |y|$ and since y was arbitrary in Ax we have

$$(4) \qquad |A_\lambda x| \leq (1-\lambda\omega)^{-1} |A^\circ x| \qquad \text{for every } x \in D(A) .$$

Since $x = J_\lambda x + \lambda A_\lambda x$ it follows from (4) that $\lim_{\lambda \to o} J_\lambda x = x$ for every $x \in D(A)$ and from (1) it follows that this is true every $x \in \overline{D(A)}$ thus

$$(5) \qquad \lim_{\lambda \to o} J_\lambda x = x \qquad \text{for every } x \in \overline{D(A)} .$$

A. Pazy

Finally using (4), (5) and lemma 1 we have

$$(6) \qquad \lim_{\lambda \to 0} A_\lambda x = A_\lambda x = A^{\circ}x \qquad \text{for every} \quad x \in D(A).$$

Perturbations.

We conclude this lecture with some remarks on perturbations of maximal monotone sets.

Let A and B be monotone sets then clearly A+B is monotone. If however A and B are maximal monotone sets A+B need not be maximal monotone. We state without proof the following perturbation theorems.

Theorem 3.

Let A be maximal monotone and let B be a function such that :

   i) A+B is monotone

  ii) $D(A) \subset D(B)$ and there exist constants k , $0 \le k < 1$ and K such that

$$(7) \qquad |Bx_1 - Bx_2| \le k \ d(Ax_1, Ax_2) + K |x_1 - x_2| \quad \text{for every} \quad x_1, x_2 \in D(A)$$

A. Pazy

then   A+B   is maximal monotone.

Here $d(Ax_1, Ax_2)$ is the distance between the sets Ax_1
and Ax_2 i.e. $d(Ax_1, Ax_2) = \inf \left\{ |\xi_1 - \xi_2| : \xi_1 \in Ax_1, \xi_2 \in Ax_2 \right\}$

Theorem 4.

Let   A   and   B   be maximal sets if $int(D(A)) \cap D(B) \neq \emptyset$
then   A+B   is maximal monotone.

Theorem 5.

Let   A   and   B   be maximal monotone. If $D(B) \supset D(A)$ and
for every   r > 0   there are constants   k(r) < 1   and   K(r)
such that

(8) $|B^o x| \leq k(r) |A^o x| + K(r)$ for every $x \in D(A)$ with $|x| \leq$

then   A+B   is maximal monotone.

Remark:

Proofs of theorem 3 and 5 are given in :

M. Crandall e A. Pazy, Semi-groups of nonlinear contractions
and dissipative sets , Jour. Func. Anal. 3 (1969) 376-418.

A. Pazy

A simple proof of theorem 4 may be found in

H. Brezis e A. Pazy, Semigroups of nonlinear contractions
on convex sets, Jour. Func. Anal. (to appear).

Theorem 4 is due to R.T. Rockafellar.

A. Pazy

Lecture 5. - Semi-groups of type ω .

Definition 1.

Let D be a subset of H . A semi-group of type ω on D is a function S with domain $[0,\infty) \times D$ and range in D satisfying the following conditions :

   i) $S(0)x = x$ and $S(t_1+t_2)x = S(t_1)S(t_2)x$ for every

      $t_1, t_2 \geq 0$ and every $x \in D$.

  ii) $|S(t)x - S(t)y| \leq e^{\omega t} |x-y|$ for every $t \geq 0$ and $x,y \in D$

 iii) For every $x \in D$, $S(t)x$ is strongly continuous in $t \geq 0$

A semi-group of type 0 is a semi-group of contractions. By the strong continuity (property (iii)) any semi-group of type-ω

   can be extended in a unique way to the closure of D and therefore we shall assume without loss of generality that D is always closed.

Definition 2.

Let S be a semi-group of type-ω on D . For every

A. Pazy

$h > 0$ let

$$A(h)x = h^{-1}(x-S(h)x) \ .$$

Let $D(A')$ be the set of all $x \in D$ for which

$\lim\limits_{h \to 0} A(h)x$ exists and let

$$A'x = \lim\limits_{h \to 0} A(h)x \qquad \text{for every} \ \ x \in D(A')$$

we shall call A' the generator of the semi-group S .

Remark:

In the usual terminology $- A'$ is called the generator

of S but in these lectures we find it more convenient to

call A' the generator of S .

Lemma 1.

Let $S(t)$ be a semi-group of type ω on D . Let $x_0 \in D$

if $|A(h)x_0|$ is bounded as $h \to 0$ then $S(t)x_0$ is

lipschitz on every interval $[0,T]$, $T > 0$ with the lipschitz

constant $L(T)$ given by :

$$L(T) = e^{\omega T} \lim\limits_{h \to 0} |A(h)x_0| \ .$$

A. Pazy

Proof:

Consider the interval $[0, T]$. It is sufficient to show that if $\{t_k\}$ is a monotonically decreasing sequence of positive numbers tending to zero and $L_1 < \infty$ such that

$$| S(t_k)x_o - x_o| \leq L_1 t_k$$

then $\quad | S(t+\delta)x_o - S(t)x_o| \leq e^{\omega T} L_1 \delta \qquad$ for $t+\delta \leq T$.

Let τ, t be positive then it is easy to show that

$$| S(\tau +nt)x-x| \leq e^{n\omega t}|S(\tau)x-x| + \sum_{j=0}^{n-1} e^{j\omega t} |S(t)x-x|.$$

For every k let n_k be a nonnegative integer such that $0 \leq \delta - n_k t_k < t_k$, then

$$| S(t+\delta)x_o -S(t)x_o| \leq e^{\omega t}|S(\delta)x_o-x_o| = e^{\omega t}|S(\delta-n_k t_k+n_k t_k)x_o - x_o| \leq$$

$$\leq e^{\omega t}\left[e^{\omega n_k t_k}|S(\delta-n_k t_k)x_o-x_o| + n_k e^{\omega n_k t_k}|S(t_k)x_o-:\right.$$

$$\leq e^{\omega t}\left[\varepsilon_k + \delta e^{\omega \delta} L_1\right] \leq L_1 e^{\omega(t+\delta)}.\delta + \varepsilon_k e^{\omega T_-}$$

$$\leq L_1 e^{\omega T}.\delta + \varepsilon_k e^{\omega T}$$

A. Pazy

by the strong continuity of $S(t)x_0$, $\varepsilon_k \longrightarrow 0$ as $k \longrightarrow \infty$ and the result follows.

Our next result is well known.

Theorem 1.*

Let H be a Hilbert space and let $f(t)$ be a lipschitz continuous function from $\mathbb{R}^1$ to H, then $f(t)$ is strongly differentiable for almost all t and it can be expressed as the indefinite integral of its derivative i.e.

$$f(t) = f(0) + \int_0^t f^1(s)\,ds .$$

From this theorem we obtain easily the following corollary.

Corollary 1.

Let $S(t)$ be a semi-group of type ω on $D \subset H$. If for some $x_0 \in D$ $\;|A(h)x_0|$ is bounded as $h \longrightarrow 0$ then $x_0 \in \overline{D(A^1)}$ where A^1 is the generator of $S(t)$.

---

* A proof of theorem 1 can be found in : Y. Kōmura, Nonlinear Semi-groups in Hilbert Space, J. Math. Soc. of Japan 19 (1967) 493–507 .

A. Pazy

Proof:

By lemma 1 $S(t)x_0$ is lipschitz on $[0,1]$ and therefore by theorem 1 there exists a sequence $t_k \longrightarrow 0$ for which $S(t_k)x_0$ is strongly differentiable i.e. $S(t_k)x_0 \in D(A^1)$. But $S(t_k)x_0 \longrightarrow x_0$ which implies $x_0 \in \overline{D(A^1)}$.

We shall see later that under the conditions of corollary 1 $x_0 \in D(A^1)$ and not only $x_0 \in \overline{D(A^1)}$.

Lemma 2.

Let $S(t)$ be a semi-group of type ω on $D \subset H$ and let A^1 be its generator then A^1 is a $\mathcal{M}(\omega)$ function i.e.   for every $x_1, x_2 \in D(A^1)$

$$(A^1 x_1 - A^1 x_2, x_1 - x_2) \geq - \omega |x_1 - x_2|^2 .$$

Proof:

$$(A(h)x_1 - A(h)x_2, x_1 - x_2) = \frac{1}{h}\left\{|x_1 - x_2|^2 - (S(h)x_1 - S(h)x_2, x_1 - x_2\right.$$

$$\geq \frac{1}{h}\left\{|x_1 - x_2|^2 - e^{\omega h}|x_1 - x_2|^2\right\} .$$

Letting h tend to zero we obtain the result.

A. Pazy

Lemma 3.

Let $S(t)$ be a semi-group of type ω on $D \subset H$ and let A be any $\mathfrak{M}(\omega)$ extension of the generator A' of $S(t)$ then :

$$
(1) \quad
\begin{aligned}
&\frac{1}{2} e^{-2\omega t} |u-S(t)x|^2 - \frac{1}{2}|u-x|^2 \leq \\
&\leq |v| \int_0^t e^{-2\omega\tau} |u-S(\tau)x| \, d\tau \qquad \forall x \in \overline{D(A')}, \quad [u,v] \in A
\end{aligned}
$$

$$
(2) \quad
\begin{aligned}
\left(\frac{x-S(t)x}{t} - v, \; x-u \right) &\geq -\frac{1}{t} \int_0^t (v, \; x-S(\tau)x) \, d\tau - \\
&\quad - \frac{\omega}{t} \int_0^t |u-S(\tau)x|^2 \, d\tau \qquad \forall x \in \overline{D(A')}, \; [u,v] \in A \; .
\end{aligned}
$$

Proof:

Assume first that $x \in D(A')$ then

$$
(3) \quad \frac{1}{2} \frac{d}{d\tau} |u-S(\tau)x|^2 = (A'S(\tau)x, \; u-S(\tau)x) \leq (v, u-S(\tau)x) + \omega |u-S(\tau)x|^2
$$

and therefore

$$
\frac{1}{2} e^{2\omega\tau} \frac{d}{d\tau} \left\{ e^{-2\omega\tau} |u-S(\tau)x|^2 \right\} \leq (v, u-S(\tau)x) \; .
$$

Integrating the last inequality between 0 and t we

A. Pazy

obtain (1) for $x \in D(A^1)$.

Since $S(t)x$ satisfies $|S(t)x-S(t)y| \leq e^{\omega t}|x-y|$ the

result is true for all $x \in \overline{D(A^1)}$. Integrating (3) from 0

to t we obtain

$$(x-S(t)x, u-x) \leq \frac{1}{2}|u-S(t)x|^2 - \frac{1}{2}|u-x|^2 \leq$$

$$\leq \int_0^t (v,u-S(\tau)x)d\tau + \omega \int_0^t |u-S(\tau)x|^2 d\tau$$

and therefore

$$\left(\frac{x-S(t)x}{t} - v, x-u\right) \geq -\frac{1}{t}\int_0^t (v,x-S(\tau)x)d\tau - \frac{\omega}{t}\int_0^t |u-S(\tau)x|^2 d\tau \quad \forall x \in D(A$$

and again since $|S(t)x-S(t)y| \leq e^{\omega t}|x-y|$ this is true for

every $x \in \overline{D(A^1)}$.

Theorem 2.

Let $S(t)$ be a semi-group of type ω on $D \subset H$. Let A^1

be the generator of $S(t)$ and let A be a maximal $\mathfrak{M}(\omega)$

extension of A^1 then :

a) $A^\circ \supset A^1$

b) $D(A^1) = D(A) \cap \overline{D(A^1)}$

c) If $\overline{D(A)} = \overline{D(A^1)}$ then $A^1 = A^\circ$.

Proof:

a) Let $x \in D(A^1)$ then $t^{-1}(x-S(t)x) \longrightarrow A^1x \subset Ax$ as $t \longrightarrow 0$.
Using (1) with $u = x$ and $v = A^\circ x$ we obtain:

(4)
$$\frac{1}{2} e^{-2\omega t}|x-S(t)x|^2 \leq |A^\circ x| \int_0^t e^{-2\omega\tau}|x-S(\tau)x|\,d\tau \leq$$

$$\leq |A^\circ x| \int_0^t e^{-\omega\tau}|x-S(\tau)x|\,d\tau$$

this implies

(5)
$$\left|\frac{x-S(t)x}{t}\right| \leq e^{\omega t}|A^\circ x|$$

and therefore $|A^1x| \leq |A^\circ x|$. Since $A^1x \in Ax$ we have
$A^1x = A^\circ x$ and $A^\circ \supset A^1$.

b) It is clear that $D(A^1) \subset D(A) \cap \overline{D(A^1)}$ since $D(A^1) \subset D(A)$.
Let $x \in D(A) \cap \overline{D(A^1)}$ for this x both (4) and (5) hold.
Let $t_n \longrightarrow 0$ such that $t_n^{-1}(x-S(t_n)x \longrightarrow \eta$. From (2) and
the maximality of A it follows that $\eta \in Ax$ and from (5)
it follows that $|\eta| \leq |A^\circ x|$ thus $\eta = A^\circ x$ and
$|t_n^{-1}(x-S(t_n)x)| \longrightarrow |A^\circ x|$. Therefore $t_n^{-1}(x-S(t_n)x) \longrightarrow A^\circ x$.
By the uniqueness of the limit it follows that $t^{-1}(x-S(t)x) \longrightarrow A^\circ x$
as $t \longrightarrow 0$ i.e. $x \in D(A^1)$ and $D(A^1) \supset D(A) \cap \overline{D(A^1)}$.

Finally (c) is an immediate consequence of (a) and (b) .

Theorem 3.

Let $S(t)$ be a semi-group of type ω on $D \subset H$. Let A' be the generator of $S(t)$ and $x_0 \in D$. Then $x_0 \in D(A')$ if and only if $|A(h) x_0|$ is bounded as $h \rightarrow 0$.

Proof:

If $x_0 \in D(A')$ the limit $\lim\limits_{h \rightarrow o} A(h)x_0 = A'x_0$ exists and clearly $|A(h)x_0|$ is bounded as $h \rightarrow 0$. Assume now that $|A(h)x_0|$ is bounded. Let A be a maximal $\mathfrak{M}(\omega)$ extension of A'. Since $|A(h)x_0|$ is bounded there exists a sequence $h_n \rightarrow 0$ such that $A(h_n)x_0 \rightarrow \eta$. From (2) and the maximality of A it follows that $[x_0, \eta] \in A$. Since $x \in \overline{D(A')}$ by corollary 1 it follows that $x_0 \in D(A) \cap \overline{D(A')}$ and from theorem 2(b) we deduce $x_0 \in D(A')$.

Remark:

From theorem 3 it follows that if we would have defined the weak generator A'_w of the semi-group $S(t)$ by

$$A'_w x = \underset{h \rightarrow o}{w\text{-lim}} \, h^{-1}(x - S(h)x)$$

A. Pazy

whenever the weak limit exists, we would have $A_w^1 = A^1$ since

the existence of the weak limit implies the boundedness of

$\left| A(h)x_o \right|$.

A. Pazy

Lecture 6. - Semi-groups of type ω (continued).

Theorem 1.

Let $S(t)$ be a semi-group of type ω on $D \subset H$. Let A^{ι} be the generator of $S(t)$. If $x_0 \in D(A^{\iota})$ then:

a) $S(t)x_0 \in D(A^{\iota})$ for every $t \geq 0$.

b) $\dfrac{d^+ S(t)x_0}{dt} = - A^{\iota} S(t)x_0$ and $A^{\iota} S(t)x_0$ is continuous from the right.

c) $e^{-\omega t}|A^{\iota} S(t)x_0|$ is a monotonic non-increasing function and $\dfrac{dS(t)x_0}{dt}$ exists and is continuous at the continuity points of $e^{-\omega t}|A^{\iota} S(t)x_0|$ which are the continuity points of $|A^{\iota} S(t)x_0|$.

Proof:

Since $S(t)$ is of type ω we have for every $t \geq r \geq 0$

(1) $\qquad e^{-t\omega}|A(h)S(t)x_0| \leq e^{-r\omega}|A(h) S(r)x_0| \leq |A(h)x_0|$.

Since $x_0 \in D(A^{\iota})$, $|A(h)x_0|$ is bounded as $h \longrightarrow 0$ by

A. Pazy

theorem 3 of the previous lecture. From (1) it then follows that $|A(h)S(t)x_0|$ is bounded as $h \to 0$ and therefore by the same theorem $S(t)x_0 \in D(A^1)$ for every $t \geq 0$.

Since $S(t)x_0 \in D(A^1)$, $\dfrac{d^+S(t)x_0}{dt}$ exists and $\dfrac{d^+S(t)x_0}{dt} =$

$= - A^1 S(t)x_0$. Let A be any maximal $\mathfrak{M}(\omega)$ extension of A^1. From theorem 2 of the previous lecture $A^0 \supset A^1$. Passing to the limit as $h \to 0$ in (1) we see that $e^{-t\omega}|A^1S(t)x_0|$ is a monotone non-increasing function of t. Let $\{t_n\}$ be a decreasing sequence of positive numbers $t_n \downarrow t$ (or if t is a continuity point of $|A^1S(t)x_0|$ then $\{t_n\}$ is any sequence satisfying $t_n \to t$), then

$$\varlimsup_{n \to \infty} e^{-t_n\omega}|A^1S(t_n)x_0| \leq e^{-t\omega}|A^1S(t)x_0| = e^{-t\omega}|A^0S(t)x_0|$$

and therefore

$$\varlimsup_{n \to \infty} |A^1S(t_n)x_0| \leq |A^0S(t)x_0|.$$

From lemma 1 (b) in lecture 4 we then have $A^1S(t_n)x_0 \to$ $\to A^0S(t)x_0 = A^1S(t)x_0$. This establishes the right continuity of $A^1S(t)x_0$ and the continuity where $|A^1S(t)x_0|$ is continuous. Since by theorem 1 of the previous lecture $S(t)x_0$ is the

A. Pazy

integral of its derivative which exists and coincides with

$\frac{d^+S(t)x_0}{dt}$ a.e. in t we have

$$S(t)x_0 - S(t+h)x_0 = \int_0^h A^1 S(\tau+t)x_0 d\tau \; .$$

Dividing by h and letting $h \longrightarrow 0$ we obtain the existence
of the derivative of $S(t)x_0$ at points of continuity of
$A^1 S(t)x_0$ which are the points of continuity of $|A^1 S(t)x_0|$.

Remark:

From theorem 1 it follows that $\frac{dS(t)x_0}{dt}$ exists for all
$t \geq 0$ except for a denumerable set $\{\tau_n\}$. As can be seen from
examples $\frac{dS(t)x_0}{dt}$ does not usually exist for all $t \geq 0$.
The characterization of the generators of $C^1[0,\infty)$ semi-groups
is still an open problem.

Our next theorem will be proved in the appendix.

Theorem 2.

Let C be a closed convex subset of H and let $S(t)$ be
a semi-group of type ω on C . Let A^1 be the generator of

A. Pazy

$S(t)$ then $\overline{D(A^1)} = C$.

From theorem 2 it follows that to every semi-group of type ω on a closed convex subset C of H we can associate a unique maximal $\mathfrak{M}(\omega)$ set which can be regarded as its generator.

Theorem 3.

Let $C \subset H$ be closed and convex and let $S(t)$ be a semi-group of type ω on C . There exists a unique maximal $\mathfrak{M}(\omega)$ set A with the following properties:

1) $\overline{D(A)} = C$

2) $A^o = A^1$

where A^1 is the generator of $S(t)$.

Proof:

Let A^1 be the generator of $S(t)$ then A^1 is a $\mathfrak{M}(\omega)$ set and by theorem 2 $\overline{D(A^1)} = C$. Let A be a maximal $\mathfrak{M}(\omega)$ extension of A^1 such that $\overline{D(A)} = C$. Such an extension, exists by theorem 1 of lecture 4 . From theorem 2 of lecture 5

A. Pazy

we deduce $A^1 = A^0$ and the uniqueness of A follows from theorem 2 of lecture 4 .

Now we start with a maximal $\mathfrak{M}(\omega)$ set A and we try to associate with it a unique semi-group of type ω which is generated by A .

Let A be a maximal $\mathfrak{M}(\omega)$ set. We saw in lecture 4 that for $\lambda\omega < 1/2$, A_λ is lipschitz with constant $1/\lambda$, defined on H and $A_\lambda \in \mathfrak{M}\left(\frac{\omega}{1-\lambda\omega}\right)$. Consider the initial value problem

(1)
$$\begin{cases} \dfrac{du_\lambda}{dt} + A_\lambda u_\lambda = 0 \\[2em] u_\lambda(0) = x \end{cases}$$

This initial value problem has a unique solution $u_\lambda(t)$ for every $x \in H$ since A_λ is everywhere defined and lipschitz on H we denote the solution $u_\lambda(t)$ of (1) by $S_\lambda(t)x$.

Lemma 1.

$S_\lambda(t)$ is a $C^1[0, \infty)^*$ semi-group of type $\dfrac{\omega}{1-\lambda\omega}$ on H . A_λ is the generator of S_λ .

---

* A semi-group $S(t)$ is a $C^1[0,\infty)$ semi-group if for every $x\in]$ $S(t)x$ is continuosly differentiable for $t \geq 0$.

A. Pazy

Proof:

By the uniqueness of the solution of (1) it follows that $S_\lambda(t)$ satisfies the semi-group property. From (1) it also follows that $S_\lambda(t)x$ is continuously differentiable for every $x \in H$. Now,

$$|S_\lambda(t)x - S_\lambda(t)y|^2 \leq |x-y|^2 - 2 \int_0^t (A_\lambda S_\lambda(\tau)x - A_\lambda S_\lambda(\tau)y, \ S_\lambda(\tau)x - S_\lambda(\tau)y)d\tau \leq$$

$$\leq |x-y|^2 + 2\left(\frac{\omega}{1-\lambda\omega}\right) \int_0^t |S_\lambda(\tau)x - S_\lambda(\tau)y|^2 \ d\tau$$

and by Gronwall inequality

$$|S_\lambda(t)x - S_\lambda(t)y| \leq \exp\left(\frac{\omega}{1-\lambda\omega}\cdot t\right)|x-y|$$

i.e. $S_\lambda(t)$ is of type $\frac{\omega}{1-\lambda\omega}$. This completes the proof of this lemma.

From (4) in lecture 4 and theorem 1 we have :

$$(2) \quad |A_\lambda S_\lambda(t)x| \leq \exp\left(\frac{\omega}{1-\lambda\omega}\cdot t\right)|A_\lambda x| \leq (1-\lambda\omega)^{-1} \exp\left(\frac{\omega}{1-\lambda\omega}\cdot t\right)|A^\circ x|$$

for every $x \in D(A)$.

For simplicity we shall henceforth assume $\lambda < 1/2\omega$ and

A. Pazy

then (2) reduces to

(3) $\quad |A_\lambda S_\lambda(t)x| \leq 2\, e^{2\omega t}\, |A^\circ x|$ for every $x \in D(A)$, $\lambda\omega < 1/2$.

Lemma 2.

Let A be a maximal $\mathfrak{M}(\omega)$ set and let $S_\lambda(t)$ be the semi-group introduced in lemma 1 then

a) For every $x \in \overline{D(A)}$, $\lim\limits_{\lambda \to 0} S_\lambda(t)$ exists. The limit is uniform in t on every bounded interval. We denote this limit by $S(t)$.

b) $S(t)$ is a semi-group of type ω on $D(A)$.

Proof:

Let $[0,T]$ be a fixed interval and let $x \in D(A)$. Further let $u_\lambda(t) = S_\lambda(t)x$. Since $A_\lambda u_\lambda \in AJ_\lambda u_\lambda$ we have for $0 \leq t \leq T$.

$$|u_\lambda(t)-u_\mu(t)|^2 = -2 \int_0^t (A_\lambda u_\lambda - A_\mu u_\mu, u_\lambda - u_\mu)\,d\tau =$$

$$= -2 \int_0^t (A_\lambda u_\lambda - A_\mu u_\mu, u_\lambda - J_\lambda u_\lambda - u_\mu + J_\mu u_\mu)\,d\tau -$$

$$- 2 \int_0^t (A_\lambda u_\lambda - A_\mu u_\mu, I_\lambda u_\lambda - J_\mu u_\mu)\,d\tau \leq$$

$$\leq -2 \int_0^t (A_\lambda u_\lambda - A_\mu u_\mu, \lambda A_\lambda u_\lambda - \mu A_\mu u_\mu)\,d\tau + 2\omega \int_0^t |J_\lambda u_\lambda - J_\mu u_\mu|^2\,d\tau$$

$$\leq \int_0^t (\mu |A_\lambda u_\lambda|^2 + \lambda |A_\mu u_\mu|^2)d\tau + 2\omega \int_0^t |J_\lambda u_\lambda - J_\mu u_\mu|^2 \, d\tau$$

$$\leq 4e^{4\omega T} \cdot T |A^\circ x|^2 (\lambda + \mu) + 2\omega \int_0^t |J_\lambda u_\lambda - J_\mu u_\mu|^2 \, d\tau \quad .$$

But

$$|J_\lambda u_\lambda - J_\mu u_\mu| \leq |J_\lambda u_\lambda - u_\lambda| + |u_\lambda - u_\mu| + |u_\mu - J_\mu u_\mu| = \lambda |A_\lambda u_\lambda| + |u_\lambda - u_\mu| + \mu |A_\mu u_\mu|$$

therefore

$$|J_\lambda u_\lambda - J_\mu u_\mu|^2 \leq 3\lambda^2 |A_\lambda u_\lambda|^2 + 3|u_\lambda - u_\mu|^2 + 2\mu^2 |A_\mu u_\mu|^2$$

and thus

$$|u_\lambda - u_\mu|^2 \leq 4e^{4\omega T} \cdot T \cdot |A^\circ x|^2 (\lambda + \mu + 6\omega(\lambda^2 + \mu^2)) + 6\omega \int_0^t |u_\lambda - u_\mu| \, d\tau$$

and from the Gronwall inequality we obtain

$$|u_\lambda(t) - u_\mu(t)| \leq 2e^{5\omega T} \sqrt{T} |A^\circ x| (\lambda + \mu + 6\omega(\lambda^2 + \mu^2))^{1/2} .$$

Therefore $S_\lambda(t)x$ is a Cauchy sequence for every $x \in D(A)$, uniformly in t on $[0,T]$. Since for every $x,y \in \overline{D(A)}$

$$(4) \quad |S_\lambda(t)x - S_\lambda(t)y| \leq e^{t \frac{\omega}{1-\omega}} |x-y| \leq e^{2\omega t} |x-y| \quad \text{for} \quad \lambda < 1/2\omega$$

it follows that $S_\lambda(t)x$ converges uniformly in t on $[0,T]$ for every $x \in \overline{D(A)}$. Clearly the limit $S(t)x$ is a semi-group

A. Pazy

on $\overline{D(A)}$ and passing to the limit as $\lambda \longrightarrow 0$ in (4) yields

$$|S(t)x-S(t)y| \leq e^{\omega t}|x-y|$$

i.e. $S(t)$ is a semi-group of type ω and the proof is complete.

Theorem 4.

Let A be a maximal $\mathfrak{M}(\omega)$ set. There exists a unique semigroup $S(t)$ of type ω on $\overline{D(A)}$ such that A^o is the generator of $S(t)$.

Proof:

We shall prove that the semi-group $S(t)$ which was constructed in lemma 2 is the desired semi-group.

Let A be the generator of $S(t)$ and let $x \in D(A^1)$, $u \in D(A)$. Using lemma 3 of the previous lecture for

$$\left(\frac{x-S_\lambda(t)x}{t} - A_\lambda u , x-u\right) \geq -\frac{1}{t}\int_o^t (A_\lambda u, x-S_\lambda(\tau)x)d\tau - \frac{\omega}{1-\lambda\omega}\cdot\frac{1}{t}\int_o^t |u-S_\lambda(\tau)x|^2$$

and passing to the limit as $\lambda \longrightarrow 0$ we obtain

A. Pazy

$$\left(\frac{x-S(t)x}{t} - A^{\circ}u, \ x-u\right) \geq -\frac{1}{t} \int_{0}^{t} (A^{\circ}u, \ x-S(\tau)x)d\tau - \frac{\omega}{t} \int_{0}^{t} |u-S(\tau)x|^{2} \ d\tau$$

Letting $t \longrightarrow 0$ yields

$$(A^{1}x - A^{\circ}u, \ x-u) \geq -\omega|x-u|^{2} \quad \text{for every} \quad u \in D(A) \ .$$

Since $x \in \overline{D(A)}$ and A° is a principal section of A it follows that $\left[x, \ A^{1}x\right] \in A$. Thus A is a maximal $\mathcal{M}(\omega)$ extension of A^{1} with $\overline{D(A)} = \overline{D(A^{1})}$ and therefore $A^{1} = A^{\circ}$ by theorem 2 of the previous lecture. The uniqueness of $S(t)$ follows directly from the fact that

$$\frac{dS(t)x}{dt} + A^{\circ}S(t)x = 0 \quad \text{a.e.} \quad \text{for every} \quad x \in D(A)$$

and the continuity of $S(t)x$ in t , and the lipschitz continuity of $S(t)$ on bounded intervals.

Combining theorem 3 and theorem 4 we obtain the following two important corollaries.

Corollary 2.

There is a one to one correspondence between maximal $\mathcal{M}(\omega)$

A. Pazy

sets in H×H and semi-groups of type ω on closed convex
subsets of H .

Corollary 2 enables us to consider maximal $\mathcal{M}(\omega)$ sets as
generators of semi-groups of type ω . This remark is extremely
important for the development of a perturbation theory for
generators of semi-groups of type – ω .

Corollary 3 (The characterization of generators)

An operator A^1 is a generator of a strongly continuous
semi-group of type ω on a closed convex subset C of H if
and only if there exists a maximal $\mathcal{M}(\omega)$ set A such that
$\overline{D(A)} = C$ and $A^1 = A^0$. In particular A^1 is a generator of a
semi-group of contractions if and only if there exists a maximal
monotone set A such that $A = A^0$.

A. Pazy

Lecture 7. - Convergence theorems

Lemma 1.

Let $C \subset H$ be closed and convex and let T be lipschitz with constant $\alpha \geq 1$ such that $T : C \longrightarrow C$. Then $I - T$ generates a semi-group of type $\alpha - 1$ on C.

Proof:

The initial value problem :

(1)
$$\begin{cases} \dfrac{du}{dt} + (I-T)u = 0 \\[2mm] u(0) = x \end{cases}$$

is equivalent to

(2)
$$u(t) = e^{-t}x + \int_0^t e^{s-t}\, T(u(s))ds \; .$$

Let $0 \leq t \leq T_0$ where T_0 is determined by $\alpha(1-e^{-T_0}) < 1$, and consider the map :

$$\Phi_x u = e^{-t}\, x + \int_0^t e^{s-t}\, T(u(s))ds$$

$\tilde{\Phi}_x$ maps the closed convex set

$$\left\{u \mid u \in C(0,T_o: H), \ u(t) \in C \ \text{for} \ 0 \le t \le T_o\right\}$$

into itself and $\tilde{\Phi}_x$ is a contraction with the constant $\alpha \, (1-e^{-T_o})$

therefore it has a unique fixed point, i.e. (2) has a unique

solution $u(t)$ for every $x \in C$ and $0 \le t \le T_o$. This solution

is continued by the usual procedure to a solution for all $t \ge 0$.

Let

$$v(t) = e^{-t}y + \int_0^t e^{s-t} \, T(v(s))ds$$

then $\qquad e^t |u(t)-v(t)| \ \le \ |x-y| + \alpha \int_0^t e^s |u(s)-v(s)| ds$

and by the Gronwall inequality

$$|u(t)-v(t)| \ \le \ e^{(\alpha-1)t} |x-y| \quad \text{for every} \ x,y \in C \ .$$

Lemma 2.[*]

Let $C \subset H$ be closed and convex and let T be lipschitz

---

[*] A proof of lemma 2 can be found in :

I Miyadera e S. Oharu - Approximation of semi-groups of

nonlinear operators, Tôhoku Math. J. 22 (1970) 24-47 .

A. Pazy

with constant α , $\alpha \geqslant 1$, $T : C \longrightarrow C$. If $S(t)$ is the semi-group generated by $I - T$ on C then

(3) $\quad |S(n)x - T^n x| \leq \alpha^{n-1} e^{(\alpha-1)n} \left[n^2(1-\alpha)^2 + n\alpha \right]^{1/2} |x-Tx| \quad \forall x \in C \quad n > 0$

Theorem 1.

Let A be a maximal $\mathcal{M}(\omega)$ set and let A^ρ be maximal $\mathcal{M}(\omega_\rho)$ sets such that $D(A^\rho) \supset D(A)$ and $0 \leq \omega_\rho < \alpha$ for some $\alpha > 0$. If $J_\lambda^\rho x \longrightarrow J_\lambda x$ for every $x \in \overline{D(A)}$ and every $0 < \lambda \leq \lambda_0$ then $\omega \leq \underline{\lim}\, \omega_\rho$ and $S^\rho(t)x \longrightarrow S(t)x$ for every $x \in \overline{D(A)}$ as $\rho \longrightarrow 0$. The limit is uniform in t on every bounded interval .

Proof:

Since $A_\lambda = \lambda^{-1}(I - J_\lambda)$ it follows that $A_\lambda^\rho x \longrightarrow A_\lambda x$ as $\rho \longrightarrow 0$ for every $x \in \overline{D(A)}$. The limit is uniform on compact sets of x for $0 < \lambda < \min (\frac{1}{2\omega} , \frac{1}{2\alpha})$ since then A_λ and A_λ^ρ are lipschitz with a constant $1/\lambda$ which is independent of ρ . Also $J_\lambda^\rho x \longrightarrow J_\lambda x$ implies directly that $\underline{\lim}\, \omega_\rho \geq \omega$.

Let $0 \leq t \leq T_0$, $x \in D(A)$ and λ small enough, then

(4) $\quad |S(t)x - S^\rho(t)x| \leq |S(t)x - S_\lambda(t)x| + |S_\lambda(t)x - S_\lambda^\rho(t)x| + |S_\lambda^\rho(t)x - S^\rho(t)$

A. Pazy

where S_λ^ρ is the semi-group generated by A_λ^ρ and S_λ is the semi-group generated by A_λ. From the proof of lemma 2 in the previous lecture, assuming that $0 < \lambda < \min(1, \frac{1}{2\omega})$ we obtain

$$(5) \qquad |S(t)x - S_\lambda(t)x| \le 2e^{5\omega T_0}\sqrt{T_0}|A^\circ x|(\lambda+\sqrt{\lambda}) \le C_0|A^\circ x|\sqrt{\lambda}$$

where C_0 depends only on ω and T_0 and is bounded uniformly for $\omega_\rho \le \alpha$. Now,

$$|S_\lambda(t)x - S_\lambda^\rho(t)x| \le \int_0^t |A_\lambda S_\lambda(\tau)x - A_\lambda^\rho S_\lambda^\rho(\tau)x|\,d\tau \le$$

$$\le \int_0^t |A_\lambda S_\lambda(\tau)x - A_\lambda^\rho S_\lambda(\tau)x|\,d\tau + \int_0^t |A_\lambda^\rho S_\lambda^\rho(\tau)x - A_\lambda^\rho S_\lambda(\tau)x|\,d\tau$$

$$\le M_\lambda^\rho(x)T_0 + \frac{1}{\lambda}\int_0^t |S_\lambda(\tau)x - S_\lambda^\rho(\tau)x|\,d\tau$$

where

$$M_\lambda^\rho(x) = \sup_{0 \le \tau \le T_0} |A_\lambda S_\lambda(\tau)x - A_\lambda^\rho S_\lambda(\tau)x|$$

and by Gronwall inequality

$$(6) \qquad |S_\lambda(t)x - S_\lambda^\rho(t)x| \le T_0 \cdot M_\lambda^\rho(x)e^{T_0/\lambda}.$$

Finally

A. Pazy

$$\left| S^\rho(t)x - S_\lambda^\rho(t)x \right| \le \left| S^\rho(t)J_\lambda^\rho x - S_\lambda^\rho(t)J_\lambda^\rho x \right| + 2e^{\alpha t}\left| x - J_\lambda^\rho x \right| \le$$

$$\le C_o \left| (A^\rho)^\circ J_\lambda^\rho x \right| \cdot \sqrt{\lambda} + C_1 \cdot \lambda \left| A_\lambda^\rho x \right|$$

$$\le (C_o + C_1\sqrt{\lambda})\sqrt{\lambda}\left| A_\lambda^\rho x \right| \le C_2 \sqrt{\lambda} \left| A_\lambda^\rho x \right| \le C_2\sqrt{\lambda}(\left| A_\lambda^\rho x - A_\lambda x \right| + \left| A_\lambda x \right|$$

and therefore

(7) $$\left| S^\rho(t)x - S_\lambda^\rho(t)x \right| \le C_2 \sqrt{\lambda}(M_\lambda^\rho(x) + \left| A^\circ x \right|).$$

The constants C_o, C_1, C_2 depend only on T_o and α . Combining (4), (5), (6) and (7) we obtain

(8) $$\left| S(t)x - S^\rho(t)x \right| \le C_3 \left| A^\circ x \right| \sqrt{\lambda} + M_\lambda^\rho(x)\left[T_o e^{\frac{T_o}{\lambda}} + C_2\sqrt{\lambda} \right].$$

Given $\epsilon > 0$ we first choose λ so that the first term on the right hand side is less than $\epsilon/2$. Having fixed λ we choose ρ so that the second term is less than $\epsilon/2$. This is possible since for any fixed λ $M_\lambda^\circ(x) \to 0$ as $\rho \to 0$. Thus for $x \in D(A)$, $S^\rho(t)x \to S(t)x$ uniformly in $t \in [0, T_o]$. Since S^ρ and S are lipschitz with bounded constants we conclude that $S^\rho(t) \to S(t)y$ for every $y \in \overline{D(A)}$ uniformly in $t \in [0, T_o]$.

A. Pazy

<u>Lemma 3.</u>

Let A^ρ be maximal $\mathfrak{M}(\omega_\rho)$ sets. If $\overline{\lim_{\rho \to 0}} \omega_\rho = \omega$ and

for some $\lambda_0 < 1/\omega$ $J^\rho_{\lambda_0} x$ converges for every $x \in H$ as

$\rho \longrightarrow 0$ to a limit denoted by $J_{\lambda_0} x$. Then there exists a

unique maximal $\mathfrak{M}(\omega)$ set A with domain $\overline{D(A)} = \overline{R(J_{\lambda_0})}$

such that $J^\rho_\lambda x \longrightarrow (I+\lambda A)^{-1} x = J_\lambda x$ for every $0 < \lambda < 1/\omega$ and

every $x \in H$.

<u>Proof:</u>

It is easy to verify the following simple identities:

(9) $(x_1 - x_2, J^\rho_\lambda x_1 - J^\rho_\lambda x_2) \geq (1 - \omega_\rho \lambda) |J^\rho_\lambda x_1 - J^\rho_\lambda x_2|^2$ $0 < \lambda < 1/\omega_\rho$ $x_1, x_2 \in H$

(10) $J^\rho_\lambda x = J^\rho_\mu (\frac{\mu}{\lambda} x + (1 - \frac{\mu}{\lambda}) J^\rho_\lambda x)$ $0 < \lambda, \mu < 1/\omega_\rho$ $x \in H$

Also

$|J^\rho_{\lambda_0} x - J^\rho_\mu (\frac{\mu}{\lambda_0} x + (1 - \frac{\mu}{\lambda_0}) J_{\lambda_0} x)| = |J^\rho_\mu (\frac{\mu}{\lambda_0} x + (1 - \frac{\mu}{\lambda_0}) J^\rho_{\lambda_0} x) - J^\rho_\mu (\frac{\mu}{\lambda_0} x + (1 - \frac{\mu}{\lambda_0}) J_{\lambda_0}$

$\leq \frac{1}{1 - \mu \omega_0} |1 - \frac{\mu}{\lambda_0}| |J^\rho_{\lambda_0} x - J_{\lambda_0} x| \longrightarrow 0$

as $\rho \longrightarrow 0$ and $\mu < 1/\omega$.

A. Pazy

Therefore $\lim_{\rho \to 0} J_{\mu}^{\rho} z$ exists for every $0 < \mu < 1/\omega$ and z

of the form $z = \frac{\mu}{\lambda_0} x + (1 - \frac{\mu}{\lambda_0}) J_{\lambda_0} x$ or $\frac{\lambda_0}{\mu} z = x + (\frac{\lambda_0}{\mu} - 1) J_{\lambda_0} x$.

But for $0 < \mu < 1/\omega$ the mapping $x \longrightarrow x + (\frac{\lambda_0}{\mu} - 1) J_{\lambda_0} x$ is

onto. Indeed if $0 < \mu < \lambda_0$ then $(\frac{\lambda_0}{\mu} - 1) J_{\lambda_0}$ is maximal

monotone and hence $R(I + (\frac{\lambda_0}{\mu} - 1) J_{\lambda_0}) = H$. If $\lambda_0 \leq \mu < 1/\omega$

then $(1 - \frac{\lambda_0}{\mu}) J_{\lambda_0}$ is a strict contraction and the mapping

$x \longrightarrow \frac{\lambda_0}{\mu} z + (1 - \frac{\lambda_0}{\mu}) J_{\lambda_0} x$ has a unique fixed point. Thus $J_{\lambda}^{\rho} x$

converges as $\rho \longrightarrow 0$ for every $x \in H$ and $0 < \lambda < 1/\omega$.

Passing to the limit in (9) and (10) we obtain

(11) $\qquad (x_1 - x_2, J_{\lambda} x_1 - J_{\lambda} x_2) \geq (1 - \omega\lambda) |J_{\lambda} x_1 - J_{\lambda} x_2|^2 \qquad 0 < \lambda < 1/\omega$

and

(12) $\qquad J_{\lambda} x = J_{\mu}(\frac{\mu}{\lambda} x + (1 - \frac{\mu}{\lambda}) J_{\lambda} x) \qquad 0 < \lambda, \mu < 1/\omega$.

From (12) we deduce that $R(J_{\mu}) = R(J_{\lambda_0})$ for every

$0 < \mu < 1/\omega$. Now let

$$A(\lambda) = \left\{ \left[J_{\lambda} x, \frac{x - J_{\lambda} x}{\lambda} \right] : x \in H \right\}$$

$A(\lambda) \in \mathcal{M}(\omega)$ by (11). From (12) it follows that $A(\lambda) = A(\mu)$

A. Pazy

for every $0 < \lambda$, $\mu < 1/\omega$. Let $A = A(\lambda)$. Obviously $\overline{D(A)} =$ $= \overline{R(J_{\lambda_0})}$ and since $R(I+\lambda A(\lambda)) = R(I+\lambda A) = H$ for every $0 < \lambda < 1/\omega$, A is a maximal $\mathcal{M}(\omega)$ set. Also since $x \in J_\lambda x + \lambda A(\lambda) J_\lambda x = J_\lambda x + \lambda A J_\lambda x$ we have $J_\lambda x = (I+\lambda A)^{-1} x$.

To prove the uniqueness of A let A_1 and A_2 satisfy the requiered properties then $\overline{D(A_1)} = \overline{D(A_2)} = \overline{R(J_{\lambda_0})}$ and $(I+\lambda A_1)^{-1} x = (I+\lambda A_2)^{-1} x \quad \forall x \in H$. Therefore $A_{1\lambda} x = A_{2\lambda} x$ for every $x \in H$ which implies by the results of lecture 4 that $D(A_1) = D(A_2)$ and $A_1^o = A_2^o$ thus $A_1 = A_2$.

Using lemma 3 and theorem 1 we obtains;

Corollary 1.

Let A^o , $\rho > 0$, be a family of maximal $\mathcal{M}(\omega_\rho)$ sets such that $\overline{\lim_{\rho \to 0}} \omega_\rho = \omega < \infty$. If

i) There exists $0 < \lambda_0 < 1/\omega$ such that $J_{\lambda_0}^\rho x$ converges

   as $\rho \to 0$ for every $x \in H$ to a limit denoted by

   $J_{\lambda_0} x$

ii) $\overline{D(A^\rho)} \supset \overline{R(J_{\lambda_0})}$ for every $\rho > 0$.

A. Pazy

Then there exists a unique maximal $\mathcal{M}(\omega)$ set A with $\overline{D(A)} = \overline{R(J_{\lambda_0})}$ such that $J_\lambda^\rho x \longrightarrow (I+\lambda A)^{-1}x$ for every $x \in H$ and $0 < \lambda < 1/_\omega$. Consequently $S^\rho(t)x \longrightarrow S(t)x$ for every $x \in \overline{D(A)}$ uniformly in t on bounded intervals.

Remark:

If $\overline{R(J_{\lambda_0})} = H$ in the previous corollary we obtain, for Hilbert space, a nonlinear extension of a theorem of H.F. Trotter.

Lemma 4.

Let A be a maximal $\mathcal{M}(\omega)$ set and let A' be a principal section of A. For every $\rho > 0$ let A^ρ be a $\mathcal{M}(\omega_\rho)$ set satisfying :

i) $J_\lambda^\rho = (I+\lambda A^\rho)^{-1}$: $\overline{D(A)} \longrightarrow \overline{D(A)}$ for every $\lambda < 1/\omega_\rho$

ii) For every $u \in D(A)$ there exists $y_\rho \in A^\rho u$ such that $y_\rho \longrightarrow A'u$ as $\rho \longrightarrow 0$.

iii) $\lim_{\rho \to 0} \omega_\rho = \omega$.

Then $J_\lambda^\rho x \longrightarrow J_\lambda x$ as $\rho \longrightarrow 0$ for every $x \in \overline{D(A)}$ and $0 < \lambda < 1/_\omega$.

A. Pazy

Proof:

Let $x_\rho = (I+\lambda A^\rho)^{-1}x = J_\lambda^\rho x$ with $x \in \overline{D(A)}$ and let $u \in D(A$

$y_\rho \in A^\rho u$ such that $y_\rho \to A'u$ as $\rho \to 0$. Then

(13) $$\left(\frac{x-x_\rho}{\lambda} - y_\rho, \; x_\rho-u\right) \geq -\omega_\rho |x_\rho-u|^2 \; .$$

For $\lambda\omega_\rho < \delta < 1$ this implies that $|x_\rho|$ is bounded as

$\rho \to 0$. Let $x_{\rho_n} \to \xi$, then $\xi \in \overline{D(A)}$ by (i) and the fact

that $\overline{D(A)}$ is convex. From (13) we obtain as $\rho_n \to 0$

$$\left(\frac{x-\xi}{\lambda} - A'u, \; \xi-u\right) \geq -\omega|\xi-u|^2$$

and since A' is a principal section of, A this implies

$\xi \in D(A)$ and $\frac{x-\xi}{\lambda} \in A\xi$, or $\xi = (I+\lambda A)^{-1}x$. In addition it

follows from (13) that

$$\left(\frac{1}{\lambda} - \omega\right) \overline{\lim_{\rho_n \to 0}} \, |x_{\rho_n}|^2 \leq \left(\frac{x}{\lambda}, \xi-u\right) + \left(\frac{\xi}{\lambda}, u\right) - (A'u, \xi-u) - 2\omega(\xi, u)$$

for every $u \in D(A)$. Substituting $u = \xi$ we obtain

$$\overline{\lim_{\rho_n \to 0}} \, |x_{\rho_n}|^2 \leq |\xi|^2$$

A. Pazy

which implies $x_{\rho_n} \longrightarrow \xi$. Since the limit is unique we have

$x_\rho \longrightarrow \xi$ as $\rho \longrightarrow 0$.

From theorem 1 and lemma 4 we obtain:

Corollary 2.

Let A be a maximal $\mathfrak{M}(\omega)$ set and let A^ρ , $\rho > 0$, be maximal $\mathfrak{M}(\omega_\rho)$ sets such that $D(A^\rho) \supset D(A)$ for every $\rho > 0$. Let $S(t)$ and $S^\rho(t)$ be the semi-groups generated by A and A^ρ respectively. If

i) $\lim_{\rho \to 0} \omega_\rho = \omega$

ii) $J_\lambda^\rho : \overline{D(A)} \longrightarrow \overline{D(A)}$

iii) $(A^\rho)^{\circ} x \longrightarrow A^{\circ} x$ for every $x \in D(A)$.

Then $S^\rho(t)x \longrightarrow S(t)x$ for every $x \in \overline{D(A)}$ uniformly in t on every bounded interval.

A. Pazy

Lecture 8. - Representation theorems.

Lemma 1.

Let $C \subset H$ be closed and convex and let $F : C \longrightarrow C$ be Lipschitz with constant $\alpha \geq 1$. Then,

i) $(I+\lambda(I-F))^{-1}$ exists on C for every $0 < \lambda < (\alpha-1)^{-1}$ and $(I+\lambda(I-F))^{-1} : C \longrightarrow C$

ii) $\rho^{-1}(I-F)$ is lipschitz and $\rho^{-1}(I-F) \in \mathfrak{M}(\rho^{-1}(\alpha-1))$.

Proof:

i) Let $x \in C$ and let $G(y) = \frac{1}{1+\lambda}x + \frac{\lambda}{1+\lambda} Fy$. Clearly G is a strict contraction for $0 < \lambda < (\alpha-1)^{-1}$ and G maps C into C . Therefore G has a unique fixed point $y \in C$. Thus $(I+\lambda(I-F))^{-1}$ is defined on C and maps C into C .

ii) Clearly $\rho^{-1}(I-F)$ is lipschitz with constant $\leq \rho^{-1}(\alpha+1)$ and $\rho^{-1}((I-F)x_1 - (I-F)x_2, x_1-x_2) = \rho^{-1}|x_1-x_2|^2 - \rho^{-1}(Fx_1 - Fx_2, x_1 - x_2) \geq \rho^{-1}(1-\alpha) |x_1 - x_2|^2$ thus $\rho^{-1}(I-F) \in \mathfrak{M}(\rho^{-1}(\alpha-1))$

A. Pazy

Theorem 1.

Let A be a maximal $\mathfrak{M}(\omega)$ set. Let $F(t)$ $t \geq 0$ be lipschitz with constant $\alpha(t)$. If

i) $\alpha(t) = 1+\omega t+o(t)$

ii) $F(t): \overline{D(A)} \longrightarrow \overline{D(A)}$

iii) $\dfrac{x-F(t)x}{t} \longrightarrow A^{\circ}x$ for every $x \in D(A)$

then

(1) $\qquad \lim_{n\to\infty} F(\dfrac{t}{n})^{n} x = S(t)x \qquad$ for every $x \in \overline{D(A)}$

and the limit is uniform in t on every bounded interval.

Proof:

Let $A^{\rho} = \rho^{-1}(I-F(\rho))$. By lemma 1 $A^{\rho} \in \mathfrak{M}(\rho^{-1}(\alpha(\rho)-1))$ and $J_{\lambda}^{\rho}x = (I+\dfrac{\lambda}{\rho}(I-F(\rho))^{-1}x$ exists for $0 < \lambda < \dfrac{\rho}{\alpha(\rho)-1}$. Since A^{ρ} is everywhere defined and lipschitz the range of $I+\lambda A^{\rho}$ is all of H i.e. A^{ρ} is a maximal $\mathfrak{M}(\rho^{-1}(\alpha(\rho)-1))$ set. Moreover by lemma 1 $J_{\lambda}^{\rho} : \overline{D(A)} \longrightarrow \overline{D(A)}$ and by lemma 1 of the previous lecture taking $C = H$ A^{ρ} generates a semi-group $S^{\rho}(t)$ on H . From corollary 2 of the previous lecture it then follows

A. Pazy

that $S^\rho(t)x \longrightarrow S(t)x$ for every $x \in \overline{D(A)}$ uniformly in t on every bounded interval. Let $t \in [0, T_o]$ and denote by $\tilde{S}(t)$ the semi-group generated by $I - F(\rho)$. Then $S^\rho(t) = \tilde{S}(\frac{t}{\rho})$. Let $x \in D(A)$, from lemma 2 of the previous lecture we have:

$$|\tilde{S}(n)x - F(\rho)^n x| = |S^\rho(n\rho)x - F(\rho)^n x| \leq$$

$$\leq \alpha(\rho)^{n-1} e^{(\alpha(\rho)-1)n} \left[n^2(\alpha(\rho)-1)^2 + n(\alpha(\rho)-1) + n \right]^{1/2} \rho|$$

$$= H(\rho,n) \, \rho |A^\rho x|$$

substituting $\rho = \frac{t}{n}$ we obtain:

$$|S^{t/n}(t)x - F(\frac{t}{n})^n x| \leq H(\frac{t}{n},n) \frac{T_o}{n} |A^{t/n}x| \leq 2H(\frac{t}{n},n) \frac{T_o}{n} |A^o x| \quad .$$

From our assumptions about $\alpha(t)$ it follows that $\alpha(\frac{t}{n})^n$ and $e^{n(\alpha(\frac{t}{n})-1)}$ are bounded as $n \longrightarrow \infty$ and that $\frac{1}{n} H(\frac{t}{n},n) \longrightarrow 0$ uniformly in $t \in [0, T_o]$ as $n \longrightarrow \infty$. Thus $|S^{t/n}(t)x - F(\frac{t}{n})^n x| \longrightarrow$ uniformly in $t \in [0, T_o]$ as $n \longrightarrow \infty$ for every $x \in D(A)$. Since $F(\frac{t}{n})^n$ and $S(t)$ are lipschitz with bounded constants we have $F(\frac{t}{n})^n x \longrightarrow S(t)x$ for every $x \in \overline{D(A)}$ uniformly in $t \in [0, T_o]$.

A. Pazy

Remark

By a slight modification of the proof (see e.g. proof of corollary 3) one obtains the results of theorem 1 assuming that $F(t)$ is only defined on $\overline{D(A)}$ and not on all of H .

Corollary 1 (Exponential Formula)

Let A be a maximal $\mathcal{M}(\omega)$ set and let $S(t)$ be the semi-group generated by A then

$$(2) \qquad S(t)x = \lim_{n \to \infty} (I + \frac{t}{n} A)^{-n} x = \lim_{n \to \infty} J^n_{t/n} x \qquad \forall\, x \in \overline{D(A)} .$$

Proof:

We check that the conditions of theorem 1 are satisfied. In lecture 4 we have proved that J_t is lipschitz with constant $\alpha(t) = \dfrac{1}{1-\omega t} = 1 + \omega t + o(t)$. Moreover $J_t : H \longrightarrow D(A)$ therefore clearly $J_t : \overline{D(A)} \longrightarrow \overline{D(A)}$ and $\dfrac{x - J_t x}{t} = A_t x \longrightarrow A^o x$ for every $x \in D(A)$. Therefore (2) holds by theorem 1 .

Corollary 2.

Let A be a maximal $\mathcal{M}(\omega)$ set and let $S(t)$ be the semi-

A. Pazy

group generated by A. Let $C \subset \overline{D(A)}$ be closed and convex

then $S(t) : C \longrightarrow C$ if and only if $J_t : C \longrightarrow C$ for all

$0 < t < 1/\omega$.

Proof:

Corollary 1 implies that if $J_t : C \longrightarrow C$ then $S(t) : C \longrightarrow C$

for all $t > 0$. If $S(t) : C \longrightarrow C$ then by lemma 1

$(I + \frac{\lambda}{h}(I - S(h)))^{-1} : C \longrightarrow C$. From lemma 4 of the previous

lecture we deduce that $(I + \frac{\lambda}{h}(I - S(h)))^{-1}x \longrightarrow (I+\lambda A)^{-1}x$ for

all $0 < \lambda < 1/\omega$ and therefore $J_\lambda : C \longrightarrow C$.

Using Corollary 2 of the previous lecture we obtain the

following representation theorem:

Corollary 3.

Let A be a maximal $\mathcal{M}(\omega)$ set and let $S(t)$ be the semi-

group generated by A . Let $A(h) = h^{-1}(I-S(h))$ then $A(h)$

generates a semi-group of type $h^{-1}(e^{\omega h} - 1)$ on $\overline{D(A)}$ which

we denote by $S^h(t)$. For every $x \in \overline{D(A)}$ we have,

$$(3) \qquad S(t)x = \lim_{h \to 0} S^h(t)x .$$

A. Pazty

Proof:

By lemma 1 of the previous lecture $I-S(h)$ generates a
semi-group $\tilde{S}(t)$ on $\overline{D(A)}$. Therefore $h^{-1}(I-S(h))$ generates
the semi-group $S^h(t) = \tilde{S}(\frac{t}{h})$ on $\overline{D(A)}$. It is easy to check
that $A(h) \in \mathcal{M}(\frac{e^{\omega h}-1}{h})$ and therefore $S^h(t)$ is of type
$\omega_h = \frac{e^{\omega h}-1}{h}$. This implies $\lim_{h \to 0} \omega_h = \omega$. Now, $(I + \frac{\lambda}{h}(I-S(h)))^{-1}$:

$: \overline{D(A)} \longrightarrow \overline{D(A)}$ for $0 < \lambda < \frac{h}{e^{\omega h}-1}$ by lemma 1 . Let A^h be
any maximal $\mathcal{M}(\frac{e^{\omega h}-1}{h})$ extension of $A(h)$ with $\overline{D(A^h)} = \overline{D(A)}$.
It is easy to check that on $\overline{D(A)}$ $(I+\lambda A^h)^{-1} = (I+\frac{\lambda}{h}(I-S(h)))^{-1}$
thus $J_\lambda^h : \overline{D(A)} \longrightarrow \overline{D(A)}$ and finally $(A^h)^o = A(h)$ (by theorem
2 of lecture 5). Thus for every $x \in D(A)$, $\lim_{h \to 0} A(h)x = \lim_{h \to 0} (A^h)^o x =$
$= A^o x$ and (3) follows from corollary 2 of the previous lecture.

We conclude this lecture with a result related to Trotter's
product formula.

Theorem 2.

Let A and B single valued maximal monotone operators

A. Pazy

such that $\overline{D(A) \cap D(B)} = H$. Let $S_A(t)$ and $S_B(t)$ be the
semi-groups generated by A and B respectively. If A+B is
maximal monotone and $S_{A+B}(t)$ is the semi-group generated by
it then

$$(4) \qquad S(t)x = \lim_{h \to \infty} \left[S_A(\tfrac{t}{n}) \, S_B(\tfrac{t}{n}) \right]^n x \quad \text{for every } x \in H$$

and the limit is uniform in t on every finite interval.

Proof.

We substitute $F(t) = S_A(t) \cdot S_B(t)$ and take A+B instead
of A in theorem 1. Clearly $S_A(t) \, S_B(t)$ is a contraction so
(i) is satisfied. (ii) is trivial in this case so in order to
prove the result it is sufficient to prove that

$$\lim_{t \to 0} \frac{x - S_A(t) \, S_B(t)}{t} = (A+B)x \qquad \forall \, x \in D(A) \cap D(B) .$$

Now,

$$t^{-1}(x - S_A(t)S_B(t)x) = t^{-1}(x - S_A(t)x) + t^{-1}(S_A(t)x - S_A(t)S_B(t)x).$$

Choosing $x \in D(A) \cap D(b)$ the first term clearly tends to
Ax as $t \to 0$. Denoting the second term by y_t it is therefore

A. Pazy

sufficient to prove that $y_t \longrightarrow Bx$ is $t \longrightarrow 0$. Clearly $|y_t| \leq t^{-1}|x-S_B(t)x| \leq |Bx|$. Using the monotonicity of $I-S_A(t)$ we obtain

$$(a-S_A(t)a - S_B(t)x + S_A(t)S_B(t)x, \ a - S_B(t)x) \geq 0$$

therefore

$$\left(\frac{a-S_A(t)a}{t} - \frac{x-S_A(t)x}{t} - \frac{x-S_B(t)x}{t} - y_t \ , \ a - S_B(t)x \right) \geq 0 \ .$$

Choosing a sequence $t_n \longrightarrow 0$ such that $y_{t_n} \longrightarrow y$ and passing to the limit yields

$$(Aa - Ax + Bx - y, \ a-x) \geq 0 \qquad \forall \ a \in D(A).$$

Since A is maximal monotone this implies $y = Bx$. Therefore we also have $y_{t_n} \longrightarrow Bx$ and since the limit is unique $y_t \longrightarrow Bx$.

A. Pazy

Appendix

We shall prove theorem 2 of lecture 6 following some lemmas.
The proof presented here is a modification by M.G. Crandall of a
proof for the monotone case due to T. Kato.

Lemma 1.

Let $S(t)$ be a semi-group of type ω on C (C closed
convex) then

a) For every $t > 0$ and every $0 < \lambda < 1/_\omega$ $\quad R(I+\lambda A(t)) \supset C$

b) For every $x \in H$, λ and t as in (a) $\quad (I+\lambda A(t))^{-1}x$ is
    a continuous function of λ and t.

Proof:

Let $x \in C$ and define $Gy = \frac{t}{t+\lambda} x + \frac{\lambda}{t+\lambda} S(t)y$, for
$0 < \lambda < 1/_\omega$ we have $\lambda e^{\omega t}(\lambda+t)^{-1} < 1$ and therefore G is a
strict contraction, $G : C \longrightarrow C$. Therefore G has a unique
fixed point in C which we denote by $y_{\lambda,t}$. We then have
$(I+\lambda A(t))y_{\lambda,t} = x$. The continuous dependence of $y_{\lambda,t}$ on t

A. Pazy

and λ follows from the strong continuity of G in λ and t and from the fact that G is a strict contraction for all $t > 0$ and $0 < \lambda < 1/\omega$.

Lemma 2.

For every $\epsilon > 0$ there exists a δ , $0 < \delta < \dfrac{1}{4\omega}$ such that if $t \in (0,\delta]$, $s = nt \in (0,\delta]$ (n is a positive integer) then

(1) $$\left| y_{\lambda,t} - y_{\lambda,s} \right|^2 \leq \epsilon(\left| y_{\lambda,t} - x \right| + \left| y_{\lambda,t} - x \right|^2)$$

for every $0 < \lambda < \dfrac{1}{4\omega} - \delta$.

Proof:

Without loss of generality we way assume $x = 0$. (This amounts to a shift of the origin in H). Also since λ is fixed we shall write y_t instead of $y_{\lambda,t}$. Thus $y_t = (I + \lambda A(t))^{-1} 0$ which is equivalent to $S(t) y_t = (1 + \dfrac{t}{\lambda}) y_t$ for every $t > 0$.

For every positive integer m we have:

(2) $$e^{2\omega t} \left| y_t - S((m-1)t) y_s \right|^2 \geq \left| S(t) y_t - S(mt) y_s \right|^2 = \left| (1 + \dfrac{t}{\lambda}) y_t - S(mt) y_s \right|^2 \geq$$

$$\geq \left| y_t - S(mt) y_s \right|^2 + \dfrac{2t}{\lambda} (y_t ; y_t - S(mt) y_s)$$

A. Pazy

From (2) it follows by induction that

$$(3) \quad |y_t-y_s|^2 \geq e^{-2\omega n t} \, |y_t-S(nt)y_s|^2 + \frac{2t}{\lambda} \sum_{k=1}^{n} e^{-2\omega(k-1)t}(y_t,y_t-S(kt))$$

but,

$$|y_t-S(nt)y_s|^2 = |y_t-S(s)y_s|^2 = |y_t-(1+\frac{s}{\lambda})y_s|^2 \geq$$

$$\geq |y_t - y_s|^2 + \frac{2s}{\lambda} (y_s,y_s - y_t)$$

substituting this into (3) and using the inequality
$\sum_{k=1}^{n} e^{-2\omega(k-1)t} \geq ne^{-2\omega s}$ we obtain after rearrangement

$$(e^{-2\omega s}+ \frac{s}{\lambda}e^{-2\omega s}-1)|y_t-y_s|^2 + \frac{s}{\lambda}e^{-2\omega s}(|y_t|^2+|y_s|^2) \leq$$

$$(4) \qquad\qquad \leq \frac{2t}{\lambda} \, e^{2\omega t} \sum_{k=1}^{n} e^{-2\omega k t}(y_t,S(kt)y_s).$$

Now,

$$|S(kt)y_s| \leq |S(kt)y_s - S(kt)0| + |S(kt)0| \leq e^{\omega k t}|y_s| + E(s)$$

where

$$E(s) = \max_{0 \leq t \leq s} |S(t)0|$$

therefore

A. Pazy

$$(e^{-2\omega s}+ \frac{s}{\lambda}e^{-2\omega s}- 1)|y_t-y_s|^2+ \frac{s}{\lambda} e^{-2\omega s}(|y_t|^2+|y_s|^2) \leq$$

$$\leq \frac{2s}{\lambda} e^{2\omega t}(|y_s| |y_t| + E(s)) \leq \frac{s}{\lambda} e^{-2\omega s}|y_s|^2+ \frac{s}{\lambda} e^{2\omega s}|y_t|^2 + \frac{2s}{\lambda} e^{2\omega t} E(s)$$

or

$$(5) \quad (e^{-2\omega s}-2\lambda\omega(\frac{1-e^{-2\omega s}}{2\omega s}))|y_t-y_s|^2 \leq 2e^{\omega t}\cdot E(s)|y_t| + (e^{2\omega s}-e^{-2\omega s})|y_t|^2 \; .$$

Choosing δ such that $\delta < \frac{1}{4\omega}$, $2e^{\omega\delta} E(\delta) < \epsilon/2$ and

$e^{2\omega\delta} - e^{-2\omega\delta} < \epsilon/2$ the right hand side of (5) is smaller that

$\frac{\epsilon}{2} (|y_t|+ |y_t|^2)$. Finally for $\lambda < 1/4\omega$ we have

$$(e^{-2\omega s} - 2\lambda\omega(\frac{1-e^{-2\omega s}}{2\omega s})) \geq \frac{1}{2}$$

and therefore

$$|y_t - y_s|^2 \leq \epsilon (|y_t|+ |y_t|^2)$$

which is the desired result for $x = 0$.

Lemma 3.

Let $\epsilon < 1/2$ and δ be as in lemma 2 then for every $t \in (0,\delta]$

A. Pazy

(6) $\qquad |y_{\lambda,t} - x| \leq 2\varepsilon\ (1 + \frac{8\lambda}{\delta})$ for every $0 < \lambda < \frac{1}{4\omega} - \delta$.

Proof:

Again without loss of generality we assume $x = 0$ and denot $y_{\lambda,t}$ by y_t . Let $t \geq \delta/2$ then

$$|y_t| = \frac{\lambda}{\lambda+t}|S(t)y_t| \leq \frac{\lambda}{\lambda+t}|S(t)y_t - S(t)0| + \frac{\lambda}{\lambda+t}\ |S(t)\cdot0| \leq$$

$$\leq \frac{\lambda e^{\omega t}}{\lambda+t}\ |y_t| + \frac{\lambda}{\lambda+t}\ |S(t)0|$$

hence,

$$|y_t| \leq \frac{\lambda\varepsilon}{t-\lambda(e^{\omega t}-1)} \leq \frac{2\lambda\varepsilon}{\delta-2\lambda(e^{\omega\delta}-1)} \leq \frac{4\lambda}{\delta}\ \varepsilon\ ,$$

since for $\lambda < \frac{1}{4\omega} - \delta$ we have $2\lambda(e^{\omega\delta} - 1) < \frac{\delta}{2}$.

Now let $t < \delta/2$ then there exists $s = nt \in [\delta/2, \delta]$ and

$$|y_t|^2 - 2(y_t, y_s) + |y_s|^2 = |y_t - y_s|^2 \leq \varepsilon(|y_t|^2 + |y_t|)$$

therefore,

$$(1-\varepsilon)|y_t|^2 \leq \varepsilon|y_t| + 2|y_t|\ |y_s| \leq \varepsilon|y_t| + \frac{8\lambda}{\delta}\ \varepsilon|y_t|$$

Since $\varepsilon < 1/2$ this implies

$$\left| y_t \right| \le 2\epsilon \left(1 + \frac{8\lambda}{\delta} \right)$$

which is the desired result for $x = 0$.

Lemma 4.

For every $x \in C$ and $\lambda < 1/4\omega$ we have

(6) $$\lim_{t \to 0} y_{\lambda, t} = y_\lambda \in C .$$

Proof:

Let $\lambda_0 < 1/4\omega$ be given. Reduce (if necessary) the δ of lemma 2 and lemma 3 so that $\lambda_0 < \frac{1}{4\omega} - \delta$. From lemma 3 we then have $\left| y_{\lambda_0, t} - x \right| \le M$ for every $t \in (0, \delta]$. Let $M_0^2 = M + M^2$. By lemma 2 we have

(7) $$\left| y_{\lambda_0, t} - y_{\lambda_0, s} \right| \le (M_0^2 \epsilon)^{1/2} = M_0 \epsilon^{1/2} .$$

Now if s and s' are in the interval $(o, \delta]$ and s/s' is rational then there exists $t > 0$ such that $s = nt$ $s' = n't$ with n and n' positive integers. From (7) we deduce

A. Pazy

$$(8) \qquad \left| y_{\lambda_0,s} - y_{\lambda_0,s'} \right| \leq 2M_0 \, \epsilon^{1/2}$$

Since $y_{\lambda_0,s}$ is a continuous function of s by lemma 1, (8) holds for every $s, s' \in (0, \delta]$. Since ϵ is arbitrary $y_{\lambda_0,t}$ is a Cauchy sequence, and from $y_{\lambda_0,t} \in C$ and the closedness of C it follows that the limit $y_{\lambda_0} \in C$.

Proof of theorem 2.

From theorem 3 in lecture 5 it follows that it is sufficien to prove the existence of a dense subset of C for which $|A(s)x|$ is bounded as $s \to 0$. We shall prove that the set

$$C' = \left\{ y_\lambda = \lim_{t \to 0} y_{\lambda,t} \; \middle| \; 0 < \lambda < 1/4\omega, \; x \in C \right\}$$

has these properties.

C' is dense in C since for every $x \in C$ and every $\epsilon > 0$ there exists a δ such that $\left| y_{\lambda,t} - x \right| \leq 2\epsilon(1 + \frac{8\lambda}{\delta})$ by lemma 3. Choosing $\lambda < \delta/16$ we have $\left| y_{\lambda,t} - x \right| < 3\epsilon$ and passing to the limit as $t \to 0$ we obtain $\left| y_\lambda - x \right| < 3\epsilon$.

Now, let $s = nt$ and $\lambda < 1/4\omega$ then

A. Pazy

$$s \left| A(s) y_{\lambda,t} \right| \leq \sum_{k=1}^{n} \left| S((k-1)t) y_{\lambda,t} - S(kt) y_{\lambda,t} \right| \leq \sum_{k=1}^{n} e^{\omega(k-1)t} \left| y_{\lambda,t} - S(t) y_{\lambda,t} \right|$$

$$= \sum_{k=1}^{n} t \, e^{\omega(k-1)t} \left| A(t) y_{\lambda,t} \right| .$$

Let s be fixed and let $n \longrightarrow \infty$. From lemma 4 it follows that $y_{\lambda,t} \longrightarrow y_\lambda$ and therefore

$$A(t) y_{\lambda,t} = \frac{1}{\lambda} (y_{\lambda,t} - x) \longrightarrow \frac{1}{\lambda} (y_\lambda - x)$$

thus

$$\left| A(s) y_\lambda \right| \leq \lim_{n \to \infty} \left(\frac{1}{n} \sum_{k=1}^{n} e^{\omega(k-1)s/n} \right) \frac{1}{\lambda} \left| y_\lambda - x \right| = \frac{1}{\lambda} \left| y_\lambda - x \right|$$

and therefore for every $y_\lambda \in C'$ $\left| A(s) y_\lambda \right|$ is bounded as $s \longrightarrow 0$.

Q.E.D.

A. Pazy

General - References

1. H. Brezis, M.G. Crandall and A. Pazy, Perturbation of
    non linear maximal monotone sets in Banach
    space, Comm. Pure Appl. Math. 23 (1970) 123-144.

2. H. Brezis and A. Pazy, Semigroups of nonlinear contractions
    on Convex sets. Jour. Func. Anal. 6 (1970).

3. P. Chernoff, Note on product formulas for operator
    semigroups. Jour. Func. Anal. 2 (1968), 238-242.

4. M.G. Crandall, Differential equations on convex sets,
    J. Math. Soc. Japan (to appear).

5. M.G. Crandall and A. Pazy, Semigroups of Nonlinear
    contractions and dissipative sets. Jour. Func.
    Anal. 3 (1969) 367-418.

6. M.G. Crandall and A. Pazy, On accretive sets in Banach
    spaces, J. Func. Anal. 5 (1970) 204-217.

7. J.R. Dorroh, A nonlinear Hille-Yosida-Phillips theorem,
    J. Func. Anal. 3 (1969) 345-353.

8. E. Hille and R.S. Phillips, Functional Analysis and Semigroups
    Amer. Math. Soc. Colloq. Publ. Vol. 31, Amer.
    Math. Soc. Providence (1957).

A. Pazy

9. T. Kato, On the generators of nonlinear semi-groups, Proc.
    Summer Inst. Global Anal. Berkeley 1968 AMS
    (to appear).

10. Y. Kōmura, Nonlinear semigroups in Hilbert space, J. Math.
    Soc. Japan 19 (1967) 493-507 .

11. Y. Kōmura, Differentiability of nonlinear semi-groups, J.
    Math. Soc. Japan 21 (1969) 375-402 .

12. G.I. Minty, Monotone (nonlinear) operators in Hilbert space,
    Duke Math. J. 29 (1962) 341-346.

13. I. Myadera and S. Oharu, Approximation of semigroups of
    nonlinear operators, Tōhoku Math. Jour. 22 (1970)
    24-47 .

14. J.W. Neuberger, An exponential formula for one parameter
    semigroups of nonlinear transformations, J. Math.
    Soc. Japan 18 (1966) 154-157.

15. H.F. Trotter, Approximation of semigroups of operators,
    Pacific J. Math. 8 (1958) 887 - 919 .

16. H.F. Trotter, On the product of semigroups of operators,
    Proc. Amer. Math. Soc. 10 (1959) 541-551.

17. K. Yosida, "Functional Analysis" Springer Verlag New York
    (London 1966).

CENTRO INTERNAZIONALE MATEMATICO ESTIVO

(C.I.M.E.)

R. TEMAM

ÉQUATIONS AUX DÉRIVÉES PARTIELLES STOCHASTIQUES

Corso tenuto a Varenna dal 20 al 29 Agosto 1970

ÉQUATIONS AUX DÉRIVÉES PARTIELLES STOCHASTIQUES

par

R. TEMAM

(Paris - Orsay)

Introduction.

Dans ce qui suit nous développons un résultat d'existence et d'unicité de solution pour des équations aux derivées partielles stochastiques non linéaires; ce résultat est extrait d'un travail à paraitre (cf. Bensaussan-Temam [2]) où l'on tente de généraliser aux équations aux dérivées partielles non linéaires (*) la théorie d'Ito [4] des équations differentielle stochastiques.

On considère une équation différentielle opérationnelle

(0.1) $\qquad \dfrac{du(t)}{dt} + Au(t) = g(t)$,

(0.2) $\qquad u(0) = u_0$,

et l'on envisage le cas où les données u_0 , g , et la fonction inconnue u sont "stochastiques" ce qui se traduit par une certaine dépendance par rapport à un paramètre ω à valeurs dans un espace $\underline{\omega}$ (l'espace des épreuves).

(*) Pour le cas linéaire cf. [1] .

R. Temam

Le résultat principal concerne en fait le cas plus délicat
où l'équation (0.1) est remplacée par

$$(0.1') \qquad \frac{du}{dt}(t) + Au(t) = g(t) + \frac{df}{dt}(t) \quad ,$$

f étant un processus de Wiener, $\frac{df}{dt}$ représentant un "bruit
blanc".

Nous nous limiterons ici au cas simple d'un opérateur A
monotone, indépendant de t , renvoyant à [ ] pour des résul
tats beaucoup plus généraux.

Nous commencerons par un rappel sur le système détermini-
ste (0.1) (0.2) et nous donnerons des résultats relatifs au système stochastique (0.1) (0.2)(§ 1). Nous introduisons ensuite
le processus de Weiner f et nous traitons l'équation stochasti
que (0.1') (0.2). Nous admettrons plusieurs résultats auxiliaires de démonstration longue et technique.

Plan

1. Un résultat simple.

2. Théorème d'existence et d'unicité.

R. Temam

1. Un résultat simple.

Dans ce paragraphe nous précisons les hypothèses et notations du cadre deterministe puis du cadre stochastique et nous
donnons ensuite un premier résultat.

1.1. Le cadre déterministe

Soit V un espace de Banach réel reflexif séparable, et
H un espace de Hilbert réel, avec $V \subset H$, l'injection étant
continue et V étant dense dans H. On identifie H et son
dual; V' désignant le dual de V, on a alors

$$(1.i) \qquad V \subset H \subset V' \ ,$$

les injections étant continues, chaque espace étant dense dans
le suivant.

On note $|\cdot|_V$ ou $|\cdot|$ la norme de V, $|\cdot|_H$ ou $|\cdot|$
celle de H, et $(.\,,\,.)$ le produit scalaire dans H, $<\cdot,\cdot>$
le produit scalaire dans la dualité entre V et V'.

Soit $T > 0$ fixé; si X est un espace de Banach, on note
$L^p([0,T] ; X)$ ou plus simplement $L^p(X)$ l'espace des fonctions
L^p sur $[0,T]$ pour la mesure de Lebesgue dt, à valeurs dans
X ; $L^\infty([0,T] ; X)$ ou encore $L^\infty(X)$ est l'espace des fonctions
mesurables, essentiellement bornées, sur $[0,T]$ à valeurs dans

R. Temam

X ; C([O,T] ; X) ou plus simplement C(X) désigne l'espace des fonctions continues sur [O,T] à valeurs dans X. Ces espaces sont de Banach pour les normes usuelles.

On se donne un opérateur A (non linéaire) de $V \rightarrow V'$, vérifiant

(1.ii) A est monotone, hémicontinu, borné,

(1.iii) $< Ay,y > \geqslant \alpha \|y\|^p$, $\forall y \in V$,

$\alpha > 0$, $\lambda \geqslant 0$, $p \geqslant 2$,

(1.iv) Si $y(\cdot) \in L^p(V)$, alors $\{t \rightarrow Ay(t)\} \in L^{p'}(V')$, $1/p + 1/p' = 1$ et l'application $y(\cdot) \rightarrow Ay(\cdot)$ de $L^p(V) \rightarrow L^{p'}(V')$ est bornée hémicontinue.

Nous rappelons le résultat bien connu suivant (cf. Browder [3] , Lions [5]).

Théorème 1.1.

Sous les hypothèses (1.i) - (1.iii) , pour y_o donné dans H et g donné dans $L^1(H) + L^{p'}(V')$, il existe y unique qui vérifie

(1.1) $y \in L^p(V) \cap L^{\infty}(H)$,

(1.2) $y' \in L^{p'}(V') + L^1(H)$,

(1.3) $y' + Ay = g$,

R. Temam

(1.4) $\qquad y(0) = y_0$

La fonction y est en autre (p.p. égale à) une fonction continue de $[0,T] \rightarrow H$ et satisfait l'égalité d'énergie

(1.5) $|y(t)|^2 + 2 \int_0^t \langle Ay(s), y(s) \rangle \, ds = |y_0|^2 + \int_0^t \langle g(s), y(s) \rangle \, ds, \ \forall t \geq 0.$

Remarque 1.1. Dans le cas stochastique les égalités d'énergie joueront (comme d'habitude....) un rôle important; nous verrons qu'elles sont plus difficiles à établir et présentent un aspect différent.

1.2. Le cadre stochastique.

Soit $(\underline{\omega}, \mu)$ un espace de probabilité: on supposera avec P.A. MEYER [6] et L. SCHWARTZ [7], que $\underline{\omega}$ est un espace topologique complétement régulier muni de sa tribu borélienne $\mathcal{C}$ et que μ est une mesure de Radon (c.à.d. une mesure abstraite sur $\mathcal{C}$ intérieurement réguliere; cf. [6] [7]).

On introduit les espaces

(1.6) $\qquad \mathcal{H} = L^2(\underline{\omega}, \mu ; H) ,$

(1.7) $\qquad \mathcal{V} = L^p(\underline{\omega}, \mu ; V) .$

L'espace $\mathcal{H}$ est de Hilbert pour le produit scalaire

(1.8) $\quad (y,z)_{\mathcal{H}} = E(y,z)_H = \int_{\underline{\omega}} (y(\omega), z(\omega))_H \, d\mu(\omega) .$

R. Temam

L'espace $\mathcal{V}$ est de Banach pour la norme

$$(1.9) \qquad \|y\|_{\mathcal{V}} = \left\{ \int_{\underline{\omega}} \|y(\omega)\|_{V}^{p} \, d\mu(\omega) \right\}^{1/p}$$

Le dual de $\mathcal{V}$ est biensur l'espace

$$(1.10) \qquad \mathcal{V}' = L^{p'}(\underline{\omega}, \mu; V') \quad,$$

de Banach pour la norme

$$(1.11) \qquad \|y\|_{\mathcal{V}'} = \left\{ \int_{\underline{\omega}} \|y(\omega)\|_{V'}^{p'} \, d\mu(\omega) \right\}^{1/p'} \quad.$$

Il résulte de (1.i) et de ce que $p \geqslant 2$, que

$$(1.12) \qquad \mathcal{V} \subset \mathcal{H} \subset \mathcal{V}',$$

les injections étant continues et chaque espace étant dense dans le suivant. On note

$$(1.13) \qquad \langle y, z \rangle_{\mathcal{V}, \mathcal{V}'} = E \langle y, z \rangle = \int_{\underline{\omega}} \langle y(\omega), z(\omega) \rangle \, d\mu(\omega) \quad;$$

la dualité entre $\mathcal{V}$ et $\mathcal{V}'$.

On sera amené à considérer les espaces $L^{2}(\mathcal{H}) = L^{2}([0,T]; \mathcal{H})$, $L^{p}(\mathcal{V}) = L^{p}([0,T]; \mathcal{V})$, $L^{p'}(\mathcal{V}') = L^{p'}([0,T]; \mathcal{V}')$; remarquons, qu'en tant qu'espaces normés ces espaces sont évidemment isomorphes aux espaces respectifs,

$$L^{2}(\underline{\omega} \times [0,T], d\mu \otimes dt; H),$$

$$L^{p'}(\underline{\omega} \times [0,T], d\mu \otimes dt, V'),$$

R. Temam

$$L^{p'}(\underline{\omega} \times [0,\underline{T}] , d\mu \otimes dt , V') .$$

L'opérateur $\mathcal{A}$.

Nous allons associer à A un opérateur $\mathcal{A}$ de $\mathcal{V} \to \mathcal{V}'$. En effet si $y \in \mathcal{V} = L^p(\underline{\omega},\mu;V)$, alors $\omega \longrightarrow Ay(\omega)$ est une application de $\underline{\omega} \to V'$. En raison de (1.ii), l'opérateur A est une application continue de V fort dans V' faible (cf. [5]) et alors $\{\omega \to Ay(\omega)\}$ est faiblement mesurable à valeurs dans V' et donc mesurable puisque V' est séparable. On note $\mathcal{A}y$ $\in$ $\underline{M}es(\underline{\omega},\mu,V')$ cette application $\{\omega \to Ay(\omega)\}$, et nous allons montrer le

Lemme 1.1.

On suppose que l'opérateur A vérifie (1.ii)-(1.iv) et en autre

$(1.v)$ $\begin{cases} \|Ay\|_{V'} \leqslant \theta(\|y\|) , \quad \forall y \in V , \\ \theta \text{ fonction continue de } \mathbb{R}_+ \longrightarrow \mathbb{R}_+ , \theta(s) = O(s^{p-1}), \underline{\text{pour}} \ s \to o \end{cases}$

Alors l'operateur $\mathcal{A}$ envoie $\mathcal{V}$ dans $\mathcal{V}'$ et vérifie

(1.14) $\mathcal{A}$ est monotone hémicontinue et borné (de $\mathcal{V} \to \mathcal{V}'$),

(1.15) $<\mathcal{A}y,y>_{\mathcal{V},\mathcal{V}'} \geqslant \alpha \|y\|_{\mathcal{V}}^p \quad \forall y \in \mathcal{V} ,$

mêmes constantes α ,p , qu'en (1.iii) ,

(1.16) <u>Si</u> $y(\cdot) \in L^P(\mathcal{V})$ <u>alors</u> $\{t \to \mathcal{A}y(t)\} \in L^{P'}(\mathcal{V}')$ <u>et l'ap</u>-

<u>plication</u> $y(\cdot) \to \mathcal{A}y(\cdot)$ <u>de</u> $L^P(\mathcal{V}) \to L^{P'}(\mathcal{V}')$ <u>est bor</u>-

<u>née hemicontinue.</u>

<u>Démonstration.</u>

On vérifie aisément avec (1.v), l'existence de $d, d_1 > 0$,

tels que

(1.17) $\theta(s) \leqslant \max (d, d_1 \, s^{P-1})$, $\forall s \geqslant 0$.

A présent si $y \in \mathcal{V}$, alors on sait déjà que $\mathcal{A}y \in \text{Mes}(\underline{\omega}, \mu, V')$

et on a

$$\|Ay(\omega)\|_{V'}^{P'} \leqslant \max (d^{P'}, \ d_1^{P'} \|y(\omega)\|^P)$$

$$\|\mathcal{A}y\|_{\mathcal{V}'}^{P'} = \int_{\underline{\omega}} \|Ay(\omega)\|_{V'}^{P'} \, d\mu(\omega) \leqslant$$

$$\leqslant \max (d^{P'}, \ d_1^{P'} \int_{\underline{\omega}} \|y(\omega)\|^P \, d\mu(\omega)$$

$$= \max (d^{P'}, \ d_1^{P'} \|y\|_{\mathcal{V}}^P) < +\infty ,$$

en sorte que $\mathcal{A}y \in \mathcal{V}'$.

Montrons (1.14): la monotonie est facile car si y et $z \in \mathcal{V}$,

$$< \mathcal{A}y - \mathcal{A}z, y-z >_{\mathcal{V}, \mathcal{V}'} = \int_{\underline{\omega}} < Ay(\omega) - Az(\omega), \ y(\omega) - z(\omega) > d\mu(\omega) \geqslant 0 .$$

R. Temam

Il est immédiat aussi, en raison de (1.v) que l'opérateur $\mathcal{A}$ est borné de $\mathcal{V} \longrightarrow \mathcal{V}'$. Pour l'hémicontinuité, soient y,z et $\varphi \in \mathcal{V}$; si $\lambda \in \mathbb{R}$,

$$< \mathcal{A}(y+\lambda z), \varphi >_{\mathcal{V}, \mathcal{V}'} = \int_{\underline{\omega}} < A(y(\omega) + \lambda z(\omega)), \varphi(\omega) > \, d\mu(\omega) \quad .$$

D'après l'hémicontinuité de A , lorsque $\lambda \longrightarrow 0$,

$$< A(y(\omega) + \lambda z(\omega)), \varphi(\omega) > \longrightarrow < A y(\omega) , \varphi(\omega) >$$

pour $d\mu$ - presque tout ω . D'après (1.17), on a

$$| < A(y(\omega) + \lambda z(\omega)) , \varphi(\omega) > | \leqslant$$

$$\leqslant \| A(y(\omega) + \lambda z(\omega)) \|_{V'} \, \| \varphi(\omega) \|_{V}$$

$$\leqslant \left\{ \max(d^{P'}, d_1^{P'} \| y(\omega) + \lambda z(\omega) \|_{V}^{P-1}) \right\} \| \varphi(\omega) \|_{V}$$

$$\leqslant \left\{ \max(d^{P'}, d_1^{P'} (\| y(\omega) \|_{V}^{P-1} + \| z(\omega) \|_{V}^{P-1})) \right\} \| \varphi(\omega) \|_{V} = \zeta(\omega)$$

(en se limitant ce qui suffit, à $|\lambda| \leqslant 1$). On note que la fonction $\zeta(\omega) \in L^1(\underline{\omega}, \mu; \mathbb{R})$ et on peut donc appliquer le théorème de Lebesgue

$$\int_{\underline{\omega}} < A(y(\omega) + \lambda z(\omega)) , \varphi(\omega) > \, d\mu(\omega) \longrightarrow$$

$$\longrightarrow \int_{\underline{\omega}} < Ay(\omega) , \varphi(\omega) > \, d\mu(\omega) \quad ,$$

et l'hémicontinuité en résulte.

R. Temam

La propriété (1.15) résulte immédiatement de (1.iii) par intégration sur ω .

On démontre enfin (1.16) à partir de (1.14) par un raisonnement semblable au précédent.

1.3. Un premier théorème d'existence et d'unicité.

Théorème 1.2.

Sous les hypothèses (1.i) à (1.v), pour y_0 donné dans $\mathcal{H}$ et g donné dans $L^{p'}(\mathcal{V}') + L^1(\mathcal{H})$, il existe y unique vérifiant

$$(1.17) \qquad y \in L^p(\mathcal{V}) \cap L^\infty(\mathcal{H}) \ ,$$

$$(1.18) \qquad y' \in L^{p'}(\mathcal{V}') + L^1(\mathcal{H}) \ ,$$

$$(1.19) \qquad y' + \mathcal{A}y = g \ ,$$

$$(1.20) \qquad y(0) = v_0 \ .$$

Démonstration.

On applique seulement le théorème 1.1. avec A, V, H, V', remplacés par $\mathcal{A}, \mathcal{V}, \mathcal{H}, \mathcal{V}$.

Remarque 1.2. L'égalité (1.19) est prise au sens des distributions vectorielles sur $]0,T[$ à valeurs dans $\mathcal{V}'$.

Nous en déduisons facilement que

R. Temam

(1.21) $y'(\omega) + Ay(\omega) = g(\omega)$, $d\mu$ - p.p. ,

cette égalité étant prise au sens des distributions vectorielles

sur $]0,T[$ à valeurs V' . Plus encore, puisque y' est une fon-

ction (pour la dépendance de t),

(1.22) $y'(t,\omega) + Ay(t,\omega) = g(t,\omega)$, dt - p.p. , $d\mu$-p.p. ,

(égalité dans V').

Remarque 1.3. Comme dans le theoreme 1.1, la fonction $t \rightarrow y(t)$

est (p.p. égale à) une fonction continue de $[0,T] \rightarrow \mathcal{H}$. L'éga-

lité (1.20) s'entend donc dans $\mathcal{H}$, c'.à.d .

(1.23) $y(0,\omega) = y_o(\omega)$, $d\mu$ - p.p.

(égalité dans H).

2. Théorème d'existence et d'unicité.

2.1. Processus de Wiener Hilbertien.

On se donne un processus de Wiener $f(t)$ à valeurs dans H ,

c'est-à-dire une application de $[0,T] \rightarrow Mes[\underline{\omega} , \mu,H]$ [1] vérifiant:

(2.i) $\begin{cases} \forall t \in [0,T], \ \forall \varphi \in H , \ (f(t),\varphi) \ \text{est une variable aléatoire} \\ \text{réelle centrée gaussienne,} \end{cases}$

---

[1] Espace des variables aléatoires à valeurs dans H , i.e.
des applications $d\mu$ - mesurables de $\underline{\omega} \rightarrow H$.

R. Temam

et

$$(2.ii) \quad \begin{cases} E\left\{(f(t),\varphi)\,(f(s),\psi)\right\} = \int_0^{\min(t,s)} (Q(\tau)\varphi,\psi)d\tau \ , \\ \\ \forall t,s \in [0,T] \ , \ \forall \varphi, \ \psi \in H \ , \end{cases}$$

où $Q(\cdot) \in L^{\infty}([0,T]; \mathcal{L}(H,H))$ et

$$(2.iii) \quad \begin{cases} \text{pour presque tout} \quad t, \quad Q(t) \in \mathcal{L}(H;H) \quad \text{est aut -adjoint}, \\ \text{positif et nucléaire}, \end{cases}$$

$$(2.iv) \quad \left\{t \to \text{trace } Q(t) = \text{tr } Q(t)\right\} \in L^1([0,T] \ ; \ \mathbb{R}) \quad .$$

On peut quitte à remplacer f par une processus équivalent, supposer que $f(t)$ est presque surement continu à valeurs dans H . On vérifie sans difficulté à partir de ce qui précède que

$$(2.1) \quad \begin{cases} \forall t, \ \tau \in [0,T], \ t \leqslant \tau \ \forall \varphi, \ \psi \in H \ , \quad (f(t),\varphi) \quad \text{et} \\ (f(t)-f(\tau),\psi) \quad \text{sont des v.a. indépendantes} \end{cases}$$

$$(2.2) \quad E|f(t)|_H^2 = \int_\omega |f(t,\omega)|^2 \, d\mu(\omega) = \int_0^t \text{tr } Q(\tau) \, d\tau$$

(et donc $f(0) = 0$) ; pour cela cf. [2] .

Nous admettrons ici la proposition suivante (cf. [2])

Proposition 2.1.

Sous les hypothèses (2.i) - (2.iv) , le processus f vérifie

R. Temam

(2.3) $f \in \mathcal{C}([0,T] ; L^r(\underline{\omega} , \mu ; H)) , \forall r \geqslant 2$.

Approximation de f.

Il sera commode par la suite de disposer d'une approximation de f par des processus à valeurs dans V . Soit pour cela $\{e_i\}_{i \geqslant 1}$ une base orthonormée de H formée d'éléments de
V ([1]), et posons

(2.4) $f_N(t) = \sum_{i=1}^{N} (f(t), e_i)e_i$

(ou, en faisant apparaître la dépendance en ω ,

(2.4') $f_N(t,\omega) = \sum_{i=1}^{N} (f(t,\omega),e_i)e_i)$.

On a alors la

Proposition 2.2.

Pour tout N , f_N est un processus stochastique à valeurs
dans H qui vérifie les propriétés (2.i) - (2.iv) avec Q(t)
remplacé par $Q_N(t)$ ainsi défini:

---

[1] Une telle base existe: comme V est dense dans H , il existe une famille libre et totale de H (et nécessairement dénombrable) formée d'éléments de V . Le procédé de Schmidt
nous permet d'en déduire une base orthonormée.

R. Temam

$$(2.5) \qquad Q_N(t)\varphi = \sum_{i,j=1}^{N} (Q(s)e_i, e_j)\,(\varphi, e_i)e_j \quad,$$

$\forall \varphi \in H$, $\forall t \in [0,T]$.

En autre, pour tout $N \geqslant 1$,

$$(2.6) \qquad f_N \in \mathscr{C}([0,T] \; ; \; L^r(\omega, \mu; V)) \;, \quad \forall r \geqslant 2 \quad,$$

et, lorsque $N \longrightarrow \infty$,

$$(2.7) \qquad f_N \longrightarrow f \quad \text{dans} \quad \mathscr{C}([0,T] \; ; \; L^2(\underline{\omega}, \mu; H)) = \mathscr{C}(\mathscr{H})$$

Nous admettrons également cette proposition dont la démonstration est trés technique (cf. [2]) .

2.2. Un cas particulier.

Nous envisageons ici le cas particulier où le processus f donné par (2.i) - (2.iv) vérifie en autre

$$(2.8) \qquad f \in \mathscr{C}([0,T] \; ; \; \mathscr{V}) \quad.$$

Nous avons alors le

Théorème 2.1.

Sous les hypothèses précédentes $[(1.i) - (1.v), (2.i) - (2.iv),$ et $(2.8)]$, pour y_0 donné dans $\mathscr{H}$ et g donné dans $L^{p'}(\mathscr{V}') +$ $+ \; L^1(\mathscr{H})$ il existe y unique vérifiant

$$(2.9) \qquad y \in L^p(\mathscr{V}) \cap L^\infty(\mathscr{H}) \quad,$$

$$(2.10) \qquad y'-f' \in L^{p'}(\mathscr{V}') + L^1(\mathscr{H}) \quad,$$

R. Temam

$$(2.11) \qquad y' + \mathcal{A}y = g + f' ,$$

$$(2.12) \qquad y(0) = y_0 .$$

Remarque 2.1. Pour la signification des égalités (2.11) et (2.12)

cf. les remarques 1.2 et 1.3.

Démonstration du théorème.

Nous posons $z = y - f$. Pour tout $t \in [0,T]$, soit $\mathcal{A}_f(t)$

l'opérateur de $\mathcal{V} \to \mathcal{V}'$ donné par

$$\varphi \in \mathcal{V} \longrightarrow \mathcal{A}(\varphi + f(t)) .$$

Il est facile de voir que pour tout t , $\mathcal{A}_f(t)$ est monoto-

ne borné, hémicontinu (grâce à (1.14)). Il est facile de voir que

l'opérateur $\mathcal{A}_f : y(\cdot) \in L^p(\mathcal{V}) \longrightarrow \mathcal{A}_f(\cdot)y(\cdot) \in L^{p'}(\mathcal{V}')$ est monoto

ne borné hémicontinu (grâce à (1.16)).

Enfin grâce à (1.15), $\mathcal{A}_f$ vérifie la propriété de coercivi-

té

$$(2.13) \quad \lim_{\substack{y \in L^p(\mathcal{V}) \\ \|y\|_{L^p(\mathcal{V})} \to \infty}} \left\{ \frac{\langle \mathcal{A}_f \cdot (y-f) , y \rangle_{L^p(\mathcal{V}), L^{p'}(\mathcal{V}')}}{\|y\|_{L^p(\mathcal{V})}} \right\} = \infty , \text{(1)}$$

---

(1) Bien sûr $\mathcal{A}_f(y-f) = \mathcal{A}y$

R. Temam

Dans ces conditions une forme un peu plus générale du théorème 1.1. entraine l'existence et l'unicité de z vérifiant (cf. [5]) :

(2.14) $z \in L^P(\mathcal{V}) \cap L^\infty(\mathcal{H})$

(2.15) $z' \in L^{P'}(\mathcal{V'}) + L^1(\mathcal{H})$

(2.16) $z' + \mathcal{A}_f \, z = g$

(2.17) $z(0) = y_o$

En repassant à $y = z + f$ on retrouve (2.9)-(2.12). Evidemment la proprieté (2.8) (^{1}) est essentielle dans tout cela.

Remarque 2.2. D'apres [·] la fonction z est encore (p.p. égale à) une fonction continue de $[0,T] \longrightarrow \mathcal{H}$, et vérifie l'égalité d'énergie

(2.18) $|z(t)|^2_{\mathcal{H}} + 2 \int_o^t < \mathcal{A}_f(t) \, z(t), z(t) >_{\mathcal{V},\mathcal{V'}} dt \; =$

$$= |z(0)|^2_{\mathcal{H}} + 2 \int_o^t < g(t), \, z(t) >_{\mathcal{V},\mathcal{V'}} dt \; ,$$

$\forall t \in [0,T]$.

_____

(^{1}) Ou au moins $f \in L^P([0,T] ; \mathcal{V})$.

R. Temam

Conséquences évidentes pour $y = z + f$.

2.3 Indépendance stochastique et égalités d'énergie.

Contrairement à ce qui précède, les développements qui sui-
vent fairont appel de manière plus apparentes à des outils des
probabilités.

Nous allons donner tout d'abord une égalité d'energie parti
culiere verifiée par la solution y de (2.9)-(2.12). Mais sous
la restriction que la propriété d'indépendance suivante soit vérifiée.

$$(2.v) \begin{cases} \forall t_1, t_2 \ , \quad \text{avec} \ \ 0 \leqslant t_1 \leqslant t_2 \leqslant T \ , \ f(t_2) - f(t_1), \ \text{v.a.} \\ \text{à valeurs dans } H \ (^1) \ \text{est indépendante de} \\ \left\{ y_o \ , \ g\big|_{[0,t_1]} \ , f(t_{j1}), \ldots, f(t_{jq}) \right\} \ , \ \text{v.a. à} \\ \text{valeurs dans } H \times L^{p'}([0,t_1]; V') \times H^q \ , \ \text{pour tout} \\ q \ \text{et quels que soient} \ t_{j1}, \ldots, t_{jq} \leqslant t_1 \ (^2) \ . \end{cases}$$

(^{1}) Pour l'indépendance de v.a. , cf. [ ] . Il est trés impor-
      tant pour des notions d'indépendance, en dimension infinie,
      de préciser l'espace des valeurs des v.a. .

(^{2}) $g\big|_{[0,t_1]}$ désigne la restriction de g sur l'intervalle
      $[0,t_1]$; cette restriction appartient evidemment à $L^{p'}([0,t_1]; V')$.

R. Temam

Remarque 2.3. L'hypothèse (2.v) recouvre en particulier le fait que nous connaissions déjà (cf.(2.1)) que $f(t_2)-f(t_1)$ est indépendant de $f(\tau)$ $\forall \tau \leqslant t_1 \leqslant t_2$.

Nous avons alors la

Proposition 2.3.

Sous les hypothèses du théorème 2.1 et en autre (2.v), la solution y de (2.9)-(2.12) vérifie l'égalité suivante:

$$(2.19) \qquad E|y(t)|^2_H + 2 E \int_o^t <A y(s) , y(s)> ds =$$

$$= E |y_o|^2 + 2 E \int_o^t <g(s), y(s)> ds + E|f(t)|^2$$

$\forall t \in [0,T]$.

Principe de la démonstration

L'égalité (2 18) écrite avec $z = y-f$, donne

$$(2.20) \quad E | y(t) - f(t)|^2_H + 2 E \int_o^t <A y(s) , y(s) - f(s)> ds =$$

$$= E |y_o|^2_H + 2 E \int_o^t <g(\underline{s}), y(s) - f(s)> ds .$$

On démontre par ailleurs (moyennant (2.i)) que l'on a

R. Temam

$$(2.21) \qquad E(y(t) , f(t)) + E \int_0^t < A y(s), f(s) > ds =$$

$$= E. \int_0^t < g(s), f(s) > ds + E \mid f(t) \mid_H^2 \ ,$$

et avec (2.20) cela donne (2.19) .

La démonstration de (2.21) est trés longue: on approxime (2.9)-(2.12) par des différences finies en t , on démontre une égalité discrète analogue à (2.21) et on passe à la limite; pour cela cf. [2] .

Supposons à présent que l'on ait deux processus f_1 et f_2 vérifiant (2.i)-(2.iv) et (2.8) et que nous associons à ces processus deux fonctions y_1 et y_2 satisfaisant à (2.9)-(2.12) (même fonction g et même donnée initiale y_o).

Nous allons voir que y_1-y_2 satisfait une égalité analogue à (2.19) sous la condition qu'une propriété analogue à (2.v) soit satisfaite:

$$(2.vi) \begin{cases} \forall t_1, t_2 , \quad \text{avec } 0 \leqslant t_1 \leqslant t_2 \leqslant T , \quad \text{la v.a.} \\ \{f_1(t_2) - f_1(t_1), f_2(t_2) - f_2(t_1)\} \text{ à valeurs dans } H^2 \\ \text{est indépendante de la v.a.} \\ \{y_o, g\}_{[0,t_1]} \ f_1(t_{j_1}),\dots,f_1(t_{j_q}), f_2(t_{i_1}),\dots, \\ f_2(t_{i_r})\} \text{ à valeur dans } H \times L^{p'}([0,t_1] ; V') \times H^{q+r} , \\ \forall q, r \text{ et } t_{j_1},\dots,t_{j_q} , t_{i_1},\dots,t_{i_r} \leqslant t_1 . \end{cases}$$

R. Temam

Proposition 2.4.

Sous les hypothèses du théorème 2.1, soient y_1 et y_2 les solutions de (2.9)-(2.12) associées à deux processus f_1 et f_2 différents. Avec l'hypothèse supplémentaire (2.vi), y_1 et y_2 vérifient:

$$(2.22) \quad E|y_1(t) - y_2(t)|_H^2 + 2 \int_0^t <Ay_1(s)Ay_2(s), y_1(s)-y_2(s)> ds$$

$$= E|f_1(t) - f_2(t)|_H^2 \, , \quad \forall t \in [0,T] \, .$$

Démonstration semblable à celle de la proposition 2.1 (cf. [2]).

2.4. Cas général.

Nous voulons étudier ici l'existence et l'unicité de solution de (2.9)-(2.12) dans le cas général, c'.à.d. f ne vérifiant pas (2.8). Nous démontrerons le résult suivant:

Théorème 2.2.

Les hypotheses sont (1.i)-(1.v) et (2.i)-(2.iv). On se donne
$$(2.23) \qquad y_0 \in \mathcal{H} = L^2(\underline{\omega}, \mu; H)$$
et
$$(2.24) \quad g \in L^{p'}([0,T]; L^{p'}(\underline{\omega}, \mu; V')) + L^1([0,T]; L^2(\underline{\omega}, \mu; H))$$

et l'on suppose que y_o, g et f satisfont la propriété d'indé-pendance (2.v).

Alors, il existe y unique qui vérifie

(2.25) $y \in L^P([0,T]; L^P(\underline{\omega}, \mu; V)) \cap L^\infty ([0,T] ; L^2(\underline{\omega}, \mu; H))$

(2.26) $\dfrac{dy}{dt} + \mathcal{A}y = g + \dfrac{df}{dt}$

(2.27) $y(0) = y_o$

Le processus y est (dt - p.p. égal à) un processus conti-nu de $[0,T] \longrightarrow \mathcal{H} = L^2(\underline{\omega}, \mu;H)$ et satisfait l'égalité d'énergie

(2.28) $E \; |y(t)|^2_H + 2 E \displaystyle\int_o^t < A \, y(s) \, , \, y(s)> ds =$

$= E|y_o|^2_H + 2 E \displaystyle\int_o^t < g(s), \, y(s)> ds + E |f(t)|^2_H \, ,$

$\forall t \in [0,T]$.

Remarque 2.4. (cf. Remarque 1.2). L'égalité (2.26) est prise au sens des distributions vectorielles sur $]0,T[$ à valeurs dans $\mathcal{V}'$.

Nous déduisons facilement de (2.26) que

(2.29) $\dfrac{dy}{dt} (\omega) + Ay(\omega) = g(\omega) + \dfrac{df}{dt} (\omega)$, $d\mu$ - p.p. ,

cette égalité étant prise au sens des distributions vectorielles sur $]0,T[$ à valeurs dans V' .

R. Temam

D'aprés (2.25) et (2.26),

(2.30) $\quad \dfrac{d}{dt}(y-f) \in L^{p'}(\mathcal{V}') + L^1(\mathcal{H})$,

mais en général $\dfrac{dy}{dt}$ et $\dfrac{df}{dt}$ <u>ne sont pas des fonctions</u>.

Si l'on veut faire apparaître explicitement la dépendance en t on écrira (conséquence facile de (2.29) et (2.30)):

(2.31) $\quad \dfrac{d}{dt}(y-f)(t,\omega) + Ay(t,\omega) = g(t,\omega), d\mu$-p.p. , dt - p.p. ;

égalité dans V'.

<u>Remarque</u> 2.5.(cf. Remarque 1.3). L'égalité (2.27) s'entend dans $\mathcal{H}$, c'est-à-dire

(2.32) $\quad y(0,\omega) = y_o(\omega)$, $d\mu$-p.p. , dans H .

<u>Démonstration de l'unicité.</u>

Le raisonnement est classique. Si y_1 et y_2 sont deux solutions de (2.25)-(2.27), on obtient par différence

(2.31) $\qquad y = y_1-y_2 \in L^p(\mathcal{V}) \cap L^\infty(\mathcal{H})$,

(2.32) $\qquad \dfrac{dy}{dt} + Ay_1 - Ay_2 = 0$,

(2.33) $\qquad y(0) = 0$.

Il résulte de (2.31) et (2.32) que $\dfrac{dy}{dt} \in L^{p'}(\mathcal{V}')$ ce qui joint à (2.31) et Lions [5] entraine que

$$(2.34) \qquad < \frac{dy}{dt} , y>_{\mathcal{V}, \mathcal{V}'} = \frac{1}{2} \frac{d}{dt} |y(t)|^2_{\mathcal{H}} .$$

Utilisant (2.34) et la monotonie de $\mathcal{A}$, on déduit de (2.32)

$$\frac{d}{dt} |y(t)|^2_{\mathcal{H}} \leq 0 ,$$

et donc $y = y_1 - y_2 = 0$.

2.5. Démonstration de l'existence dans le théorème 2.2.

On utilise ici les approximations f_N de f . D'après la proposition 2.2 ces approximations vérifient (2.i)-(2.iv) et en autre (cf.(2.6)) la propriété de régularité (2.8).

a. Approximation du problème.

D'après la remarque précédente, le théorème 2.1 donne l'existence et l'unicité pour tout entier N de y_N vérifiant:

$$(2.35) \qquad y_N \in L^P(\mathcal{V}) \cap \mathcal{C}(\mathcal{H})$$

$$(2.36) \qquad \frac{dy_N}{dt} + \mathcal{A}y_N = g + \frac{df_N}{dt}$$

$$(2.37) \qquad y_N(0) = y_0$$

Pour un autre entier M on a aussi

$$(2.38) \qquad y_M \in L^P(\mathcal{V}) \cap \mathcal{C}(\mathcal{H})$$

$$(2.39) \qquad \frac{dy_M}{dt} + \mathcal{A}y_M = g + \frac{df_M}{dt}$$

$$(2.40) \qquad y_M(0) = y_0 .$$

En vue d'appliquer les propositions 2.3. et 2.4, nous allons montrer le

Lemme 2.1.

L'hypothèse (2.v) est satisfaite avec f remplacé par f_N, $\forall N \geqslant 1$.

L'hypothèse (2.vi) est satisfaite avec f_1 et f_2 rempla-cé par f_N et f_M , $\forall N,M \geqslant 1$.

Démonstration

Soit P_N le projecteur orthogonal dans H sur l'espace en-gendré par $\{e_1,\ldots,e_N\}$. On a , d'après (2.4)

(2.41) $\qquad f_N(t) = P_N \, f(t)$.

D'après (2.v) la v.a. $f(t_2)-f(t_1)$ est indépendante de la v.a. $\left\{y_0, \, g\big|_{[0,t_1]} \, , \, f(t_{j_1}),\ldots,f(t_{j_q})\right\}$ (pour $t_1,t_2,t_{j_1},\ldots,t_{j_q}$ convenables). L'opérateur P_N étant continu dans l'espace appro-prié (dans H), la v.a. $P_N(f(t_2)-f(t_1)) = f_N(t_2)-f_N(t_1)$ est in-dépendante de la v.a. $\left\{y_0, \, g\big|_{[0,t_1]} \, , \, f_N(t_{j_1}),\ldots,f_N(t_{j_q})\right\}$.

Ce qui prouve (2.v) pour f_N .

De même la v.a. $\left\{f_N(t_2)-f_N(t_1), \, f_M(t_2)-f_M(t_1)\right\}$ à valeurs dans H^2 se déduit de $f(t_2)-f(t_1)$ par une opération continue dans H; la v.a. $\left\{y_0, \, g\big|_{[0,t_1]} \, , f_N(t_{j_1}),\ldots,f_N(t_{j_q}), \, f_M(t_{i_1}), \ldots\right.$

$f_M(t_{i_r})\}$ se déduit de $\{y_o, g\big|_{[0,t_1]}$, $f(t_{j_1}), \ldots, f(t_{j_q})$,

$f(t_{i_1}), \ldots, f(t_{i_r})\}$ par une opération continue dans

$H \times L^{p'}([0,t_1]; V') \times H^{q+r}$ (en l'occurence $I \times I \times P_N^q \times P_M^r$) et

les deux premières v.a. sont indépendantes grace à (2.v).

Les propositions 2.3 et 2.4 entrainent alors le

Lemme 2.2.

Pour tout N et $M \geqslant 1$, y_N et y_M vérifient:

$$(2.42) \quad E \, |y_N(t)|_H^2 + 2 \, E \int_o^t <A\,y_N(s) \, , \, y_N(s)> \, ds =$$

$$= E \, |y_o|_H^2 + 2 \, E \int_o^t <g(s), \, y_N(s)> ds + E|f_N(t)|^2$$

$$(2.43) \quad E \, \Big|y_N(t) - y_M(t)\Big|_H^2 + 2 \, E \int_o^t <A\,y_N(s) - A\,y_M(s), y_N(s) - y_M(s)> ds$$

$$= E \, \Big| f_N(t) - f_M(t)\Big|_H^2 \quad ,$$

$\forall t \in [0,T]$.

Avec (2.42), la coercivité de A et un calcul classique on

en déduit:

$$(2.44) \quad y_N \in \text{borné de } L^p(\mathcal{V}) \cap L^\infty(\mathcal{H}) \, , \, N \longrightarrow \infty \, ;$$

d'après (2.43) et la proposition 2.2 on obtient aussi

(2.45) la suite y_N est de Cauchy dans $\mathscr{C}(\mathscr{H})$, $N \to \infty$;

(2.46) $E \int_0^T < Ay_N(s) - Ay_M(s), y_N(s)-y_M(s)> ds \to 0$, $N,M \to \infty$.

b. Passage à la limite.

Il s'agit de passer à la limite, $N \to \infty$, dans (2.35)-(2.37) en utilisant (2.44)-(2.46).

D'aprés (2.44) et (2.45), il existe y ,

(2.47) $y \in L^P(\mathcal{V}) \cap \mathscr{C}(\mathscr{H})$,

avec

(2.48) $y_N \longrightarrow y$ dans $\mathscr{C}(\mathscr{H})$,

(convergence uniforme sur [0,T] à valeurs dans $\mathscr{H}$) ,

(2.49) $y_{N_i} \longrightarrow y$ dans $L^P(\mathcal{V})$ faible.

Comme l'opérateur $\mathscr{A}: L^P(\mathcal{V}) \to L^{P'}(\mathcal{V}')$ est borné, nous avons

(2.50) $\mathscr{A}y_N \in$ borné de $L^{P'}(\mathcal{V}')$.

La sous-suite N_i de (2.49) peut être choisie en sorte que

(2.51) $\mathscr{A}y_{N_i} \longrightarrow \chi$ dans $L^{P'}(\mathcal{V}')$ faible.

Nous allons prouver le

Lemme 2.3.

(2.52) $\chi = \mathscr{A}y$.

Démonstration.

D'après (2.46) pour $\varepsilon > 0$, il existe $N(\varepsilon)$ tel que $M \geqslant N(\varepsilon)$, $N \geqslant N(\varepsilon)$ entrainent:

$$0 \leqslant \int_0^T < \mathcal{A}y_N(s) - \mathcal{A}y_M(s),\ y_N(s) - y_M(s)>_{\mathcal{V},\mathcal{V}'} ds \leqslant \varepsilon$$

Fixons $N \geqslant N(\varepsilon)$ et faisons tendre M vers l'infini, avec $M = N_j$, $j \to \infty$. En raison de (2.49) et (2.51) on a alors

$$0 \leqslant \limsup_{N_j \to \infty} \int_0^T < \mathcal{A}y_{N_j}(s),\ y_{N_j}(s)>_{\mathcal{V},\mathcal{V}'} ds + \int_0^T < \mathcal{A}y_N(s), y_N(s)>_{\mathcal{V},\mathcal{V}'} ds -$$

$$- \int_0^T < \chi(s),\ y_N(s)>_{\mathcal{V},\mathcal{V}'} ds - \int_0^T < \mathcal{A}y_N(s), y(s)>_{\mathcal{V},\mathcal{V}'} ds \leqslant \varepsilon$$

Faisons alors $N = N_i$ et $N_i \to \infty$, il vient:

$$0 \leqslant 2 \limsup_{N_j \to \infty} \int_0^T < \mathcal{A}y_{N_j}(s) ,\ y_{N_j}(s)>_{\mathcal{V},\mathcal{V}'} ds$$

$$- 2 \int_0^T < \chi(s),\ y(s)>_{\mathcal{V},\mathcal{V}'} ds \leqslant \varepsilon \quad .$$

Comme $\varepsilon > 0$ est arbitrairement petit, on a en fait

$$(2.53) \quad \limsup_{N_j \to \infty} \int_0^T < \mathcal{A}y_{N_j}(s),\ y_{N_j}(s)>_{\mathcal{V},\mathcal{V}'} ds =$$

$$= \int_0^T < \chi(s) ,\ y(s)>_{\mathcal{V},\mathcal{V}'} ds .$$

R. Temam

Un raisonnement standard sur les opérateurs monotones nous permet d'obtenir (2.52).

Le passage à la limite dans (2.35)-(2.37) est alors trivial (pour la suite N_i) et donne (2.25)-(2.27).

Comme d'abitude l'unicité de solution de (2.25)-(2.27) fait que les convergences (2.49) et (2.51) sont aussi vraies pour la suite $\hat{N}$ elle même.

Le théorème 2.2 sera completement démontré si l'on établit l'égalité d'énergie (2.28). Or, d'aprés (2.53) et (2.52)

$$\limsup_{N \to \infty} \int_o^T < \mathcal{A}y_N(s), y_N(s)>_{\mathcal{V},\mathcal{V}'} ds = \int_o^T < \mathcal{A}y(s), y(s)>_{\mathcal{V},\mathcal{V}'} ds \quad .$$

Il suffit donc de passer à la limite supérieure dans l'égalité d'énergie (2.42) correspondante à y_N .

Remarque 2.6. Dans les exemples g et y_o seront en général des processus gaussiens non corrélés avec f et la propriété (2.v) sera automatique.

Le modèl peut s'appliquer à des équations aux derivées partielles non linéaires du type monotone (cf. Lions [ ]). Renvoyons encore à [ ] pour les exemples, pour des développements analo

R. Temam

gues pour les équations non linéaires relevant des méthodes de

compacité, et aussi dans le cadre des semi-groupes non linéai-

res de contraction.

R. Temam

BIBLIOGRAPHIE

[1] A. Bensaussan, Filtrage optimal des systèmes linéaires,
    Dunod, Paris, à paraître.

[2] A. Bensoussan et R. Temam, Équations aux dérivées partiel-
    les stochastiques et applications, à paraître.

[3] F.E. Browder, Strongly non linear boundary value problem,
    Am. J. Math., 86, 1964 p. 339-357.

[4] K. Ito, Stochastic Integrals, Conférence du Tata Institute,
    Bombay.

[5] J.L. Lions, Quelques methodes de résolution des problèmes
    aux limites non linéaires, Dunod-Gauthier-Villars,
    Paris 1969.

[6] P.A. Meyer, Probabilités et potentiels, Hermann, Paris, 1966.

[7] L. Schwartz, Radon measures on arbitrary topological spaces,
    à paraître, Tata Institute of fundamental research,
    Bombay.

CENTRO INTERNAZIONALE MATEMATICO ESTIVO

(C. I. M. E.)

M. M. VAINBERG

LE PROBLÈME DE LA MINIMISATION DES FONCTIONNELLES

NON LINÉAIRES

Corso inviato per il 4º ciclo 1970

M. M. V A I N B E R G

(Université de Moscou)

LE PROBLÈME DE LA MINIMISATION DES FONCTIONNELLES

NON LINÉAIRES.

M. M. Vainberg

I. LA SEMICONTINUITÉ FAIBLE DES FONCTIONNELLES.

1.1 Critère de semi-continuité faible des fonctionnelles.

On va considérer ici les propositions fondamentales sur la semi-continuité faible des fonctionnelles, nécessaires pour étudier la question du minimum des fonctionnelles.

Définition 1.1. La fonctionnelle réelle $f(x)$ donnée sur l'ensem ble U d'un espace normé est dite faiblement semicontinue inférieurement (supérieurement) au point $x_0 \in U$ si pour chaque suite $\{x_n\} \subset U$ faiblement convergente vers $x_0 (x_n \to x_0)$ on a l'i négalité

$$f(x_0) \leqslant \varliminf_{n \to \infty} f(x_n) \qquad \left(f(x_0) \geqslant \varlimsup_{n \to \infty} f(x_n) \right).$$

De cette définition, comme dans le cas classique, il suit que la somme de fonctionnelles faiblement semi-continues inférieurement est faiblement semi-continue inférieurement.

Définition 1.2. On dit que l'ensemble σ dans un espace normé est faiblement fermé (ou faiblement complet par suites) s'il est faiblement fermé par suites c'est-à-dire, pour chaque suite $\{x_n\} \subset \sigma$ faiblement convergente vers x_0 il suit $x_0 \in \sigma$.

Remarque 1.1. Notre définition d'ensemble faiblement fermé d'un espace normé E ne coincide pas avec celle ordinaire.

Selon celle-ci on dit que l'ensemble $x \subset E$ est faiblement fermé lorsq'il est fermé dans la topologie faible de E.

Les voisinages de la topologie faible de l'espace E sont défi-

M. M. Vainberg

nis de la façon suivante. Pour $x_o \in E$ soit ε positif quelconque et soient $1_1, 1_2, \ldots, 1_n$ des fonctionnelles linéaires continues n arbitraire. Un voisinage de x_o est alors

$$U(x_o; 1_1, 1_2, \ldots, 1_n, \varepsilon) = \left\{ x \in E : | 1_k(x - x_o) | < \varepsilon, \ k = 1, 2, \ldots, n \right\},$$

La famille de tous ces voisinages définit la topologie faible de l'espace E. L'ensemble de tous les voisinages $U(x_o; 1_1, 1_2, \ldots, 1_n, \varepsilon)$ forment un système fondamental de voisinages du point x_o. La fermeture faible d'un ensemble dans cette topologie a le défaut suivant. Il existe des suites ayant un seul point limite dans cette topologie et qui pourtant ne contiennent aucune sous-suite qui converge faiblement vers ce point. (Un tel exemple a été donné par J. von Neumann [I]). On note toutefois que si l'ensemble σ est la fermeture de $X \subset E$ au sens de la déf. 1.2 et $\bar{\sigma}$ est la fermeture de $X \subset C$ au sens de la topologie faible de l'espace E, on a $\bar{\sigma} \supset \sigma$. C'est à cause de cette inclusion que l'on peut utiliser plusieurs résultats sur les ensembles de E fermés dans la topologie faible. Par exemple le théorème connu sur la fermeture faible des ensembles convexes (fortement) fermés d'un espace de Banach reste valable lorsqu'on considère la fermeture faible au sens de la déf. 1.2.

Théorème 1.1. Afin que la fonctionnelle $f(x)$ donnée sur un espace normé soit faiblement semi-continue inférieurement il est nécessaire et suffisant que la condition suivante soit satisfaite:

M. M. Vainberg

pour chaque nombre réel c l'ensemble

$$E_c = \left\{ x : f(x) \leqslant c \right\} \tag{1.1}$$

est faiblement fermé.

<u>Démonstration</u> de la suffisance. La condition (1.1) soit satisfaite. Démontrons que pour une suite quelconque $x_n \rightarrow x_0$ on a

$$f(x_0) \leqslant \varliminf_{n \to \infty} f(x_n) = b \tag{1.2}$$

Supposons, au contraire, que $f(x_0) > b$. Alors il existe $\varepsilon > 0$ tel que

$$f(x_c) > b + \varepsilon \tag{1.3}$$

Comme

$$b = \varliminf_{n \to \infty} f(x_n)$$

il existe une sous-suite $\left\{ x_{n_k} \right\}$ pour laquelle

$$f(x_{n_k}) \leqslant b + \varepsilon/2 \tag{1.4}$$

Mais comme l'ensemble

$$E_{c_1} = \left\{ x ; f(x) \leqslant b + \varepsilon/2 \right\}, \quad c_1 = b + \varepsilon/2$$

est faiblement fermé et $x_{n_k} \rightarrow x_0$ pour $k \to \infty$, on déduit de l'inégalité (1.4) $\left\{ x_{n_k} \right\} \subset E_{c_1}$, donc $x_0 \in E_{c_1}$, c'est-à-dire $f(x_0) \leqslant b + \varepsilon/2$, ce qui contredit l'inégalité (1.3). La contradiction prouve la suffisance de la condition.

<u>Démonstration</u> de la nécessité. La fonctionnelle $f(x)$ soit faiblement semi-continue inférieurement. Démontrons que l'ensemble E_c est faiblement fermé. Considérons une suite $\left\{ x_n \right\} \subset E_c$ qui soit faiblement fondamentale (de Cauchy) et faiblement convergen

M. M. Vainberg

te vers un element x_o de l'espace. Si l'espace est semicomplet
(faiblement complet par suites) il n'est pas nécessaire de deman-
der que $x_n \to x_o$). Comme la fonctionnelle $f(x)$ est faiblement
semi-continue inférieurement au point x_o on a la relation (1.2).
Mais $f(x_n) \leqslant c$ pour chaque n , donc $b \leqslant c$. Pourtant $f(x_o) \leqslant$
$\leqslant b \leqslant c$, c'est-à-dire $x_o \in E_c$. Le théorème est ainsi démontré.
Remarquons que le théorème s'étend aux espaces localement conve-
xes.

1.2 Condition de semi-continuité faible pour les fonctionnelles convexes.

Définition 1.3. La fonctionnelle réelle $f(x)$ donnée sur un ensemble convexe ω d'un espace linéaire, est dite convexe sur ω
si pour chaque x_1, $x_2 \in \omega$ et pour $\lambda \in (0,1)$ arbitraire on a l'i-
négalité

$$f(\lambda x_1 + (1 - \lambda) x_2) \leqslant \lambda f(x_1) + (1 - \lambda) f(x_2) \, ,$$

strictement convexe si l'égalité ne vaut pas pour $x_1 \neq x_2$.
Définition 1.4. La fonctionnelle réelle $f(x)$ donnée sur un ensemble ouvert U d'un espace normé, est dite semicontinue inférieurement (supérieurement) au point $x_o \in U$, si pour chaque suite $\{x_n\} \subset U$ convergente vers x_o $(x_n \to x_o)$ on a l'inégalité

$$f(x_o) \leqslant \varliminf_{n \to \infty} f(x_n) \qquad \left(f(x_o) \geqslant \varlimsup_{n \to \infty} f(x_n) \right).$$

M. M. Vainberg

Lemma 1.1. Si la fonctionnelle réelle $f(x)$ est semi-continue inférieurement, alors l'ensemble $E_c = \{x : f(x) \leqslant c\}$, est fermé pour c réel quelconque.

Démonstration. Considérons une suite fondamentale quelconque $\{x_n\} \subset E_c$. Soit $x_o = \lim_{n \to \infty} x_n$. Supposons que $x_o \notin E_c$, c'est-à-dire que $f(x_o) > c$. Il existe alors $\varepsilon > 0$ tel que $f(x_o) > c + \varepsilon$.

De la semicontinuité inférieure de la fonctionnelle $f(x)$ au point x_o on a un nombre naturel $n_o = n_o(\varepsilon)$, tel que si $n \geqslant n_o$ il suit

$$f(x_n) > f(x_o) - \varepsilon/2 > c + \varepsilon/2 .$$

Cette inégalité contredit le fait que $\{x_n\} \subset E_c$. La contradiction démontre le lemma.

Lemma 1.2. Si la fonctionnelle convexe et semi-continue inférieurement $f(x)$ est donnée dans un espace de Banach, alors l'ensemble $E_c = \{x : f(x) \leqslant c\}$ est faiblement fermé pour chaque c réel.

Démonstration. De la convexité de $f(x)$ il suit la convexité de l'ensemble E_c et de la semi-continuité inférieure de $f(x)$ il suit en vertu du Lemma 1.1. que E_c est fermé. Donc E_c, étant un ensemble convexe et fermé d'un espace de Banach est aussi faiblement fermé (Cfr. E. Hille - R. Phillips [I] , 2.9.3).

Du lemma et du théorème 1.1. on obtient le

M. M. Vainberg

Théorème 1.2. Si E est un espace de Banach alors chaque fonctionnelle donnée sur E , convexe et semi-continue inférieurement est aussi faiblement semi-continue inférieurement.

Dans la suite on démontrera des propositions plus fortes (Cfr.
Théorème 1.10).

1.3 Conditions pour la semi-continuité faible et la convexité de fonctionnelles différentiables.

Dans le cas où la fonctionnelle réelle non linéaire est différen
tiable au sens de Gateaux on a des conditions simples pour sa con
vexité et sa semi-continuité inférieure. Nous en donnerons quelques unes.

Théorème 1.3. La fonctionnelle réelle f(x), donnée dans un espa
ce normé, soit différentiable au sens de Gateaux et soit $D f(x,h)$
la différentielle de Gateaux, linéaire en h. Si au point x_0 on
a l'inégalité

$$f(x) - f(x_0) - D f(x_0, x - x_0) \geqslant 0 \qquad (1.5)$$

pour chaque $x \in U$, où U est un certain voisinage du point
x_0, alors f(x) est faiblement semi-continue inférieurement au
point x_0 .

Si l'inégalité (1.5) est vérifiée pour chaque x, x_0 appartenant à un ensemble convexe et ouvert U , alors f(x) est faiblement semi-continue inférieurement et convexe dans U (stricte
ment convexe si dans (1.5) l'égalité ne vaut pas pour $x \neq x_0$).

M. M. Vainberg

<u>Démonstration</u>. Considérons une suite $\{x_n\} \subset U$ quelconque, convergente vers x_0 faiblement. Comme $D\, f(x_0, x_n - x_0)$ est une fonctionnelle linéaire continue par rapport à $x_n - x_0$ et $x_n \to x_0$ on a

$$\lim_{n \to \infty} D\, f(x_0, x_n - x_0) = 0 .$$

De (1.5) on tire

$$\lim_{n \to \infty} \left[f(x_n) - f(x_0) \right] \geqslant \lim_{n \to \infty} D\, f(x_0, x_n - x_0) =$$

$$= \lim_{n \to \infty} D\, f(x_0, x_n - x_0) = 0 ,$$

donc

$$f(x_0) \leqslant \lim_{n \to \infty} f(x_n) .$$

Cela prouve la première partie du théorème sur la semi-continuité faible au point x_0 ou bien sur U . Pour démontrer la deuxième partie sur la convexité de $f(x)$ on écrit l'inégalité (1.5) pour deux vecteurs quelconques (différents) $x_1, x_2 \in U$:

$$f(x_1) \geqslant f(x_0) + D\, f(x_0, x_1 - x_0)$$

$$f(x_2) \geqslant f(x_0) + D\, f(x_0, x_2 - x_0)$$

Multipliant ces deux inégalités par t et $1 - t$ avec $0 < t < 1$ et posant $x_0 = t\, x_1 + (1 - t)\, x_2$ on obtient

$$t\, f(x_1) + (1 - t)\, f(x_2) \geqslant f(x_0)$$

c'est-à-dire

$$t\, f(x_1) + (1 - t)\, f(x_2) \geqslant f(t\, x_1 + (1-t)\, x_2) .$$

Cette inégalité montre la convexité ou la convexité stricte. Le

M. M. Vainberg

théorème est démontré.

Il faut remarquer que ce théorème reste valable pour une classe

étendue d'espaces vectoriels topologiques, mais pour la démonstra

tion de la première partie il est nécessaire de considérer des

suites généralisées.

On connait plusieurs conditions suffisantes pour que l'inégalité

(1.5) soit vérifiée. En voici quelques unes.

Lemma 1.3. Sur l'ensemble convexe σ d'un espace normé la fon-

ctionnelle $f(x)$ soit donnée avec une différentielle de Gateaux

$D f(x,h)$ linéaire et continue par rapport à h , qui satisfait

la condition

$$D f(x, x - x_o) - D d(x_\sigma, x - x_o) \geqslant 0 , (x,x_o \in \sigma) \qquad (1.6) .$$

Alors on a l'inégalité (1.5).

Démonstration. En appliquant la formule de Lagrange (Cfr. M.M.

Vainberg [2]) on obtient

$$f(x) - f(x_o) = D f(x_o + \tau(x - x_o), x - x_o) =$$

$$= D f(x_o,x - x_o) + \left[D f(x_o+ \tau(x - x_o),x - x_o) - D f(x_o,x - x_o) \right.$$

$$(0 < \tau < 1)$$

D'ici et de l'inégalité (1.6) on tire

$$f(x) - f(x_o) \geqslant D f(x_o, x - x_o) .$$

Le lemma est démontré.

Remarquons que l'inégalité (1.6) peut être écrite d'une autre

façon. En effet, comme la différentielle $D f(x,h)$ est continue

en h , on aura selon la définition de gradient (Cfr. M.M. Vain

M. M. Vainberg

berg [2])

$$D f(x,h) = \langle F(x), h \rangle \; ; \; F(x) = \text{grad } f(x) \; ,$$

où l'on denote par $\langle y, x \rangle$ la valeur de la fonctionnelle linéaire $y \in E^*$ au point x de l'espace normé E . Avec cette notation l'inégalité (1.6) devient

$$\langle F(x) - F(x_o) . \; x - x_o \rangle \geqslant 0$$

Cela signifie que $F(x)$ est un opérateur monotone.

Pourtant du lemma 1.3 et du théorème 1.3 on obtient le

Théorème 1.4. Pour la semi-continuité inférieure faible de la fonctionnelle réelle il est suffisant que son gradient soit un opérateur monotone.

Remarquons que ce théorème s'étend aux espaces vectoriels topologiques localement convexes séparables.

Lemma 1.4. Sur l'ensemble convexe σ d'un espace normé la fonctionnelle $f(x)$ soit deux fois différentiable au sens de Gateaux avec

$$D^2 f(x, h, h) \geqslant 0 \; , \quad (x \in \sigma) \tag{1.8}.$$

Alors l'inégalité (1.5) est vérifiée.

Démonstration. Au moyen de la formule de Lagrange et de l'inégalité (1.8) on peut écrire

$$D (x_o + \tau(x - x_o), \; x - x_o) - D f(x_o, \; x - x_o) =$$

$$= \tau D^2 f(x_o + \tau_1(x - x_o) \; ; \; x - x_o, \; x - x_o) \geqslant 0 \; .$$

D'ici et de l'inégalité (1.7) on tire l'inégalité (1.5) .

M. M. Vainberg

De ce lemma et du théorème 1.3 il suit le

Théorème 1.5. Si le gradient $F(x)$ de la fonctionnelle réelle

$f(x)$ admet la dérivée de Gateaux $F'(x)$, et si

$$\langle F'(x) h, h \rangle \geqslant 0$$

alors la fonctionnelle $f(x)$ est faiblement semicontinue infé-

rieurement.

En effet, dans les conditions de ce théorème l'inégalité préce-

dente coincide avec l'inégalité (1.8).

Démontrons encore une proposition sur la semicontinuité faible

des fonctionnelles.

Théorème 1.6. Si la fonctionnelle convexe $f(x)$ donnée sur un

ensemble convexe ouvert U d'un espace normé V est différentiable au

sens de Gateaux et si pour chaque $x \in U$ fixé la différentielle

$D f(x,h)$ est continue en h , alors $f(x)$ est faiblement semi-

continue dans U .

Démonstration. Soient $F(x) = \text{grad } f(x)$; x, x_0 deux vecteurs

(différents) quelconques de U et $h = x - x_0$. Considérons la

fonction réelle

$$\varphi(t) = f(x_0 + t h) , \quad 0 \leq t \leq 1 .$$

Cette fonction est convexe puisque la fonctionnelle $f(x)$ est

convexe, et la dérivée $\varphi'(t) = \langle F(x_0 + t h), h \rangle$ ne décroit pas.

Donc

$$f(x) - f(x_0) = \varphi(1) - \varphi(0) = \varphi'(\tau) \geq \varphi'(0) = \langle F(x_0), h \rangle$$

M. M. Vainberg

puisque $0 < \tau < 1$. Par conséquent

$$f(x) - f(x_o) - D\, f(x_o, x - x_o) \geqslant 0$$

c'est-à-dire l'inégalité (1.5) est satisfaite et du théorème 1.3 il suit la semi-continuité inférieure faible de la fonctionnelle $f(x)$.

Il fait remarquer que le théorème 1.6 suit immédiatement du théorème 1.3 et comme on verra (Cfr. Théorème 1.10) un résultat plus fort est valable.

Des théorèmes 1.3 et 1.6 on obtient le

Corollaire 1.1. Si on donne sur l'ensemble convexe et ouvert U une fonctionnelle $f(x)$ différentiable au sens de Gateaux et si pour chaque $x \in U$ la différentielle $D\, f(x,h)$ est continue en h , alors l'inégalité (1.5) est nécessaire et suffisante pour la convexité de la fonctionnelle $f(x)$ sur U . Si dans (1.5) l'égalité n'est pas valable pour $x \neq x_o$ alors (1.5) est nécessaire et suffisante pour la convexité stricte.

Du théorème 1.3 e du lemma 1.4 on a le

Corollaire 1.2. Si le gradient $F(x)$ de la fonctionnelle réelle $f(x)$ donnée sur l'ensemble convexe et ouvert U d'un espace normé est différentiable au sens de Gateaux, alors pour la convexité de $f(x)$ il est nécessaire et suffisant que l'on ait $\langle F'(x)\, h, h \rangle \geqslant 0$ pour chaque $x, h \in U$.

Si, en outre, dans l'inégalité précedente l'égalité ne vaut pas pour $h \neq 0$ alors $f(x)$ est strictement convexe.

1.4 Relation avec les fonctions d'appui.

Soit E un espace normé et soit E^* son dual.

Définition 1.5. La fonctionnelle linéaire $y_o \in E^*$ est dite fonction d'appui de la fonctionnelle $f(x)$ au point x_o si

$$f(x) - f(x_o) \geqslant \langle y_o, x - x_o \rangle \qquad (1.9)$$

où $\langle y_o, h \rangle$ est la valeur de la fonctionnelle linéaire continue y_o au vecteur h.

Comme si l'on remplace l'inégalité (1.5) par (1.9) le théorème 1.3 et sa démonstration restent valables on a la proposition suivante

Théorème 1.7. Si la fonctionnelle réelle définie sur l'ensemble ouvert et convexe $U \subset E$ admet une function d'appui à chaque point de U, alors $f(x)$ est convexe et faiblement semicontinue inférieurement dans U.

On voit que le critère de convexité d'une fonctionnelle réelle donnée dans U sert à assurer l'existence de sa fonction d'appui à chaque point de U car il vaut le

Théorème 1.8. Si la fonctionnelle convexe finie $f(x)$ est donnée sur l'ensemble ouvert et convexe $U \subset E$ elle admet une fonction d'appui à chaque point $x \in U$.

Démonstration. Considérons deux vecteurs $x, x + h \in U$ quelconques et la fonction réelle

M. M. Vainberg

$$\varphi(t) = f(x + t\,h) , \quad (-\varepsilon < t < 1 + \varepsilon , \ \varepsilon > 0) .$$

Pour x, h fixés cette fonction est convexe sur l'intervalle $(-\varepsilon , 1 + \varepsilon)$. En effet, soient $\alpha \geqslant 0$, $\beta \geqslant 0$, $\alpha + \beta = 1$. Alors de la convexité de la fonctionnelle $f(x)$ on a

$$\varphi(\alpha\, t_1 + \beta\, t_2) = f(\alpha(x + t_1 h) + \beta(x + t_2 h)) \leqslant \alpha\, \varphi(t_1) + \beta\, \varphi(t_2) .$$

Mais pour une fonction covexe il existe la dérivée gauche $\varphi'_-(t)$, la dérivée droite $\varphi'_+(t)$, et l'on a $\varphi'_-(t) \leqslant \varphi'_+(t)$. En particulier $\varphi'_-(0) \leqslant \varphi'_+(0)$, donc

$$V_- f(x,h) \leqslant V_+ f(x,h) , \qquad (1.10)$$

où

$$V_- f(x,h) = -V_+ f(x,h), \quad V_+ f(x,h) = \lim_{t \to +\infty} \frac{f(x+th) - f(x)}{t}$$

La variation $V_+ f(x,h)$ est, par définition, un opérateur positivement homogène en h . En outre elle est sub-additive en h car

$$V_+ f(x, h_1 + h_2) = \lim_{t \to +0} \frac{f(x + t\,(h_1 + h_2)/2) - f(x)}{t/2} =$$

$$= \lim_{t \to +0} \frac{2\, f\left(\dfrac{x + t\,h_1}{2} + \dfrac{x + t\,h_2}{2}\right) - 2\, f(x)}{t} \leqslant$$

$$\leqslant \lim_{t \to +0} \frac{f(x + t\,h_1) - f(x)}{t} + \lim_{t \to +0} \frac{f(x + t\,h_2) - f(x)}{t} =$$

$$= V_+ f(x , h_1) + V_+ f(x , h_2) .$$

Pourtant $V_+ f(x,h)$ est une fonctionnelle sublinéaire, d'où l'on a (1.10) encore une fois (voir (1.10') après).

Nous allons utiliser maintenant la proposition suivante:

Lemma 1.5. Si $f(x)$ est une fonctionnelle convexe finie sur l'en semble ouvert et convexe U, alors pour chaque $x \in U$ fixe il existe au moins une fonctionnelle linéaire $y \in E^*$ telle que

$$V_-f(x,h) < \langle y.h \rangle < V f(x,h) \qquad (1.11)$$

Démonstration. Ayant fixé le vecteur h_o différent de zéro on con sidère la droite $z = \lambda h_o$ $(-\infty < \lambda < +\infty)$ dans l'espace E. On définit après une fonctionnelle linéaire y_o telle que

$$y_o(h_o) = V_+f(x, h_o).$$

Comme $V_+f(x,h)$ est positivament homogène en h on tire de l'é galité précedente

$$y_o(\lambda h_o) = V_+f(x,\lambda h_o) \quad , \quad \lambda \geqslant 0 .$$

Et encore, pour la sublinéarité de $V_+f(x,h)$

$$0 = V_+f(x,h - h) \leq V_+f(x,h) + V_+f(x, -h)$$

ou

$$- V_+f(x,h) \leq V_+f(x, -h) \qquad (1.10')$$

D'ici pour $\lambda < 0$ on a

$$y_o(\lambda h_o) = \lambda V_+f(x,n_o) = -|\lambda| V_+f(x,h_o) \leq$$

$$< |\lambda| V_+f(x,-h_o) = V_+f(x,\lambda h_o) .$$

Pourtant pour chaque valeur de $z = \lambda h_o$ on a

$$y_o(z) \leq V_+f(x,z) \qquad (1.12)$$

M. M. Vainberg

La fonctionnelle linéaire y_0 étant donnée sur une droite et satisfaisant l'inégalité (1.12) peut être étendue (pour le théorème de Hahn-Banach) à tout l'espace donnant lieu à une fonctionnel$\underline{le}$
le $y \in E^*$ qui satisfait les conditions

$$\langle y, h \rangle \leqslant V_+ f(x,h) \quad , \quad \forall h \in E$$

$$\langle y, n_0 \rangle = V_+ f(x,h_0) .$$

De l'inégalité précédente on a $\langle y,-h \rangle \leqslant V_+ f(x,-h)$ ou

$$\langle y,h \rangle = -\langle y, -h \rangle \geqslant - V_+ f(x, -h) = V_- f(x,h)$$

et enfin

$$V_- f(x,h) \leqslant \langle y, h \rangle \leqslant V_+ f(x,h)$$

Le lemma est démontré.

Pour achever la démonstration du théorème considérons encore la
fonction $\varphi(t) = f(x + th)$ et utilisons la propriété suivante

$$\varphi'_+(0) \leqslant \frac{\varphi(1) - \varphi(0)}{1 - 0} < \varphi'_-(1)$$

D'ici $\varphi(1) - \varphi(0) \geqslant \varphi'_+(0)$ ou

$$f(x + h) - f(x) > V_+ f(x,h)$$

et de l'inégalité (1.11) on tire

$$f(x + h) - f(x) \geqslant \langle y, h \rangle$$

c'est-à-dire y est une fonction d'appui de la fonctionnelle
$f(x)$ au point x . Le théorème est démontré.

Des théorèmes 1.7 et 1.8 on déduit les propositions suivantes.

<u>Théorème</u> 1.9 Afin qu'une fonctionnelle réelle finie donnée sur
un ensemble ouvert et convexe $U \subseteq E$ soit convexe sur U il est

M. M. Vainberg

nécessaire et suffisant qu'elle admette une fonction d'appui à chaque point $x \in U$.

Théorème 1.10. Chaque fonctionnelle convexe finie donnée sur un ensemble ouvert et convexe de E est faiblement semi-continue inférieurement.

1.5 Sur la semi-continuité faible des fonctionnelles quasi-convexes.

Définition 1.6. La fonctionnelle réelle $f(x)$ donnée sur un ensemble convexe ω d'un espace linéaire est dite quasi-convexe sur ω si pour x_1, $x_2 \in \omega$ quelconques et $\lambda \in (0,1)$ arbitraire on a l'inégalité

$$f(\lambda x_1 + (1 - \lambda) x_2) \leq \max \left[f(x_1) , f(x_2) \right] ,$$

strictement quasi-convexe si l'égalité ne vaut pas pour $x_1 \neq x_2$.

Lemma 1.6. Si la fonctionnelle $f(x)$ définie sur un espace de Banach est quasi-convexe et semi-continue inférieurement alors l'ensemble $E_c = \left\{ x : f(x) \leq c \right\}$ est faiblement fermé pour chaque c réel.

Démonstration. De la définition de quasi-convexité il suit que l'ensemble E_c est convexe et de la semicontinuité inférieure E_c est fermé grâce au lemma 1.1. Donc E_c étant un ensemble convexe et fermé d'un espace de Banach est aussi faiblement fermé. (Cfr. E. Hille - R. Phillips I , Th. 2.9.3).

De ce lemma et du théorème 1.1. on obtient

M. M. Vainberg

Theorème. 1.2. Si E est un espace de Banach alors chaque fonctionnelle quasi-convexe et semi-continue inférieurement sur E est aussi faiblement semi-continue inférieurement.

Il faut remarquer que dans les conditions de ce théorème on ne peut pas supprimer celle de la semi-continuité inférieure, comme on le voit de l'exemple suivant de la fonction réelle d'une variable réelle

$$f(x) = \begin{cases} e^x & , \quad x < 1 \\ e^2 & , \quad 1 \leqslant x \leqslant 2 \\ e^x & , \quad x > 2 \end{cases}$$

Cette fonction est quasi-convexe, mais non pas faiblement semi-continue inférieurement.

1.6 Exemples de fonctionnelles faiblement semicontinues inférieurement.

Exemple 1.1. Dans un espace normé la fonctionnelle $f(x) = \|x\|$ est convexe, donc faiblement semicontinue inférieurement en vertu du théorème 1.10.

Exemple 1.2. Soit A un opérateur borné autoadjoint positif dans un espace de Hilbert réel. Alors

$$A x = \frac{1}{2} \operatorname{grad} (A x , x)$$

et comme A est monotone, $(A x - A y , x - y) \geqslant 0$, la fonctionnelle $f(x) = (A x , x)$ est faiblement semicontinue inférieurement grâce au théorème 1.4.

M. M. Vainberg

Exemple 1.3. Dans un espace de Hilbert réel on donne un opérateur borné autoadjoint T dont le spectre a une partie positive arbitraire tandis que la partie negative est fermée d'un nombre fini d'autovaleurs, chacun de multiplicité finie. Alors la fonctionnelle f(x) = (T x , x) est faiblement semicontinue inférieurement.

Exemple 1.4. (Cfr. M.M. Vainberg [2 , 3]). La fonction $g(u,x)$ soit la génératrice d'un opérateur de Nemytzkii de l'espace de Lebesgue L^p dans l'espace $L^q(p > 1,\ p^{-1} + q^{-1} = 1)$, c'est-à-dire

$$|g(u,x)| \leqslant a(x) + b|u|^{p-1}, \quad a(x) \in L^q, \quad b > 0$$

et $g(u,x)$ ne décroit pas par rapport à u pour presque tous les x ∈ G , G étant un ensemble mésurable de l'espace euclidien Alors l'opérateur potentiel de Nemytzkii

$$f(u) = \int_G d\,y \int_0^{u(y)} g(v,y)\,d\,v$$

est faiblement semi-continu inférieurement dans l'espace $L^p(G)$. En effet grad f(u) = g(u(x),x) = h u et comme g(u,x) ne décroit pas par rapport à u pour presque tous les x ∈ G , on a

$$\langle h\,u - h\,v,\ u - v\rangle = \int_G \big[g(u(x),x) - g(v(x),x)\big]\big[u(x) - v(x)\big]\,dx \geqslant 0$$

donc le gradient de la fonctionnelle f(u) est un opérateur monotone et pourtant la fonctionnelle f(u) est faiblement semicontinue inférieurement dans L^p grâce au théorème 1.4.

De plus, soit T un opérateur linéaire borné de L^2 dans L^p .

M. M. Vainberg

Pourtant il transforme chque suite $u_n \to u_o (u_n, u_o \in L^2)$ dans u-

ne suite $T u_n \to T u_o (T u_n, T u_o \in L^p)$ et pourtant la fonctio̲n

nelle

$$\varphi(u) = f(T u) = \int_G d y \int_0^{Tu} g(v,y) d v$$

est faiblement semicontinue dans l'espace $L^2(G)$.

Exemple 1.5. Soit $g(u,x)$ la fonction génératrice d'un opérateur

de Nemytzkii de Lebesgue $L(G)$ en soi, c'est-à-dire (Cfr. M.M.

Vainberg [2] , Teorema 19.1)

$$|g(u,x)| \leqslant a(x) + b|u| , a(x) \in L(G) , b > 0$$

et $g(u,x)$ est convexe en u pour presque tous les $x \in G$.

Alors la fonctionnelle

$$\psi(u) = \int_G g(u(x),x) dx$$

est faiblement semicontinue inférieurement dans L étant une fo̲n

ctionnelle convexe. Il faut remarquer que la fonctionnelle $\Upsilon(u)$

est continue en L car l'opérateur de Nemytzkii $h u = g(u(x),x)$

est continu.

Exemple 1.6. Soit B un opérateur linéaire borné positif de l'e̲

space de Banach réel réflexif E dans le dual E^*. Considérons

la fonctionnelle quadratique

$$f(x) = \langle B x , x \rangle .$$

On a $\operatorname{grad} f(x) = B x + B^* x$, donc

$$\langle \operatorname{grad} f(x) - \operatorname{grad} f(y), x - y \rangle = 2 \langle B (x - y), x - y \rangle \geqslant 0$$

M. M. Vainberg

Donc grad f(x) est un opérateur monotone et en vertu du théorè

me 1.4 la fonctionnelle quadratique $\langle B x , x \rangle$ est faiblement se

micontinue inférieurement. Il faut remarquer que l'opérateur

$B + B^*$ est un potentiel et comme il est monotone son potentiel,

c'est-à-dire $\langle B x, x \rangle$ est une fonctionnelle convexe en vertu du

Corollaire 1.2.

Exemple 1.7. Soit F(x) un opérateur potentiel monotone, positi

vement homogène de dégré k > 0 , donné sur l'espace linéaire

E . Alors son potentiel $\varphi(x) = \frac{1}{k+1} \langle F(x),x \rangle$ est une fonction-

nelle convexe et faiblement semicontinue inférieurement.

Si F(x) est un opérateur dissipatif (c'est-à-dire (-1) F(x) est

un opérateur monotone) alors $\varphi(x)$ est une fonctionnelle faible-

ment semicontinue supérieurement.

Enfin si F(x) est un opérateur compact donné sur un espace nor-

mé, alors $\varphi(x)$ est une fonctionnelle faiblement continue.

1.7 Functionnelles semi-convexes.

Soit ω un ensemble convexe d'un espace normé.

Définition 1.7. La fonctionnelle réelle f(x) donnée sur l'ensem

ble convexe ω est dite semi-convexe, s'il existe une fonctionnel

le g(x) faiblement continue sur ω telle que f(x) + g(x) soit

convexe sur ω .

La fonctionnelle considerée dans l'exemple 1.3 est semi-convexe.

Evidéntement chaque fonctionnelle convexe est semi-convexe, mais

M. M. Vainberg

le contraire n'est pas vrai comme le montre l'exemple 1.3.

En outre si $f(x)$ est une fonctionnelle semi-convexe et $g(x)$ est une fonctionnelle faiblement continue quelconque alors $f(x)+$ $+ g(x)$ est semi-convexe.

Il faut remarquer enfin qu'une fonctionnelle finie semi-convexe $f(x)$ est faiblement semi-continue inférieurement. En effet selon la définition il existe une fonctionnelle faiblement continue $g(x)$ telle que $f(x) + g(x)$ est convexe et pourtant (Cfr. Théorème 1.10) faiblement semicontinue inférieurement. Par conséquent

$$f(x) = \left[f(x) + g(x) \right] + \left[- g(x) \right]$$

est faiblement semicontinue inférieurement étant la somme de deux telles fonctionnelles.

RÉFÉRENCES

N. Dunford - J. Schwartz [I] ; E. Hille - R. S. Phillips [I] ;
J. Neumann [I] ; M.M. Vainberg [2, 3, 8, 9] .

M. M. Vainberg

II. THÉORÈMES D'EXISTENCE ET D'UNICITÉ DU MINIMUM.

On considère ici essentiellement des fonctionnelles réelles données sur un espace de Banach réel réflexif. Rappelons que dans
un tel espace chaque boule fermée est faiblement compacte.

2.1 Points d'extremum des fonctionnelles et théorème de Weierstrass
généralisé.

Soit $f(x)$ une fonctionnelle réelle donnée sur un espace normé
E . Le point $x_o \in$ E est dit point d' e x t r e m u m
d e l a fonctionnelle $f(x)$ si dans un certain voisinage
$U(x_o)$ de ce point une des conditions suivantes est satisfaite

$$\text{I) } f(x) \leq f(x_o) \; ; \; 2) \; f(x) \geq f(x_o) \; .$$

Si la deuxième inégalité est satisfaite pour $x \in$ E quelconque
alors x_o est dit point de minimum absolu de la fonctionnelle
$f(x)$.

Si en outre $f(x)$ est différentiable au sens de Gateaux au point
x_o on dit que x_o est un point critique pour $f(x)$ si la condition $D f(x_o,h) = 0$ est satisfaite.

De cette égalité entre la différentielle et zéro il suit la continuité de la différentielle même, donc l'égalité précedente peut
s'écrire

$$\langle \text{grad } f(x_o) \, , \, h \rangle = 0$$

et comme celle-ci est satisfaite pour n'importe quel $h \in$ E on

M. M. Vainberg

peut dire que x_o est un point critique de $f(x)$ si

$$\text{grad } f(x_o) = 0 .$$

Théorème 2.1. Soit $f(x)$ une fonctionnelle donnée dans un domaine ω de l'espace normé E et soit x_o un point intérieur au domaine ω où il existe une différentielle de Gateaux linéaire. Alors on a le propriétés suivantes: I. Afin que x_o soit un point d'extremum il est nécessaire qu'il soit un point critique, c'est-à-dire

$$\text{grad } f(x_o) = 0 \qquad\qquad (2.1)$$

2. Si de plus la fonctionnelle $f(x)$ est convexe dans un certain voisinage convexe $U(x_o) \subset \omega$ (ou grad $f(x)$ est un opérateur monotone) alors l'égalité (2.1) est nécessaire et suffisante afin que x_o soit un point de minimum.

Démonstration. Soit h un vecteur quelconque fixé de l'espace E. Alors $f(x_o + th)$ est une fonction réelle définie dans un certain voisinage du point $t = 0$ et elle a une dérivée au point $t = 0$. D'ici si x_o est un point d'extremum pour la fonctionnelle $f(x)$, $t = 0$ est un point d'extremum pour la fonction $f(x_o+th)$ et pourtant

$$\frac{d}{dt} f(x_o + th)\Big|_{t=0} = 0$$

donc $D f(x_o,h) = 0$, ou $\langle \text{grad } f(x_o),h \rangle = 0$. D'ici, grâce à l'arbitrariété de h on a grad $f(x_o) = 0$. La première affirma

M. M. Vainberg

tion est démontrée. Pour prouver la seconde considérons la fonctionnelle $\varphi(z) = f(z + x_0) - f(x_0)$. Pour la convexité il existe la différentielle de Gateaux au point $z = 0$ où l'on a $\varphi(0) = 0$. Si l'on suppose que x_0 soit un point de minimum on a grad $\varphi(0) = 0$ de ce qu'on a déjà démontré. Supposons que cette égalité, qui coincide avec (2.1), ait lieu et montrons que $z = 0$ est un point de minimum de la fonctionnelle $\varphi(z)$, d'où il suit que x_0 est un point de minimum de $f(x)$.

La demonstration se fait par contrédiction. Supposons que pour un certain vecteur $z_1 \in (U(x_0) \setminus x_0)$ on ait $\varphi(z_1) = \alpha < 0$.

Alors, grâce à la convexité

$$\varphi(t z_1 + (1 - t) 0) \leqslant t \varphi(z_1) = t\alpha, \quad (0 < t < 1)$$

Par conséquent

$$[\varphi(0 + t z_1) - \varphi(0)]/t \leqslant \alpha$$

et pourtant

$$\langle \text{grad } \varphi(0), z_1 \rangle = \lim_{t \to 0} \frac{\varphi(0 + t z_1) - \varphi(0)}{t} \leqslant \alpha < 0$$

c'est-à-dire grad $\varphi(0) \neq 0$. La contrédiction démontre le théorème.

Démonstrons encore la proposition suivante (Cfr. M.M. Vainberg [2], Teorema 9.2):

Théorème 2.2. (Premier théorème de Weierstrass généralisé).

Si dans un ensemble borné faiblement fermé σ d'un espace de Banach réflexif E on donne une fonctionnelle faiblement semi-con-

M. M. Vainberg

tinue inférieurement, elle est inférieurement bornée et atteint

son minimum dans σ .

On montre que dans ce théorème on peut supprimer l'hypothèse de

réflexivité en la remplaçant avec l'hypothèse que E soit le dual

d'un espace normé séparable. Pour cela il faut **rappeler** quelques

notions bien connues de l'analyse fonctionnelle (Cfr. E. Hille-

R. Phillips [1] , 2.10) .

Soit E un espace normé et soit E^* son dual. On dit que la sui

te de fonctionnelles $\{\varphi_n\} \subset E^*$ converge E-faiblement vers

$\varphi \in E^*$ (pour cette convergence on écrira $\varphi_n \xrightarrow{E} \varphi$) , si pour

chaque vecteur $x \in E$ on a $\lim_{n \to \infty} \varphi_n(x) = \varphi(x)$. On voit que

par rapport à ce type de convergence faible l'espace E^* est fai

blement complet et, si E est séparable, chaque sous-ensemble

borné de E est faiblement compact.

On dit que la fonctionnelle non lineaire f( ) donnee sur E^*

est E-faiblement semi-continue inferieurement au point φ_o si

pour chaque suite $\varphi_n \xrightarrow{E} \varphi_o$ on a l'inégalité

$$f(\varphi_o) \leqslant \varliminf_{n \to \infty} f(\varphi_n) .$$

L'ensemble $\sigma \subset E^*$ est dit E-faiblement fermé si pour chaque

suite $\{\varphi_n\} \subset E^*$ qui converge E-faiblement vers φ_o on a $\varphi_o \in \sigma$.

Un exemple d'ensemble E-faiblement fermé est donné par la boule

$\|\varphi\| \leqslant a$ de l'espace E^* .

Il faut remarquer que de la semi-continuité inférieure faible

M. M. Vainberg

d'une fonctionnelle il suit sa semi-continuité inférieure E-faible, car de la convergence faible d'une suite d'éléments il suit la convergence E-faible de la même suite.

Théorème 2.3. (Deuxième théorème de Weierstrass généralisé).

Soit E un espace normé séparable et soit σ un ensamble borné E-faiblement fermé de l'espace dual E^*. Alors chaque fonctionnelle $f(\varphi)$ donnée dans σ et E-faiblement semicontinue inférieurement est inférieurement bornée dans σ et atteint le minimum sur σ.

Démonstration. Si l'on suppose que $f(\varphi)$ ne soit pas inférieurement bornée sur σ alors il existe une suite $\{\varphi_n\} \subset \sigma$ telle que $f(\varphi_n) < -n$. Comme par hypothèse $\|\varphi_n\| \leq c =$ constant, alors grâce à la compacité E-faible de chaque ensemble borné de E^*, il existe une sous-suite $\varphi_{n_k} \xrightarrow{E} \varphi_0$ et comme est E-faiblement fermé on a $\varphi_0 \in \sigma$. De la semi-continuité inférieure E-faible on a

$$f(\varphi_0) \leq \varliminf_{k \to \infty} f(\varphi_{n_k}) = -\infty$$

ce qui n'est pas possible. Par conséquent il existe

$$d = \inf_{\varphi \in \sigma} f(\varphi), \quad d > -\infty \tag{2.2}$$

Soit alors $\{\varphi_n\} \subset \sigma$ une suite minimisante, c'est-à-dire

$$\lim_{n \to \infty} f(\varphi_n) = d.$$

En vertu du fait que σ est faiblement fermé et E-faiblement compact on peut extraire de $\{\varphi_n\}$ une sous-suite $\{\varphi_{n_k}\}$ telle que $\varphi_{n_k} \xrightarrow{E} \varphi_0$. D'ici en vertu de la semi-continuité E-faible de $f(\varphi)$

M. M. Vainberg

sur σ on a

$$f(\varphi_o) \leqslant \lim_{k \to \infty} f(\varphi_{n_k}) = \lim_{n \to \infty} f(\varphi_n) = d$$

donc en vertu de (2.2) $f(\varphi_o) = d$. Le théorème est démontré.

Remarque 2.1. On note que des conditions des théorèmes 2.2 et 2.3
on peut supprimer l'hypothèse que l'ensemble σ soit borné si la
fonctionnelle f satisfait à la condition

$$\lim_{R \to \infty} f(z) = + \infty \qquad\qquad (R = \| z \|).$$

Cette remarque suit immédiatement de la démonstration des théorèmes 2.2 et 2.3.

2   Sur la m-propriété des fonctionnelles.

Soit ω un ensemble convexe borné et ouvert de l'espace de Ba-
bach E , ω' sa frontière et $\overline{\omega} = \omega \cup \omega'$. L'ensemble est faiblement fermé. Nous considérons tous les ensembles possibles $\overline{\omega}$
sur lesquels la fonctionnelle φ (x) est faiblement semi-continue inférieurement.

Definition 2.1. Si parmi les divers ensembles faiblement fermés
$\overline{\omega}$ sur lesquels la fonctionnelle $\varphi(x)$ est faiblement semicontinue inférieurement il en existe un au moins $\overline{\omega}_o$ tel que sur
ω'_o l'inégalité $\varphi(x) > \varphi(x_o)$ est satisfaite, où x_o est un point
quelconque de ω_o, alors on dit que la fonctionnelle $\varphi(x)$ a la
m-propriété (propriété de minimum).

En vertu du théorème de Weierstrass généralisé, si la fonctionnel

M. M. Vainberg

le $\varphi(x)$ a la m-propriété, alors elle atteint le minimum sur $\overline{\omega}_o$

et le points de minimum ne peuvent pas appartenir à ω_o, donc

ils appartiennent à ω_o. Pour cela si la fonctionnelle $\varphi(x)$ est

différentiable au sens de Gateaux on a la proposition

Théorème 2.4. Si la fonctionnelle $\varphi(x)$ donnée sur un espace de

Banach réflexif, différentiable au sens de Gateaux, a la m-pro-

priété, alors il existe au moins un point critique pour cette fon

ctionnelle, c'est-à-dire grad $\varphi(x_o) = 0$.

Le théorème subsiste si l'on remplace la condition de réflexivi-

té avec la condition que $E = X^*$, où X est normé et séparable.

Corollaire 2.1. Si dans les conditions du théorème 2.4 on rempla

ce la semi-continuité inférieure faible de la fonctionnelle $\varphi(x)$

par sa convexité stricte alors le point de minimum x_o est uni-

que.

En effet, en vertu du théorème 1.10 de la convexité de $\varphi(x)$ il

suit sa semi-continuité inférieure faible et pourtant elle admet

au moins un point de minimum.

Mais comme $\varphi(x)$ est fonctionnelle strictement convexe elle n'a

dmet plus qu'un point de minimum car si x_1 , x_2 étaient deux

points de minimum différents en posant $d_1 = \varphi(x_1)$, $d_2 = \varphi(x_2)$

avec $d_1 \leqslant d_2$ on aurait de la convexité stricte de $\varphi(x)$

$$\varphi(\lambda x_1 + (1 - \lambda)x_2) < \lambda d_1 + (1 - \lambda)d_2 \leqslant d_2 , \quad (0 < \lambda < 1) .$$

Toutefois pour un voisinage quelconque $U(x_2)$ du vecteur x_2 il

M. M. Vainberg

existe un λ suffisamment petit tel que le vecteur $\lambda x_1 +$

$+ (1-\lambda)x_2 = z \in U(x_2)$. D'ici et de ce qu'on a vu auparavant il

suit l'inégalité

$$\varphi(z) < \varphi(x_2) = d_2 \quad , \quad z \in U(x_2)$$

qui contrédit au fait que x_2 est un point de minimum pour $\varphi(x)$.

Au n° suivant on donnera des conditions suffisantes assurant la

m-propriété des fonctionnelles. Ces conditions donnent plusieurs

théorèmes d'existence et d'unicité du minimum des fonctionnelles.

2.3 Critères suffisants pour le minimum sans contraintes.

Démontrons préliminairement la suivante simple proposition.

Lemma 2.1. Si la fonctionnelle $f(x)$ strictement convexe sur

l'espace linéaire E atteint son minimum au point x_0 alors x_0

est un point de minimum absolu et il n'y pas d'autres points de

minimum car la fonction $f(x_0 + t x)$ croit avec t $(t > 0)$ pour

chaque $x \neq 0$.

Démonstration. Posons

$$\varphi(z) = f(x_0 + z) - f(x_0) .$$

La fonctionnelle $\varphi(z)$ est convexe et par hypothése elle atteint

son minimum, égal zéro, à l'origine de l'espace. Montrons que à

chaque point $z \neq 0$ on a $\varphi(z) > 0$. D'ici il suit que si $x \neq x_0$

alors $f(x) > f(x_0)$.

Remarquons d'abord que pour démontrer 2.1 il faut établir que si

une fonctionnelle strictement convexe admet le minimum ce minimum est stricte et il n'existe pas d'autres points de minimum. Pour

cela on considère un voisinage U de l'origine tel que pour

$x \in U$ et $x \neq 0$ on a $\varphi(x) > \varphi(0) = 0$. Soit maintenant $x \in E$

quelconque différent de zéro. Pour λ positif assez petit le point

$\lambda x \in U$. Mais pour la convexité stricte de $\varphi(x)$ on a

$$\varphi(\lambda x + (1 - \lambda) 0) < \lambda \varphi(x) \quad , \quad (0 < \lambda < 1) .$$

Comme $\lambda x \in U$ alors

$$0 = \varphi(0) < \varphi(\lambda x) < \lambda \varphi(x) < \varphi(x) .$$

Cela démontre la prémière partie du lemma. Pour démontrer la se-

conde on considère la fonction réelle d'une variable réelle

$$\Psi(t) = f(x_0 + t x) - f(x_0) \quad , \quad x \neq 0 , \ t \geqslant 0 \quad .$$

Cette fonction est strictement convexe et pourtant $[\Psi(t) - \Psi(0)]/t$

est une fonction croissante de t , c'est-à-dire

$$\frac{\Psi(t_2)}{t_2} > \frac{\Psi(t_1)}{t_1} \quad , \quad t_1 < t_2 .$$

D'ici, comme on a démontré que $\Psi(t) > 0$ pour $t > 0$, on a

$\Psi(t_2) > \Psi(t_1)$ et pourtant $f(x_0 + t x)$ est une fonction crois-

sante de $t \geqslant 0$. Le lemma est démontré.

Théorème 2.5. Dans l'espace de Banach réflexif E soit donné la

fonctionnelle $f(x)$ finie faiblement semicontinue inférieurement

qui satisfait à la condition

$$\overline{\lim_{R \to \infty}} \ f(x) = + \infty \quad (R = \|x\|) \tag{2.3}$$

Alors il existe un point de minimum sans contraintes de $f(x)$.

M. M. Vainberg

Démonstration. De la condition (2.3) on tire que pour chaque ve-

cteur x_o il existe une surface sphétique $S_r = \left\{ x \in E : \| x \| = \right.$

$\left. = r > 0 \right\}$ sur laquelle $f(x) > f(x_o)$. Comme la boule $D_r =$

$= \left\{ x \in E : \| x \| \leqslant r \right\}$ est faiblement fermée et $f(x)$ est faiblement semi-continue inférieurement

elle a la m-proprieté, c'est-à-dire elle admet un minimum local à

un certain point intérieur à la boule D_r. Le théorème est démon-

tré.

On note que si dans (2.3) on remplace la limite supérieure par la

limite alors le point x^* de minimum local coincide avec un point

de minimum absolu de $f(x)$ dans tout l'espace, c'est-à-dire

$f(x^*) \leqslant f(x)$ pour $x \in E$ quelconque, car dans ce cas on a $f(x) >$

$> f(x_o)$ non seulement pour $\| x \| = r$ mais pour $\| x \| > r$ aussi.

On note encore que au lieu de la réflexivité de E on peut suppo-

ser $E = X^*$ avec X séparable.

Théorème 2.6. La fonctionnelle réelle $f(x)$ donnée dans un espa-

ce de Banach réel réflexif E soit différentiable au sens de Ga-

teaux et grad $f(x) = F(x)$ satisfasse aux conditions:

1. la fonction $\langle F(tx), x \rangle$ soit continue en t sur $[0,1]$ pour

chaque $x \in E$;

2. $\langle F(x+h) - F(x), h \rangle \geqslant 0$, $(\forall x, h \in E)$;

3. $\lim\limits_{R \to \infty} \dfrac{\langle F(x), x \rangle}{R} = + \infty$ $(R = \| x \|)$.

Alors il existe un point de minimum x_o pour la fonctionnelle

M. M. Vainberg

$f(x)$ et grad $f(x_o) = 0$. Si dans la condition 2. l'égalité a

lieu seulement pour $h = 0$, c'est-à-dire si $F(x)$ est un opé-

rateur strictement monotone, alors le point de minimum de la fon

ctionnelle est unique et à ce point $f(x)$ atteint le minimum

absolu.

Démonstration. De la condition 2. en vertu du théorème 1.4 il suit

que la fonctionnelle $f(x)$ est faiblement semicontinue inférieu

rement. En outre de la formule connue (Cfr. M.M. Vainberg [2] ,

Teorema 5.1)

$$f(x) = f(x_o) + \int_0^1 \langle F(x_o + t(x - x_o)) , x - x_o \rangle \, dt$$

on a

$$f(x) = f(0) + \int_0^1 \langle F(tx), x \rangle \, dt$$

où l'intégrale existe grâce à la condition I du théorème.

D'ici pour la condition 2 on a

$$f(x) - f(0) = \langle F(0), x \rangle + \int_0^1 \langle F(tx) - F(0), x \rangle \, dt \geqslant$$

$$\geqslant \langle F(0), x \rangle + \int_{1/2}^1 \langle F(tx) - F(0), x \rangle \, dt =$$

$$= \frac{1}{2} \langle F(0), x \rangle + \frac{1}{2} \langle F(t_1 x), x \rangle$$

où $1/2 \leqslant t_1 \leqslant 1$. D'ici

$$f(x) - f(0) \geqslant \frac{1}{2} \|x\| \left(\frac{\langle F(t_1 x), t_1 x \rangle}{\|t_1 x\|} - \|F(0)\| \right)$$

car en vertu de la condition 3 on a l'égalité (2.3). Pourtant les

M. M. Vainberg

conditions du théorème 2.5 sont vérifiées, donc dans un certain

point x_o intérieur à la boule D_r la fonctionnelle $f(x)$ admet

un minimum local. Dans ce point on a grad $f(x_o) = 0$ car $f(x)$

est différentiable au sens de Gateaux. Finalement, si $F(x)$ est

un opérateur strictement monotone, alors $f(x)$ est une fonction

nelle strictement convexe et pourtant en vertu du théorème 2.1

et du lemma 2.1 x_o est le seul point de minimum où $f(x)$ atteint

le minimum absolu. Le théorème est démontré.

Remarque 2.2. Il faut remarquer que l'unicité du minimum de $f(x)$

lorsque grad $f(x) = F(x)$ est un opérateur strictement monotone

est immédiate. En effet si $f(x)$ avait le minimum dans deux

points différente x_1, $x_2 \in E$, alors $F(x_1) = 0$ et $F(x_2)=0$,

donc $\langle F(x_1) - F(x_2), x_1 - x_2 \rangle = 0$. Cette égalité contrédit la

condition de monotonicité stricte de $F(x)$.

Il faut aussi remarquer que si dans la condition 3 du théorème

on remplace la limite supérieure par la limite alors x_o est un

point de minimum absolu pour $f(x)$ dans tout l'espace (Cfr. la

remarque au théorème 2.5).

Notons enfin que l'hypothèse de la réflexivité de l'espace E

peut être remplacée par l'hypothèse que $E = X^*$ où X est un

espace normé séparable.

Théorème 2.7. L'opérateur potentiel monotone $F(x)$ donné dans

un espace de Banach réflexif E satisfasse aux conditions:

M. M. Vainberg

I. $\langle F(tx), x \rangle$ est intégrable par rapport à t sur $(0,1)$ pour chaque $x \in E$;

2. $\langle F(x), x \rangle \geqslant \|x\| \gamma (\|x\|)$, où $\gamma(t)$ est une fonction intégrable sur $[0,R]$ pour chaque $R > 0$;

3. On a

$$\overline{\lim_{R \to \infty}} \int_0^R \gamma(t) \, dt = c > 0 \ .$$

Alors le potentiel $f(x)$ de l'opérateur $F(x)$ admet un point de minimum. Ce point de minimum est unique et $f(x)$ y atteint le minimu absolu lorsque F est un opérateur strictement monotone.

Démonstration. On trouve comme auparavant que $f(x)$ est une fonctionnelle faiblement semicontinue inférieurement qui satisfait à la condition

$$f(x) - f(0) \geqslant \int_0^1 \|x\| \gamma(t \|x\|) \, dt = \int_0^{\|x\|} \gamma(z) \, dz$$

D'ici, posant $\|x\| = R$ on a

$$\overline{\lim_{R \to \infty}} f(x) \geqslant f(0) + \overline{\lim_{R \to \infty}} \int_0^R \gamma(z) \, dz = f(0) + c$$

donc pour un certain $r = R$ sur la sphère $\|x\| = r$ on a $f(x) > f(0)$, c'est-à-dire $f(x)$ a la m-propriété. Par conséquent $f(x)$ a un minimum à un certain point x_0 intérieur à la boule $\|x\| \leqslant r$ et pourtant $\operatorname{grad} f(x_0) = 0$. L'unicité du minimum et le fait que x_0 est un point de minimum absolu se démontrent comme dans la démonstration du théorème 2.6.

M. M. Vainberg

Remarquons que la condition de la réflexivité de E peut être remplacée par l'hypothèse que $E = X^*$ où X est un espace normé séparable.

Pour la suite il nous sera utile le suivant

Lemma 2.2. Soit $f(x)$ une fonctionnelle réelle différentiable au sens de Gateaux, donnée sur un espace réel normé E . Alors si grad $f(x) = F(x)$ satisfait sur la sphère $S = \{ x \in E : \|x\| = R > 0$ à la condition

$$\langle F(x), x \rangle > 0 \qquad (2.4)$$

sur cette sphère S il n'existe pas de vecteurs z tels que l'on ait l'égalité

$$\min_{\|x\| \leqslant R} f(x) = f(z)$$

si ce minimum existe.

La démonstration s'obtient facilement par contrédiction. Supposons que $\min_{\|x\| \leqslant R} f(x) = f(x_o)$, $\|x_o\| = R$. Alors la fonction réelle $\varphi(t) = f(x_o + t(-x_o))$ pour t suffisamment petit satisfait à la condition $\varphi(t) \geqslant \varphi(0)$ et pourtant $\varphi'_+(0) \geqslant 0$. Mais $\varphi'_+(0) =$

$$= \lim_{t \to +0} \frac{f(x_o + t(-x_o)) - f(x_o)}{t} = \langle F(x_o), -x_o \rangle$$

donc

$$\varphi'_+(0) = - \langle F(x_o), x_o \rangle < 0 \quad .$$

La contrédiction démontre le lemme.

M. M. Vainberg

Lemma 2.3. Dans un espace normé réel E soit donnée la fonctionnelle réelle f(x) differentiable au sens de Gateaux et soit

F(x) = grad f(x). Alors si sur la sphère $S_a = \left\{ x \in E : \| x - a \| = R > 0 \right\}$ la condition

$$\langle F(x),\ x - a \rangle > 0$$

est satisfaite, il n'y a pas sur S_a de vecteurs z tels que

$$\min_{\| x-a \| \leqslant R} f(x) = f(z)$$

si le minimum existe.

Pour la démonstration il est suffisant de considérer la fonctionnelle $\Psi(t) = f(z + th)$, h = a - z , et de repeter la démonstration du lemma précedente.

Théorème 2.8. La fonctionnelle réelle f(x) donnée sur un espace
de Banach réel réflexif E soit différentiable au sens de Gateaux,
faiblement semi-continue inférieurement et telle que

$$\langle F(x),\ x \rangle > 0 \qquad\qquad (2.4)$$

pour chaque vecteur x ∈ E tel que ∥x∥ = R > 0 , où R est un
nombre positif quelconque et F(x) = grad f(x) .

Alors il existe un point x_o intérieur à la boule ∥x∥⩽ R auquel
f(x) a un minimum local et par conséquent grad $f(x_o)$ = 0 .

Démonstration. Considérons la boule $D_R = \left\{ x \in E : \| x \| \leqslant R,\ R > 0 \right\}$.
Elle est faiblement fermé à cause de la réflexivité de E et
f(x) est faiblement semi-continue inférieurement sur elle.

Pourtant en vertu du Théorème 2.2 f(x) a un minimum sur D_R et

M. M. Vainberg

à cause du lemma 2.2. le point de minimum est intérieur à la bou-

le D_R. Le théorème est démontré.

Remarque 2.3. Dans les conditions du théorème 2.8 on peut rempla-

cer l'hypothèse de réflexivité de E par l'hypothèse $E = X^*$, où

X est un espace normé séparable. Dans ce cas pour la démonstra-

tion il faut se servir du théorème 2.3 au lieu du théorème 2.2.

Il faut aussi remarquer que en vertu du lemma 2.3 l'inégalité

(2.4) peut être remplacée par

$$\langle F(x), \ x - a \rangle > 0 \qquad\qquad (2.4')$$

sur la sphère $\|x - a\| = R > 0$. Alors il existe un point x_o inté-

rieur à la boule $\|x - a\| \leqslant R$ auquel $f(x)$ a un minimum local et

$F(x_o) = 0$.

Théorème 2.9. L'opérateur $F(x) = \mathrm{grad}\, f(x)$ strictement monotone

donné dans un espace de Banach E réflexif (ou bien dans un espa-

ce E qui est le dual d'un espace normé séparable) vérifie les

conditions suivantes: pour chaque $x \in E$ la fonction $\langle F(tx), x \rangle$

soit intégrable par rapport à t dans $[0,1]$ et

$$\langle F(x) - F(0), x \rangle \geqslant \|x\| \gamma(\|x\|)$$

où la fonction $\gamma(z)$ est intégrable dans $[0,R]$ pour chaque $R > 0$

et la fonction

$$c(R) = \int_0^R \gamma(z)\,dz$$

satisfait à l'inégalité $c(R) > R\,\|F(0)\|$ pour un certain R .

Alors la fonctionnelle $f(x)$ a un seul point de minimum où elle atteint le minimum absolu.

<u>Demonstration</u>. De la monotonicité stricte de $F(x)$ il suit la convexité stricte et la semi-continuité inférieure faible de la fonctionnelle $f(x)$. On a aussi, des conditions du théorème

$$f(x) - f(0) = \langle F(0), x \rangle + \int_0^1 \langle F(tx) - F(0), x \rangle \, dt \geqslant$$

$$\geqslant \langle F(0), x \rangle + \int_0^1 \|x\| \, \gamma(t\|x\|) dt \geqslant$$

$$\geqslant R \left(\frac{1}{R} \int_0^R \gamma(z) dz - \| F(0) \| \right)$$

où $R = \|x\|$. Pourtant pour un certain R_0

$$f(x) > f(0) \qquad (\|x\| = R_0) \tag{2.5}$$

Comme pour les hypothèses du théorème la boule $\|x\| \leqslant R_0$ est faiblement fermée et la fonctionnelle $f(x)$ est faiblement semicontinue inférieurement sur elle, alors du théorème 2.2 (ou 2.3) $f(x)$ a un point de minimum dans cette boule et pour l'inégalité (2.5) ce point de minimum est un point intérieur de la boule $\|x\| \leqslant R_0$. Enfin, pour la convexité stricte de $f(x)$ ce point de minimum est unique (Cfr. Corollaire 2.1 et Lemma 2.1) et le minimum de $f(x)$ est absolu. Le théorème est démontré.

<u>Théorème</u> 2.10. La fonctionnelle $f(x)$ deux fois différentiable au sens de Gateaux, donnée sur un espace de Banach E réflexif (où dans l'espace dual d'un espace normé séparable) satisfasse aux conditions suivantes:

M. M. Vainberg

1. $D^2f(y + tx ; x , x)$ comme fonction de t soit intégrable sur $[0,1]$ pour chaque $x, y \in E$;

2. $D^2f(x ; h, h) \geqslant \gamma(\|x\|) \|h\|^2$ pour chaque $x, h \in E$ où $\gamma(z)$ est un fonction positive de croissante intégrable sur $[0,R]$ pour $R > 0$ arbitraire, et l'on ait pour un certain $R > 0$

$$\frac{1}{R} \alpha(R) = \frac{1}{R} \int_0^R z \, \gamma(z) \, dz > \|F(0)\| \qquad (2.6)$$

où $F(x) = \text{grad } f(x)$.

Alors il existe un seul point de minimum de la fonctionnelle $f(x)$ et le minimum est absolu.

Démonstration. Comme $F(x) = \text{grad } f(x)$ alors

$$D f(x,h) = \langle F(x), h \rangle , \quad D^2f(x; h,h) = \langle D F(x,h), h \rangle$$

et procédant comme dans la démonstration du théorème 9.4 de M.M. Vainberg [2] on obtient

$$D f(x,h) = D f(0,h) + \int_0^1 D^2f(tx ; h,x) \, dt$$

ou

$$\langle F(x), h \rangle = \langle F(0), h \rangle + \int_0^1 \langle D F(tx,x), h \rangle \, dt \quad .$$

Donc

$$\langle F(x),x \rangle \geqslant \langle F(0), x \rangle + \int_0^1 \|x\|^2 \gamma(u\|x\|) \, du$$

et

M. M. Vainberg

$$f(x) - f(0) = \int_0^1 \langle F(tx) , tx \rangle t^{-1} dt \geqslant$$

$$\geqslant \int_0^1 \left[\langle F(0), tx \rangle + \int_0^1 \gamma (ut \|x\|) t^2 \|x\|^2 du \right] t^{-1} dt \geqslant$$

$$\geqslant \langle F(0), x \rangle + \int_0^1 t \gamma(t\|x\|)\|x\|^2 dt \geqslant$$

$$\geqslant - \|x\|\|F(0)\| + \int_0^{\|x\|} z \gamma(z) dz .$$

Posant $R = \|x\|$ on obtient

$$f(x) - f(0) \geqslant R \left(\frac{1}{R} \int_0^R z \gamma(z) dz - \|F(0)\| \right) \qquad (2.7)$$

D'ici en vertu de la condition du théorème relative à un certain

$R > 0$ on déduit l'inégalité $f(x) - f(0) > 0$, c'est-à-dire $f(x)$

a la m-propriété et pourtant il existe un point de minimum x_0

de la fonctionnelle $f(x)$. Comme des conditions du théorème on

déduit que pour $x \neq y$

$$\langle F(x) - F(y), x - y \rangle = \langle D F(y + t(x - y), x - y \rangle =$$

$$= D^2 f(y + t(x - y) ; x - y , x - y) \geqslant \|x - y\|^2 \gamma (\|y + t(x-y)\|) > 0$$

on a que $F(x)$ est un opérateur strictement monotone et pourtant

x_0 est le seul point de minimum absolu de la fonctionnelle $f(x)$

Le théorème est démontré.

Notons que la condition (2.6) est satisfaite si l'on a

$$\overline{\lim_{R \to + \infty}} \frac{1}{R} \alpha(R) = + \infty$$

M. M. Vainberg

4 Sur l'extremum conditionné.

Si la fonctionnelle réelle $f(x)$ faiblement semicontinue inférieurement, donnée sur un espace de Banach réflexif E n'a pas la m-propriété on ne peut pas affirmer que, quelle que soit la boule $D_r = \left\{ x \in E : \|x\| \leqslant r , r > 0 \right\}$ le point de minimum x_o de la fonctionnelle $f(x)$ sur D_r soit intérieur, c'est-à-dire que $\|x_o\| < < r$. Il peut arriver que $f(x)$ atteint le minimum sur la sphère $\|x_o\| = r$.

Par exemple si $F(x)$ est un opérateur potentiel compact positivement homogène de dégré $k > 0$, son potentiel

$$\varphi(x) = \frac{1}{k+1} \langle F(x), x \rangle , \quad F(tx) = t^k F(x)$$

a le minimum et le maximum sur la boule D_r . Mais si $\varphi(x)$ prend des valeurs négatives, alors $\varphi(x)$ en vertu de l'homogèneité atteint le minimum et le maximum sur la sphère $\|x\| = r$ pour chaque $r > 0$.

Si le minimum de $f(x)$ sur la boule D_r est atteint au point x_o de la sphère $S_r = \left\{ x \in E : \|x\| = r \right\}$ alors il existe un voisinage $U(x_o)$ du point x_o tel que

$$f(x_o) \leqslant f(x), \quad \forall x \in S_r \cap U(x_o) \tag{2.8}$$

Lorsque l'inégalité (2.8) est satisfaite on dira que x_o est un point de minimum conditionné de $f(x)$ rélatif à la sphère S_r . Cette définition s'étend à des cas plus generaux.

M. M. Vainberg

Soit E un espace vectoriel topologique réel sur lequel sont données deux fonctionnelles $f(x)$ et $\varphi(x)$. On note par $V_c(\varphi)$ la
variété $\varphi(x) = c = $ constant.

Définition 2.2. Le point $x_0 \in V_c(\varphi)$ est dit un point d'extremum conditionné de $f(x)$ rélativement à la variété $V_c(\varphi)$ s'il
existe un voisinage $U(x_0)$ du point x_0 tel que pour chaque
$x \in U(x_0) \cap V_c(\varphi)$ la différence $f(x) - f(x_0)$ ne change pas de
signe.

Supposons que les fonctionnelles $\varphi(x)$ et $f(x)$ ayent une différentielle de Gateaux linéaire continue au point x_0 et pourtant qu'ils existent grad $\varphi(x_0)$ et grad $f(x_0)$.

Définition 2.3. Le point $x_0 \in V_c(\varphi)$ est dit un point ordinaire
de la varieté $V_c(\varphi)$ si grad $\varphi(x_0) \neq 0$.

Définition 2.4. Le point $x_0 \in V_c(\varphi)$ est dit un point critique
conditionné de la fonctionnelle $f(x)$ rélativement à la variété
$V_c(\varphi)$ si

$$\text{grad } f(x_0) = \mu \text{ grad } \varphi(x_0)$$

pour un certain nombre μ .

2.5 Théorème fondamental sur l'extremum conditionné.

Supposons que E soit un espace vectoriel topologique réel avec
assez de fonctionnelles linéaires continues (comme par exemple
c'est le cas des espaces localement convexes).

On indicherd par $d\ f(x,h)$ et $d\varphi(x,h)$ respectivement les variations des fonctionnelles $f(x)$ et $\varphi(x)$, continues en h et x et on les appellera différentielles de Fréchet car dans le cas des espaces normés la différentielle de Gateaux continue en h est une différentielle de Fréchet.

Théorème 2.11(Cfr. M.M. Vainberg et Ia. L. Enghel'son [1]) .

Si le point ordinaire x_o de la variété $V_c(\varphi)$ est un point d'extremum conditionné de la fonctionnelle $f(x)$ relativement à la variété $V_c(\varphi)$, alors il existe λ tel que

$$\text{grad } f(x_o) = \lambda \text{ grad } \varphi(x_o) \qquad (2.9)$$

6 Exemples.

Exemple 2.1. Dans un espace de Banach réel réflexif E soyent données une fonctionnelle réelle $f(x)$ différentiable au sens de Fréchet faiblement semi-continue inférieurement et une fonctionnelle $\varphi(x)$ non négative strictement convexe différentiable au sens de Fréchet, avec la propriété que l'ensemble

$$D_c(\varphi) = \left\{ x \in E : \varphi(x) \leqslant c, \ c > 0 \right\}$$

est borné pour chaque c fini et, de plus, de l'égalité $\varphi(x)=0$ il suit $x = 0$. Un exemple d'une fonctionnelle de ce type est donné par $\varphi(x) = \|x\|$ si la norme dans E est différentiable au sens de Fréchet et la boule unité de l'espace est strictement convexe. Pour de telles fonctionnelles on a du théorème 2.1 et du lemma 2.1 grad $\varphi(x) \neq 0$ pour $x \neq 0$. Comme $\varphi(x)$ est une fonctionnelle strictement convexe, l'ensemble $D_c(\varphi)$ est strictement convexe et (pour la continuité de $\varphi(x)$) fermé. Par conséquent l'ensemble $D_c(\varphi)$ convexe et fermé dans un espace de Banach est faiblement fermé. Pour cela en vertu du théorème 2.2 il existe un vecteur $x_0 \in D_c(\varphi)$ tel que

$$f(x_0) = \min_{D_c(\varphi)} f(x) = m(c)$$

Montrons que si $m(c) \neq f(0)$ l'équation

$$\text{grad } f(x) = \mu \text{ grad } \varphi(x)$$

admet un continuum de solutions non nulles.

En effet si pour chaque $c > 0$ le point de minimum x_0 de la fonctionnelle $f(x)$ sur $D_c(\varphi)$ appartient à la variété

$$V_c(\varphi) = \left\{ x \in e : \varphi(x) = c \right\} \tag{2.10}$$

alors du théorème 2.11 (Cfr. Remarque 2.4) l'équation (2.10) admet une solution $x_0 \neq 0$ pour un certain μ réel et pourtant l'existence d'un continuum de solutions de l'équation (2.10) est assurée. Si pour un certain $c = r > 0$ la fonctionnelle $f(x)$ n'a pas de points critiques conditionnés relativement à la variété $V_r(\varphi)$, alors comme dans la démonstration du théorème 13.9 de M.M. Vainberg [2] on considère la fonctionnelle

$$\psi_\alpha(x) = \alpha \varphi(x) + f(x) \tag{2.11}$$

où α est un nombre quelconque appartenant à l'intervalle

M. M. Vainberg

$(0, \{f(0) - m(r)] \varphi^{-1}(x_o))$. Cette fonctionnelle est faiblement

semi-continue inférieurement et différentiable au sens de Fréchet

étant la somme de deux telles fonctionnelles. La semicontinuité

faible inférieure de la fonctionnelle $\varphi(x)$ suit du théorème 1.1

ou du théorème 1.10 . D'ici en vertu du théorème 2.2 il suit l'existence d'un point $x_\alpha \in D_r(\varphi)$ tel que

$$\min_{D_r(\varphi)} \psi_\alpha(x) = \psi_\alpha(x_\alpha)$$

Montrons que $x_\alpha \neq 0$. En effet $\psi_\alpha(0) = f(0)$ et comme par hypothèse $0 < \alpha < [\varphi(x_o)]^{-1} [f(0) - m(r)]$ on a

$$\psi_\alpha(0) = f(0) > \alpha \varphi(x_o) + m(r) = \gamma \varphi(x_o) + f(x_o) =$$

$$= \psi_\alpha(x_o) \geqslant \psi_\alpha(x)$$

donc $x_\alpha \neq 0$. Montrons que $x_\alpha \notin V_r(\varphi)$. En effet si l'on suppose que $x_\alpha \in V_r(\varphi)$ du théorème 2.11 il suit que

$$\operatorname{grad} \psi_\alpha(x_\alpha) = \lambda \operatorname{grad} \varphi(x_\alpha) .$$

Mais de (2.11) il suit

$$\operatorname{grad} \psi_\alpha(x_\alpha) = \alpha \operatorname{grad} \varphi(x_\alpha) + \operatorname{grad} f(x_\alpha)$$

donc

$$\operatorname{grad} f(x_\alpha) = (\lambda - \alpha) \operatorname{grad} \varphi(x_\alpha)$$

avec $x_\alpha \in V_r(\varphi)$, c'est-à-dire la fonctionnelle $f(x)$ a un point

critique conditionné rélativement à la variété $V_r(\varphi)$, contre

l'hypothèse. Pourtant $x_\alpha \in (D_r(\varphi) \setminus V_r(\varphi))$ et en vertu du theorème 2.1 on a $\operatorname{grad} \psi_\alpha(x) = 0$, ou

M. M. Vainberg

$$\text{grad } f(x_\alpha) = -\alpha \text{ grad } \varphi(x)$$

pour chaque $\alpha \in (0, [f(0) - m(r)] [\varphi(x_0)]^{-1})$. Donc l'équation

(2.10) admet un continuum de solutions correspondant à des valeurs

réelles du paramètre μ.

Evidentement la thèse s'étend au cas où $f(x)$ est faiblement

semi-continue supérieurement et $f(0) \neq \max\limits_{D_c(\varphi)} f(x) = M(c)$.

Enfin si la fonctionnelle est faiblement continue alors l'existen-

ce d'un continuum de solutions est assurée sans limitations pour

la valeur $f(0)$, ou bien de $m(c) = M(c)$ il suit $f(x) = $ con-

stante.

Remarquons encore que l'hypothèse de la réflexivité de E peut

être remplacée par l'hypothèse $E = X^*$ où X est un espace nor-

mé séparable.

Exemple 2.2. Dans un espace de Banach réel réflexif E (ou bien

dans le dual E d'un espace réel normé séparable) soyent données

une fonctionnelle réelle $f(x)$ différentiable au sens de Fréchet,

faiblement continue inférieurement, telle que

$$\lim\limits_{R \to \infty} f(x) = + \infty \qquad (R = \|x\|)$$

et une fonctionnelle $\varphi(x)$ strictement convexe non négative nul-

le seulement à l'origine et différentiable au sens de Fréchet.

Ici, différemment de l'exemple précedent, l'ensemble $D_c(\varphi)$ con-

sidéré dans le même exemple, peut être non borné, mais il est en-

core convexe et fermé, donc faiblement fermé. Etant $D_c(\varphi)$ faiblement fermé, en vertu du théorème 2.2 ou 2.3, compte tenu de la

remarque 2.1, pour chaque $c > 0$ il existe un vecteur $x_o \in D_c(\varphi)$

tel que

$$f(x_o) = \min_{D_c(\varphi)} f(x) = m(c) .$$

Pourtant comme dans l'exemple précedent on arrive à la conclusion

que si $f(0) \neq m(c)$ l'equation (2.10) admet un continuum de solu-

tions non nulles, chacune corréspondant à une valeur du paramètre

réel μ .

Exemple 2.3. Soit E un espace réel de Banach réflexif dans le-

quel la boule unité soit strictement convexe et la norme soit dif-

férentiable au sens de Fréchet. Considérons dans cet espace un o-

pérateur fortement potentiel compact $F(x)$ positivement homogène

de dégré $k > 0$ et une fonctionnelle $\varphi(x) = \|x\|^{\alpha+1}, \alpha > 0$.

Comme on a vu (Cfr. l'exemple 1.7) le potentiel de l'opérateur F

est faiblement continu et il est de la forme

$$f(x) = \frac{1}{k+1} < F(x), x > ,$$

tandis que $\varphi(x)$ est une fonctionnelle convexe telle que chaque

point de la variété $V_c(\varphi) = \left\{ x \in E: \varphi(x) = c, c > 0 \right\}$ est un

point ordinaire, car pour $x \neq 0$ on a

$$< \text{grad } \varphi(x), x > = (\alpha + 1) \|x\|^{\alpha+1} > 0 .$$

Supposons que $f(x)$ prend des valeurs négatives: cela arrive,

par exemple si $F(x) \neq 0$ et $F(-x) = F(x)$. Alors, à cause de son homogéneité elle atteindra le minimum dans la boule $\varphi(x) \leqslant c$ à un point de sa frontière $V_c(\varphi)$ et pourtant en vertu du théorème 2.11

$$\text{grd } f(x_o) = \mu_o \text{ grad } \varphi(x_o), \quad (x_o \in V_c(\varphi))$$

ou

$$F(x_o) = \lambda_o \|x_o\|^{\alpha} \text{ grad } \|x_o\| \ , \ \lambda_o = (1 + \alpha)\mu_o \qquad (2.12)$$

D'ici on a

$$\lambda_o = (1 + k) \|x_o\|^{-\alpha - 1} f(x_o) < 0 \ .$$

De (2.11) il suit que si $\alpha = k$ le vecteur $x = tx_o$ est une solution de l'équation

$$F(x) = \lambda \|x\|^{\alpha} \text{ grad } \|x\| \qquad (2.13)$$

pour $\lambda = \lambda_o$ et $t > 0$ quelconque, c'est-à-dire chaque vecteur tx est une solution de l'équation (2.13) pour $\lambda = \lambda_o$.

Pour $\alpha \neq k$ chaque vecteur du rayon $x = t^{-1}x_o$ $(t > 0)$ est encore une solution de l'équation (2.13) corréspondant à la valeur $\lambda = \lambda_o t^{\alpha - k}$ du paramètre. Lorsque t varie on obtient que à chaque valeur négative de λ il corréspond une solution de l'équation (2.13) collinéaire avec le vecteur x_o .

Si $f(x)$ prend des valeurs positives tout ce qu'on a vu pour λ

M. M. Vainberg

quation (2.13) reste valable en échangeant λ avec $-\lambda$.

Il faut encore remarquer que si E est un espace de Hilbert réel

et $\alpha = 1$, alors l'equation (2.13) devient

$$F(x) = \lambda x \tag{2.14}.$$

Les solutions non nulles de cette équation (pour $F(0) = 0$)

sont dites autofonctions ou autovecteurs de l'opérateur F et

les valeurs du paramètre λ corréspondantes à ces solutions sont

dites autovaleurs de l'opérateur F . Par conséquent si x_o est

un autovecteur de l'opérateur F et λ_o une.autovaleur corré-

spondante, c'est-à-dire $F(x_o) = \lambda_o x_o$, alors à cause de l'homo-

généité de F chaque vecteur du rayon tx_o est aussi un auto-

vecteur. Avec cela si $(F(x),x)$ prend des valeurs négatives (po

sitives) alors $\lambda_o < 0$ ($\lambda_o > 0$ respectivement).

De plus, si k = 1 le rayon des autovecteurs tx_o corréspond à

la valeur λ_o et si $k \neq 1$ chaque nombre négatif (respective-

ment, positif) est une autovaleur de l'operateur F .

RÉFÉRENCES

M. M. Vaiberg $[1, 2, 7, 10]$; M.M. Vainberg-Ia. L. Enghel'son $[1]$.

M. M. Vainberg

III. LA CONVERGENCE DES SUITES MINIMISANTES.

3.1 Quelques remarques preliminaires.

Soit E un espace linéaire, $f(x)$ une fonctionnelle réelle définie sur E , σ un certain ensemble de E et

$$\inf_{x \in \sigma} f(x) = d > - \infty$$

Definition 3.1. Chaque suite $\{x_n\} \subset E$ satisfaisant à la condition

$$\lim_{n \to \infty} f(x_n) = d$$

est dite minimisante.

Cette définition reste valable lorsque $d = f(x_0)$ où x_0 est
un point de minimum de $f(x)$ conditionne (relatif à l'ensemble
σ) ou un point de minimum de $f(x)$, c'est-à-dire un point de minimum locale de $f(x)$ ou absolu dans tout l'espace E .

Il nous intéresse la question de la convergence d'une suite minimisante vers un point de minimum non conditionné de la fonctionnel
le.

Pour étudier cette question un rôle particulier est joué par les
fonctions convexe satisfaisant certains conditions supplémentaires.

Rappelons qu'une fonction convexe quelconque n'a pas nécessairement un minimum non conditionné. Meme une fonction strictement
convexe peut ne pas avoir un minimum non conditionné. Par exemple,
la fonction réelle d'une variable réelle $f(t) = e^t$ n'a pas de

minimum absolu.

De plus si une fonction convexe (même strictement convexe) a un minimum une suite minimisante quelconque ne converge pas néccéssairement vers un point de minimum. Avant de donner un exemple on note que si la fonctionnelle $f(x)$ strictement convexe a un minimum au point x_0 elle croit sur chaque rayon sortant du point x_0 . En effet la fonctionnelle

$$\varphi(x) = f(x_0 + x) - f(x_0)$$

est aussi strictement convexe et elle a un minimum au point 0 . Ce minimum est stricte et unique (Cfr. Ch. II). Soit maintenant x un vecteur quelconque $\neq 0$. Pour la convexité stricte on a

$$\varphi(\lambda x + (1 - \lambda) 0) < \lambda \varphi(x) \quad , \quad (0 < \lambda < 1) .$$

D'ici en vertu du lemma 2.1 on a

$$\varphi(0) < \varphi(\lambda x) < \lambda \varphi(x) \quad , \quad 0 < \lambda < 1$$

et pourtant la fonctionnelle $\varphi(x)$ croit le long de chaque rayon joignant l'origine avec un vecteur $x \in E$ quelconque, donc $f(x)$ croit le long de chaque rayon sortant du point x_0 . En relation avec ces résultats on note que si E est un espace de dimension finie, alors à cause de la compacité de chaque boule fermée on peut affirmer que: si la fonction réelle $f(x)$ croit le long de chaque rayon sortant du point x_0 alors il existe une minorante monotone $c(t)$ pour $t \geqslant 0$, $c(0) = 0$ qu'on peut supposer continue, strictement croissante et satisfaisante à la condition

M. M. Vainberg

$$f(x) - f(x_o) \geqslant c(\|x - x_o\|) \qquad\qquad (3.1).$$

Notons que si l'inégalité (3.1) a lieu pour une fonctionnelle donnée sur un espace de dimension infinie il suit que x_o est un
point de minimum de $f(x)$ et chaque suite minimisante converge
vers ce point x_o .

Toutefois dans un espace de dimension infinie il existe des fonctionnelles strictement convexes ayant un minimum à un point x_o
et pourtant croissantes le long de chaque rayon sortant de x_o pour
lesquelles l'inégalité (3.1) n'a pas lieu. Cela est montré de
l'exemple suivant.

Exemple 3.1. Soit $\{e_k\}$ un système orthonormal complet d'un espace
de Hilbert réel H . Considérons dans H la fonctionnelle strictement convexe

$$f(x) = \sum_{k=1}^{\infty} \frac{(x,e_k)^2}{k^2} \qquad\qquad x \in H$$

qui est égale à zéro seulement pour $x = 0$, tandis que $f(x) > 0$
pour $x \neq 0$. Pour cette fonctionnelle la suite $x_n = \sqrt{n}\, e_n$
$(n = 1,2,3,\ldots)$ est minimisante car pour $n \to \infty$ on a

$$f(x_n) = 1/n \to 0$$

mais $\|x_n\| = \sqrt{n} \to \infty$.

3.2 Sur les suites minimisantes bornées.

Dans l'exemple ci-dessus la suite minimisante n'est pas bornée.
On se demande alors de donner des conditions afin que toute suite minimisante soit bornée. Pour cela il faut la

Definition 3.2. La fonctionnelle réelle finie $f(x)$ donnée dans

un espace normé E est dite croissante si pour chaque vecteur

$y \in E$ il existe $r > 0$ tel quel si $\|x\| > r$ on a $f(x) > f(y)$.

Remarquons que si

$$\lim_{R \to \infty} f(x) = +\infty \quad (R = \|x\|)$$

alors $f(x)$ est une fonctionnelle croissante.

Lemma 3.1. La fonctionnelle réelle croissante $f(x)$ donnée dans

l'espace normé E soit bornée inférieurement avec

$$d = \inf_{x \in E} f(x)$$

ou bien x_0 soit un point de minimum (eventuellement local) pour

$f(x)$ et $d = f(x_0)$. Alors chaque suite minimisante $\{x_n\}$ c'est-à-

dire telle que $\lim\limits_{n \to \infty} f(x_n) = d$, est bornée.

Démonstration. Si l'on suppose le contraire on aura une soussuite

$\{y_m\} \subset \{x_n\}$ telle que $\|y_m\| > m$ $(m = 1,2,3,\ldots)$.

Soit $m_0 > d$. Comme $f(x)$ est croissante il existe $r \geqslant m_0$ tel

que

$$f(y_m) > f(y_{m_0}) \geqslant m_0, \quad (m = m_1, m_1 + 1, \ldots; \; m_1 > m_0) \; .$$

D'ici il suit que

$$d = \lim_{m \to \infty} f(y_m) \geqslant m > d$$

La contrédiction démontre le lemma.

3.3 Sur la convergence faible des suite minimisantes.

Théorème 3.1. La fonctionnelle $f(x)$ croissante et strictement

convexe, donnée sur un espace de Banach réflexif E a un mini-

mum absolu à un certain point x_o, il n'y a pas d'autres points
de minimum et chaque suite minimisante converge vers le point de
minimum faiblement.

Démonstration. Comme $f(x)$ est une fonctionnelle croissante a-
lors il existe une boule $D_r = \{x \in E : \|x\| \leqslant r$ telle que sur sa
surface on a $f(x) > f(0)$. Comme de la convexité de $f(x)$ pour
le théorème 1.10 il suit sa semi-continuité inférieure faible et
comme pour la réflexivité de l'espace de Banach la boule D_r est
faiblement fermée et faiblement compacte, $f(x)$ a la m-propriété, c'est-à-dire il existe un point x_o intérieur à la boule D_r
auquel $f(x)$ atteint le minimum. Ce point de minimum est unique
et (Cfr. Lemma 2.1) le minimum est absolu.

Pour démontrer la deuxième partie du théorème considérons une
suite minimisante quelconque $\{x_n\}$, qui sera bornée grâce au lemma 3.1. Comme l'espace E est réflexif on peut extraire de $\{x_n\}$
une sous-suite $y_m \to y_o$ et de la semi-continuité inférieure faible de $f(x)$ il suit

$$f(y_o) \leqslant \lim_{m \to \infty} f(y_m) = \lim_{m \to \infty} f(y_m) = f(x_o)$$

D'ici on déduit $y_o = x_o$ car si l'on avait $y_o \neq x_o$ du lemma
2.1 on aboutirait à la contrédiction $f(y_o) > f(x_o)$.

Il reste à prouver que toute la suite minimisante $\{x_n\}$ converge
faiblement vers x_o . En effet si l'on suppose que cela n'est pas
vrai alors de $\{x_n\} \setminus \{y_m\}$ on peut extraire une suite $\{z_k\}$ qui con

verge faiblement vers $z_o \neq x_o$. Répetant pour la suite $\{z_k\}$ ce qu'on a dit pour la suite $\{y_m\}$ on obtient

$$f(z_o) \leqslant \lim_{k - \infty} f(z_k) = \lim_{k \to \infty} f(z_k) = f(x_o)$$

ce qui contrédit au lemma 2.1. Le théorème est démontré.

Remarquons que la thèse du théorème reste valable avec les varia tions suivantes. L'hypothèse de la réflexivité de E peut être remplacée par l'hypothèse que E soit le dual d'un espace normé séparable (Cfr. Théorème 2.3). L'hypothèse de la convexité stri- cte de $f(x)$ peut être remplacée par l'hypothèse que $f(x)$ soit strictement quasi-convexe et semi-continue inférieurement.

Théorème 3.2. Dans un espace de Banach E réflexif soit donnée une fonctionnelle finie $f(x)$ faiblement semi-continue inférieu- rement, satisfaisante à la condition

$$\lim_{R \to \infty} f(x) = + \infty \qquad (R = \|x\|) \qquad (3.2)$$

Alors il existe au moins un point de minimum absolu pour $f(x)$ et de chaque suite minimisante on peut extraire une sous-suite faiblement convergente vers un point de minimum absolu.

Démonstration. L'existence de points de minimum absolu pour $f(x)$ suit du théorème 2.5. Indiquons avec x_o un de ces points. De l'égalité (3.2) il suit que $f(x)$ est une fonctionnelle crois sante et pourtant en vertu du lemma 3.1 chaque suite minimisante $\{x_n\}$ est bornée, donc on peut extraire une sous-suite $y_m \to y_o$. De la semi-continuité inférieure faible de $f(x)$ on a

$$f(y_o) \leqslant \underset{m \to \infty}{\underline{\lim}} \ f(y_m) = \lim_{m \to \infty} f(y_m) = f(x_o) \ .$$

Comme $f(x_o) = \underset{x \in E}{\inf} f(x)$ on a $f(y_o) = f(x_o)$ donc y_o est un point de

minimum absolu pour $f(x)$. Le théorème est démontré. Remarquons que si

x_o est le seul point de minimum absolu pour $f(x)$ alors comme dans

la démonstration du théorème 3.1 on établit que $x_n \to x_o$.

Théorème 3.3. Les conditions du théorème 2.6 soient vérifiées. Alors

la thèse du théorème 3.2. est valable et, de plus, si dans la condi-

tion 2 le signe = a lieu seulement pour $h = 0$ alors chaque suite

minimisante converge faiblement vers le seul point de minimum absolu

de la fonctionnelle.

Démonstration. Comme dans la démonstration du théorème 2.6, des con-

ditions de ce théorème on a celles du théorème 3.2, donc la thèse aus

si du théorème 3.2. En outre si dans la condition 2 du théorème 2.6

le signe = a lieu seulement pour $h = 0$ alors $F(x)$ est un opéra-

teur strictement monotone, donc son potentiel est une fonctionnelle

strictement convexe. Comme $f(x)$ est une fonctionnelle strictement

convexe et croissante du théorème 3.1 chaque suite minimisante conver

ge faiblement vers le point de minimum. Le théorème est démontré.

3.4 Sur les problèmes de minimisation bien posés.

Définition 3.3. Le problème de minimisation d'une fonctionnelle réel

le donnée sur un certain ensemble d'un espace vectoriel est dit bien

posé s'il est résoluble, il y a une solution unique et si chaque sui

te minimisante converge vers cette solution au sens de la topologie

faible de l'espace.

Dans ce n° nous allons considérer le minimum non conditionné de la

fonctionnelle donnée sur un espace de Banach réel, différentiable au sens de Gateaux et nous donnerons des conditions pour que le pro-

blème soit bien posé.Essentiellement ces conditions suffisantes se rédui

sent à une inégalité du type (3.1) d'où, comme on verra, peut éta-

blir quand est que le problème de minimisation de la fonctionnelle

est bien posé.

<u>Théorème</u> 3.4. L'operateur $F(x) = \mathrm{grad}\, f(x)$ donné sur un espace de

Banach réflexif E satisfasse aux conditions: pour chaque $h, y \in E$ la

fonction $\langle F(y+th), h \rangle$ est intégrable par rapport à t sur $[0,1]$ et

$$\langle F(y+h) - F(y),\ h \rangle \geq \|h\|\ \gamma(\|h\|) \tag{3.3}$$

où $\gamma(t)$ est une fonction non négative, intégrable dans $[0,R]$

pour chaque $R > 0$, telle que

$$c(R) = \int_0^R \gamma(t)dt \tag{3.4}$$

est croissante et

$$c(R) > R\,\|F(0)\| \tag{3.5}$$

pour un certain R. Alors le problème de minimisation pour $f(x)$

est bien posé.

<u>Démonstration</u>. De la condition (3.3) il suit en premier lieu la

monotonicité de $F(x)$, donc la semi-continuité inférieure faible

de $f(x)$, et deuxièmement que (Cfr. la démonstration du Théorème

2.9)

$$f(x) - f(0) \geq \int_0^R \gamma(z)\ dz - R\,\|F(0)\|$$

où $R = \|x\|$. D'ici pour la condition (3.5) la fonctionnelle $f(x)$

a la m-propriété, c'est-à-dire $f(x)$ a un point de minimum x_0

et pourtant $F(x_0) = 0$. De cela on a

$$f(x) - f(x_o) = \int_o^1 \left\langle F(x_o + t(x - x_o)) - F(x_o), x - x_o \right\rangle dt \geqslant$$

$$\geqslant \int_o^1 \|x - x_o\| \, \gamma \, (t \, \| x - x_o \|) \, dt = c(\|x - x_o\|)$$

D'ici on déduit l'unicité du minimum et la convergence vers lui de chaque suite minimisante.

Il faut remarquer que pour vérifier l'inégalité (3.5) il est suffisant, par exemple, que

$$\overline{\lim_{R \to \infty}} \, \frac{c(R)}{R} = + \infty$$

Théorème 3.5. La fonctionnelle $f(x)$ deux fois différentiable au sens de Gateaux, donnée sur un espace de Banach réflexif E satisfasse aux conditions:

1. La différentielle $D^2 f(y + tx \, ; \, x, x)$ soit intégrable en $t \in [0,1]$ pour chaque x , $y \in E$;

2. $D^2 f(x \, ; \, h, h) \geqslant \|h\| \, \gamma \, (\|h\|)$, $(\forall x, h \in E)$

où $\gamma(t)$ est une fonction non négative, intégrable dans $[0,R]$ pour chaque $R > 0$ telle que la fonction

$$c(R) = \int_o^R \gamma(z) \, dz$$

soit croissante et

$$c(R) > R \, \|F(0)\|, \quad F(x) = grad \, f(x) \tag{3.6}$$

pour un certain R . Alors le problème de minimisation de $f(x)$ est bien posé.

Démonstration. Écrivons

$$f(x) - f(0) = \left\langle F(0, x) \right\rangle + \int_o^1 \left\langle F(tx) - F(0, x) \right\rangle dt$$

M. M. Vainberg

Mais $\langle F(0), x \rangle = D\ f(0,x)$, $\langle F(tx), x \rangle = D\ f(tx, x)$ et de la for-

mule de Lagrange

$$D\ f(tx, x) - D\ f(0,x) = D^2 f(\tau tx\ ;\ tx, x)$$

avec $0 < \tau < 1$. Par conséquent

$$f(x) - f(0) = \langle F(0), x \rangle + \int_0^1 D^2 f(\tau tx\ ;\ tx, x)\ dt\ .$$

D'ici

$$f(x) - f(0) \geqslant -\|x\|\ \|F(0)\| + \int_0^1 \gamma(t\ \|x\|)\|x\|\ dt\ =$$

$$= R\ (\ \frac{1}{R}\ c(R) - \|F(0)\|)$$

où $R = \|x\|$. D'ici et de (3.6) il suit $f(x) > f(0)$ pour $\|x\| =$

$= R$.

Puisque des conditions il suit la monotonicité de $F(x)$ (et pour-

tant, en vertu du théorème 1.4, la semi-continuité inférieure fai-

ble de $f(x)$) $f(x)$ a la m-propriété, c'est-à-dire il existe un

point de minimum x_0 de la fonctionnelle $f(x)$.

De plus, comme $F(x_0) = 0$, on a

$$f(x) - f(x_0) = \int_0^1 \langle F(x_0 + t(x - x_0)) - F(x_0), x - x_0 \rangle\ dt =$$

$$= \int_0^1 \langle D\ F(x_0 + \tau t(x - x_0), t(x - x_0)), x - x_0 \rangle\ dt \geqslant$$

$$\geqslant \int_0^1 \gamma(t\|x - x_0\|)\|x - x_0\|dt = \int_0^R \gamma(z)\ dz = c(R)$$

où $R = \|x - x_0\|$. Pourtant

$$f(x) - f(x_0) \geqslant c(\|x - x_0\|)\ .$$

M. M. Vainberg

De cette inégalité il suit l'unicité du minimum de $f(x)$ et la convergence vers le point de minimum x_o de toute suite minimisante. Le théorème est démontré.

<u>Théorème</u> 3.6. Admettons les conditions du théorème 2.10 et, de plus, l'hypothèse suivante: la fonction $\gamma(z)$ soit continue pour $z \geqslant 0$. Alors le problème de minimisation de $f(x)$ est bien posé.

<u>Démonstration</u>. Soit x_o le point de minimum unique de $f(x)$ dont l'existence et l'unicité a été établie par le théorème 2.10. Alors, comme $F(x) = \text{grad } f(x)$ et $F(x_o) = 0$, on a encore une fois

$$f(x) - f(x_o) = \int_0^1 D^2 f(x_o + \tau t(x - x_o); t(x - x_o), x - x_o) \, dt$$

où $0 < \tau < 1$. D'ici en vertu de la condition 2 du théorème 2.10

$$f(x) - f(x_o) \geqslant \int_0^1 \gamma(\|x_o + \tau t(x - x_o)\|) \, t \, R^2 \, dt \geqslant$$

$$\geqslant \int_0^1 \gamma(\|x_o\| + tR) \, t \, R^2 \, dt = \int_0^R \gamma(\|x_o\| + z) \, z \, dz = c(R)$$

où $R = \|x - x_o\|$. Mais $c'(R) = \gamma(\|x_o\| + R) \, R > 0$ et pourtant $c(R)$ est une fonction continue croissante qui satisfait la condition $c(0) = 0$. Par conséquent

$$f(x) - f(x_o) \geqslant c(\|x - x_o\|) \, ,$$

c'est-à-dire $f(x)$ vérifie un'inégalité du type (3.1) d'où la thèse du théorème.

R É F É R E N C E S

M. M. Vainberg [10] .

Dans la partie précedente on a étudié la question de la convergence des suites minimisantes. Dans la suite on va considérer
quelques méthodes qui permettent la construction de suites minimi
santes. Parmi les méthodes connus pour la minimisation de fonction
nelles non linéaires nous considérons seulement les trois suivan
ts: méthode de la plus grande pente, méthode de Ritz et méthode
de Newton.

IV. METHODE DE LA PLUS GRANDE PENTE.

4.1 Idée de la méthode.

Pour illustrer l'idée de cette méthode nous considérons d'abord
le cas des espaces de Hilbert.

Dans un espace de Hilbert réel H soit donné une fonctionnelle
réelle non linéaire $f(x)$ différentiable au sens de Gateaux et
bornée inférieurement. Posons

$$d = \inf_{x \in H} f(x) \quad , \quad F(x) = \text{grad } f(x)$$

Prenons un vecteur quelconque $x_1 \in H$ tel que $F(x_1) \neq 0$. Rappelons que si $f(x)$ est une fonctionnelle strictement convexe on
a vu (cfr. Lemma 2.1, Corollaire 2.1 et théorème 2.1) qu'elle
peut avoir un seul point de minimum auquel elle atteint la valeur
d, donc la condition $F(x_1) \neq 0$ signifie que x_1 n'est pas un
point de minimum pour $f(x)$. Prenons un vecteur $h \in H$ tel que
la longueur soit $\|h\| = \|F(x_1)\|$ tandis que la direction de h
est telle que la dérivée

M. M. Vainberg

$$\frac{d}{dt} f(x_1 + th) = (F(x_1 + th), h)$$

ait la valeur la plus petite pour $t = 0$, c'est-à-dire h ait

la direction de décroissance maximum de $f(x)$ au point x_1.

Pour cela on prend la direction de h de telle façon que

$(F(x_1), h)$ ait la plus grande valeur et après on change le signe

de h, de sorte que $(F(x_1), -h)$ aura la valeur la plus petite.

Comme

$$(F(x_1), h) \leqslant \| F(x_1) \| \|h\| = \| F(x_1) \|^2$$

$(F(x_1), h)$ sera maximum seulement pour $h = F(x_1)$ et minimum

pour $h = -F(x_1)$, c'est-à-dire lorsque la direction de h coïn

cide avec celle de l'anti-gradient de $f(x)$. Posons $h_1 = -F(x_1)$

et considérons la fonction réelle

$$\varphi(t) = f(x_1 + th_1), \quad t \geqslant 0 .$$

Par construction $\varphi(t)$ décroit dans un certain interval à droi-

te du point $t = 0$. Soit t_1 la plus petite valeur positive de

t pour laquelle $\varphi(t_1) = \min \varphi(t)$. Posons

$$x_2 = x_1 + t_1 h_1 = x_1 - t_1 F(x_1) .$$

Par construction

$$f(x_2) = f(x_1 + t_1 h_1) < f(x_1) .$$

Si $F(x_2) \neq 0$ on répetera le raisonnement précedent. Pourtant si

$F(x_k) \neq 0$ on est conduit au processus suivant

$$x_{n+1} = x_n - t_n F(x_n), \quad (n = 1, 2, 3, \ldots) \qquad (4.1)$$

que l'on appelle méthode de la plus grande pente ou bien méthode

M. M. Vainberg

du gradient.

Bien que la suite (4.1) ait la propriété,

$$f(x_{n+1}) < f(x_n) \quad (n = 1,2,3,\ldots) \tag{4.2}$$

elle n'est pas nécessairement minimisante.

Definition 4.1. Chaque suite $\{x_n\}$ qui vérifié l'inégalité (4.2) est dite une suite relaxante et les nombres t_n dans (4.1) sont dits multiplicateurs de relaxation.

Il faut remarquer que déjà dans le cas d'un espace de Hilbert il est difficile de définir ces multiplicateurs de relaxation. En effet seulement dans des cas bien simples on peut les calculer effectivement. Pourtant on semplace les nombres t_n par des nom bres positifs ε_n qui sont ou bien donnée apriori ou bien choisis parmi ceux compris entre des bornes fixées à l'avance.

Dans le cas où les t_n sont remplacés par des ε_n on appelle le processus (4.1) méthode du type de la plus grande pente (ou du type du gradient).

Considérons maintenant le cas plus général où la fonctionnelle réelle $f(x)$ différentiable au sans de Gateaux et bornée inférieurement est donnée dans un espace de Banach réel E.

Soit $F(x) = \text{grad } f(x)$ et soit x_1 un vecteur quelconque de E tel que $F(x_1) \neq 0$. Prenons un vecteur $h \in E$ tel que $\|h\| = \|F(x_1)\|$. Choisissons la direction de h de telle façon que la dérivée

$$\frac{d}{dt} f(x_1 + th) = \langle F(x_1 + th), h \rangle$$

ait la valeur minimum au point $t = 0$. On prend d'abord h tel que $\langle F(x_1), h \rangle$ ait la valeur maximum et après $-h$. Alors $\langle F(x_1), -h \rangle$ aura la valeur minimum. Comme

$$\langle F(x_1), h \rangle \leqslant \| F(x_1) \| \|h\| = \| F(x_1) \|^2 \tag{4.3}$$

il faut prendre le vecteur h de telle façon que $\langle F(x_1), h \rangle = $

$$= \| F(x_1) \|^2 .$$

Pour cela on considère un opérateur U de E^* dans E satisfai-sant aux conditions

$$\| U\, y \| = \|y\| \quad , \quad \langle y, U\, y \rangle = \| y \|^2 \quad (y \in E^*) .$$

Par exemple si E est réflexif et si la norme de E^* est dif-férentiable au sens de Gateaux on peut prendre comme U l'opé-rateur défini par

$$U\, y = \|y\| \, \text{grad} \, \|y\| \, , \, y \neq 0 \, , \quad U\, 0 = 0 \quad .$$

Posant $h = U\, F(x_1)$ on obtient

$$\langle F(x_1), U\, F(x_1) \rangle = \| F(x_1) \|^2$$

et en vertu de (4.3) la borne supérieure de $\langle F(x_1), h \rangle$ est at-teinte à $h = U\, F(x_1)$. Il reste alors seulement à poser $h_1 = $ $= U\, F(x_1)$. On prend maintenant $t_1 > 0$ tel que la fonction $\varphi(t) = f(x_1 + th_1)$ soit minimum pour $t = t_1$, donc $\varphi(t_1) = $ $= \min \varphi(t)$. Alors

$$f(x_2) = f(x_1 + t_1 h_1) < f(x_1)$$

avec $x_2 = x_1 - t_1 U\, F(x_1)$.

M. M. Vainberg

Continuant ce procédé on obtient

$$x_{n+1} = x_n - t_n \, U \, F(x_n) \quad (n = 1,2,3,\dots) \tag{4.4}$$

Celui-ci est la méthode du gradient. D'ordinaire on ne calcule

pas t_n et on considère le processus

$$x_{n+1} = x_n - \xi_n \, U \, F(x_n) \tag{4.5}$$

Pour ξ_n on assigne des limitations telles que la méthode du type du gradient soit convergente.

La convergence de la méthode du gradient pour les fonctionnelles différentiables au sens de Gateaux.

On commencera par donner des propositions auxiliaires.

<u>Lemma</u> 4.1. (Pour le processus de relaxation voir la note de Yu. I. Liubitch - G.D. Maistrovskii [1] où l'on trouvera aussi une bibliographie).

La fonctionnelle réelle $f(x)$ donnée dans un espace de Banach réel réflexif E soit différentiable au sens de Gateaux et son gradient $F(x)$ satisfasse à la condition

$$\langle F(x+h) - F(x), h \rangle \leqslant M(r)\|h\|^2, \; x, \, x + h \in D_r = \left\{ x \in E : \|x\| \leqslant r \right\} \tag{4.6}$$

où $M(r)$ est une fonction croissante quelconque sur le semi-axe $r \geqslant 0$ et la norme de l'espace E est différentiable au sens de Gateaux. Alors si

$$\xi_n \, M_n \leqslant 1/2$$

où

$$M_n = \max \left[1, \, M(R_n) \right] , \; R_n \geqslant \| x_n \| + \| F(x_n) \|$$

M. M. Vainberg

le procéssus (4.5) est relaxante.

Démonstration. Faisant usage de la formule de Lagrange, de (4.5)

et des propriétés de l'opérateur U on obtient

$$f(x_n) - f(x_{n+1}) = \langle F(x_n), x_n - x_{n+1} \rangle -$$

$$- \langle F(x_{n+1}) + \gamma(x_n - x_{n+1})) - F(x_n), x_n - x_{n+1} \rangle \geq$$

$$\geq \varepsilon_n \| F(x_n) \|^2 - M_n^- \| x_{n+1} - \bar{x}_n \|^2 =$$

$$= \varepsilon_n \| F(x_n) \|^2 - \varepsilon_n^2 M_n \| F(x_n) \|^2$$

ou

$$f(x_n) - f(x_{n+1}) \geq \frac{1}{2} \varepsilon_n \| F(x_n) \|^2 .$$

Le lemma est démontré.

Remarquons que afin que l'inégalité (4.6) soit satisfaite il est

suffisant que l'opérateur F(x) . satisfasse à une condition de

Lipschitz

$$\| F(x + h) - F(x) \| \leq M(r) \| h \| \tag{4.8}.$$

Les conditions du lemma 4.1 soient satisfaites et soit $\inf_{x \in E} f(x) =$

$= d > - \infty$. Posons $r_n = f(x_n) - d$.

De l'inégalité (4.7) il suit

$$r_n - r_{n+1} = f(x_n) - f(x_{n+1}) > 0$$

c'est-à-dire la suite de nombres non négatifs r_n est décroissan

te et pourtant il existe

$$\lim_{n \to \infty} r_n = r_0 \geq 0 .$$

Si l'on admet que l'ensemble

$$E_0 = \left\{ x \in E : f(x) \leq f(x_0) \right\} \tag{4.9}$$

M. M. Vainberg

soit borné alors en vertu du lemma 4.1 la suite $\{x_n\}$ est bornée.

Observons que afin que l'ensemble E_o soit borné il est suffisant par exemple que $f(x)$ soit une fonctionnelle croissante (au sens de la définition 3.2), en particulier que

$$\lim_{R \to \infty} f(x) = + \infty \qquad (R = \|x\|)$$

Lorsque la suite $\{x_n\}$ est bornée, si $F(x)$ satisfait à la condition (4.8) alors $\{\|F(x_n)\|\}$ aussi est bornée, donc $\{M_n\}$ est bornée (si l'on prend $\{R_n\}$ bornée comme il est possible), enfin $1 \leqslant M_n \leqslant M_o$. Pourtant si l'on suppose $\xi_n \geqslant \dfrac{1}{4 M_n}$ on aura $\xi_n \geqslant \dfrac{1}{4 M_o}$ et de l'inégalité (4.7) on obtient

$$\lim_{n \to \infty} F(x_n) = 0 \qquad\qquad (4.10)$$

Cela prouve le

Lemma 4.2. Les conditions suivantes soient satisfaites:

I. E soit un espace de Banach réel réflexif et E^* soit un espace dont la norme est différentiable au sens de Gateaux.

2. La fonctionnelle réelle $f(x)$ bornée inférieurement et différentiable au sens de Gateaux soit croissante (ou bien l'ensemble E_o soit borné) et son gradient satisfasse à la condition de Lipschitz (4.8).

3. Les multiplicateurs de relaxation ξ_n satisfassent aux inégalités $1/4 \leqslant \xi_n \leqslant 1/2$.

Alors le procès itératif (4.5) est relaxant et $\lim\limits_{n \to \infty} F(x_n) = 0$.

M. M. Vainberg

Remarque 4.1. Rappelons que si l'on ajoute aux conditions du lemma 4.2 l'hypothèse $f(x) - d \leqslant \|F(x)\|^{\alpha}$, $\alpha > 0$, ou $d = \inf f(x)$, alors la suite $\{x_n\}$ est minimisante.

Théorème. 4.1. Les conditions suivantes soient satisfaites:

1. E soit un espace de Banach réel réflexif et la norme de E^* soit différentiable au sens de Gateaux.

2. La fonctionnelle réelle $f(x)$ donnée sur E soit différentiable au sens de Gateaux et son gradient $F(x)$ soit tel que la fonction $\langle F(tx), x \rangle$ soit intégrable en $t \in (0,1)$ et satisfasse à l'inégalité (4.8) et

$$\langle F(x + h) - F(x), h \rangle \geqslant \|h\| \gamma(\|h\|) \tag{4.11}$$

où $M(r)$ est une fonction non négative croissante continue pour $r \geqslant 0$ et $\gamma(t)$ est une fonction croissante continue pour $t \geqslant 0$ telle que $\gamma(0) = 0$ et la fonction

$$c(R) = \frac{1}{R} \int_0^R \gamma(z) \, dz$$

soit croissante et telle que $c(R) > \|F(0)\|$ pour un certain R .

3. $1/4 \leqslant \varepsilon_n M_n \leqslant 1/2$, $M_n = \max [1, M(R_n)]$, $R_n \geqslant \|x_n\| + \|F(x_n)\|$.

   Alors la suite

$$x_{n+1} = x_n - \varepsilon_n U F(x_n) \tag{4.5}$$

est relaxante, minimisante et convergente vers le point de minimum absolu de $f(x)$.

Démonstration. De la formule pour la potentiel des opérateurs (Cfr. la démonstration du théorème 2.6) il suit que

$$f(x) - f(0) = \langle F(0), x \rangle + \int_0^1 \langle F(tx) - F(0), x \rangle \, dt$$

D'ici et de (4.11) on a

$$f(x) - f(0) \geqslant \langle F(0), x \rangle + \|x\| \int_0^1 \gamma(t\|x\|) \, dt \geqslant R(\varepsilon(R) - \|F(0)\|)$$

où $R = \|x\|$. D'ici par la condition 2 du théorème

$$\lim_{R \to \infty} f(x) = +\infty \quad (R = \|x\|)$$

donc $f(x)$ est une fonctionnelle croissante.

Comme de la condition (4.11) il suit que $f(x)$ est strictement convexe, en vertu du théorème 3.1 il existe un seul point $x_0 \in E$ auquel $f(x_0) = \inf_{x \in E} f(x) = d$ et $f(x)$ n'a pas d'autres points de minimum. En outre grâce su lemma 4.2 la suite (4.5) est relaxante et on a $\lim_{n \to \infty} F(x_n) = 0$.

Mais de la condition (4.11) il suit que

$$\|F(x_n)\| \geqslant \langle F(x_n), \frac{x_n - x_0}{\|x_n - x_0\|} \rangle = \langle F(x_n) - F(x_0), \frac{x_n - x_0}{\|x_n - x_0\|} \rangle \geqslant$$

$$\geqslant \gamma(\|x_n - x_0\|)$$

car $F(x_0) = 0$ (Cfr. Théorème 2.1). D'ici $\|x_n - x_0\| \leqslant \gamma^{-1}(\|F(x_n)\|)$ γ^{-1} etant la fonction inverse de dont l'existence est assuree par la condition 2 du theorème. De l'inégalité précedente on a $\lim_{n \to \infty} x_n = x_0$. Le théorème est démontré.

4.3 Sur la convergence de la méthode du gradient pour les fonctionnelles deux fois différentiables au sens de Gateaux.

Dans le n° précedent il n'y a pas dans les conditions ni la différentiabilité ni la continuité de $F(x)$. Maintenant nous supposons que le gradient $F(x)$ de la fonctionnelle $f(x)$ soit différentiable au sens de Gateaux.

Lemma 4.3. Les conditions suivantes soient satisfaites:

1. E soit un espace de Banach réel réflexif et la norme de E^* soit différentiable au sens de Gateaux.

2. La fonctionnelle $f(x)$ réelle, différentiable au sens de Gateaux, donnée sur E soit croissante et inférieurement bornée et son gradient $F(x)$ soit différentiable au sens de Gateaux et il satisfasse à l'inégalité

$$\langle F'(x) h_1, h_2 \rangle \leqslant M(\|x\|) \|h_1\| \|h_2\| \qquad (4.12)$$

pour chaque $x, h_1, h_2 \in E$, où $M(r)$ est une fonction positive non décroissante, donnée pour $r \geqslant 0$, bornée sur chaque intervalle de ce semi-axe, et en outre $\langle F'(x + th) h, h \rangle$ soit continue en $t \in [0,1]$.

3. Les multiplicateurs ε_n dans (4.5) satisfassent aux inégalités $1/2 \leqslant \varepsilon_n M_n \leqslant 1$ et $\varepsilon_n \leqslant 1$ où

$$M_n = \max \left[1, M(R_n) \right] \quad , \quad R_n = \|x_n\| + \|F(x_n)\| \; .$$

Alors le procès itératif (4.5) est relaxant et $\lim\limits_{n \to \infty} F(x_n) = 0$.

Démonstration. Faisant usage de la formule de Taylor (Cfr. L. Graves [1]) on peut écrire

$$- \left[f(x_{n+1}) - f(x_n) \right] = -\langle F(x_n), x_{n+1} - x_n \rangle -$$

$$- \int_0^1 \langle F'(x_n + t(x_{n+1} - x_n))(x_{n+1} - x_n), x_{n+1} - x_n \rangle (1 - t) \, dt =$$

$$= -\langle F(x_n), x_{n+1} - x_n \rangle - \frac{1}{2} \langle F'(x_n + \theta(x_{n+1} - x_n))(x_{n+1} - x_n), x_{n+1} - x_n \rangle$$

où x_n , x_{n+1} sont les vecteurs qui entrent dans (4.5) et $0 <$ $< \vartheta < 1$.

D'ici et de l'inégalité (4.12) il suit

$$f(x_n) - f(x_{n+1}) \geqslant -\langle F(x_n), x_{n+1} - x_n\rangle -$$

$$- \frac{1}{2} M(\|x_n + \theta (x_{n+1} - x_n)\|) \| x_{n+1} - x_n\|^2$$

et comme $\|x_n + \vartheta (x_{n+1} - x_n)\| \leqslant R_n$

$$f(x_n) - f(x_{n+1}) \geqslant \quad -\langle F(x_n), x_{n+1} - x_n\rangle -$$

$$- \frac{1}{2} M(R_n)\|x_{n+1} - x_n \|^2 .$$

D'ici et de (4.5), utilisant les propriétés de l'opérateur U

on obtient

$$f(x_n) - f(x_{n+1}) \geqslant \varepsilon_n \|F(x_n)\|^2 (1 - \frac{1}{2} \varepsilon_n M_n) \geqslant$$

$$\geqslant \frac{1}{2} \varepsilon_n \| F(x_n)\|^2 0 \qquad\qquad (4.7)$$

Ceci prouve la première partie du lemma. Pour démontrer la deuxiè

me affirmation posons inf $f(x) = d$ et considérons la suite

$r_n = f(x_n) - d$. Cette suite est décroissante et pourtant il exi

ste

$$\lim_{n \to \infty} r_n = r_0 \geqslant 0 \qquad\qquad (4.13)$$

Comme $f(x)$ est une fonctionnelle croissante il suit de (4.7)

que la suite $\{x_n\}$ est bornée. Montrons que la suite $\{\|F(x_n)\|\}$

est bornée. En effet pour chaque $x \neq 0$ fixé, en vertu d'un co-

rollaire du théorème de Hahn-Banach (Cfr. N. Dunford - J. Schwartz

[1] , 2.3.14) et pour la réflexivité de E il existe un seul

vecteur $h \in E$ tel que $\langle F(x) - F(0), h \rangle = \| F(x) - F(0) \|$.

D'ici pour la formule de Lagrange généralisée

$$\| F(x) - F(0) \| = \langle F'(\tau x) x, h \rangle , \qquad 0 < \tau < 1$$

d'où

$$\| F(x) \| \leqslant \| F(0) \| + \| F(x) - F(0) \| = \| F(0) \| + \langle F'(\tau x) x, n \rangle$$

et pour la condition (4.12)

$$\| F(x) \| \leqslant \| F(0) \| + M(\tau \| x \|) \| x \| \leqslant \| F(0) \| + M(\| x \|) \| x \|$$

donc la suite $\left\{ \| F(x_n) \| \right\}$ est bornée. Comme $\| x_n \|$ et $\| F(x_n) \|$ sont bornées, $R_n = \| x_n \| + \| F(x_n) \|$ et M_n sont aussi bornées, donc $1 \leqslant M_n \leqslant M_0 < + \infty$. D'ici et grâce à la condition 3 du lemma $0 < \dfrac{1}{2M_0} \leqslant \xi_n \leqslant 1$. De cette inégalité et des inégalités (4.7) et (4.13) il suit

$$\lim_{n \to \infty} F(x_n) = 0 .$$

Le lemma est démontré.

Théorème 4.2. Les conditions suivantes soient satisfaites:

1. E soit un espace de Banach réel réflexif et la norme de E^* soit différentiable au sens de Gateaux.

2. La fonctionnelle réelle $f(x)$ donnée sur E soit deux fois différentiable au sens de Gateaux, la fonction $\langle F'(y+tx)x,x \rangle$ soit continue en $t \in [0,1]$ pour chaque $x, y \in E$, et on ait l'inégalité

$$\langle F'(x) h_1, h_2 \rangle \leqslant M(\| x \|) \| h_1 \| \| h_2 \| \tag{4.12}$$

où $M(r)$ est une fonction positive non décroissante donnée

M. M. Vainberg

pour $r \geqslant 0$, bornée sur chaque intervalle du demi-axe, et

pour chaque $x, y \in E$

$$\langle F'(y + tx) x, x \rangle \geqslant \|x\| \, \gamma(\|x\|) \qquad (4.14)$$

où $\gamma(t)$ est une fonction continue croissante pour $t \geqslant 0$,

$\gamma(0) = 0$ et la fonction croissante

$$c(R) = \frac{1}{R} \int_0^R \gamma(z) \, dz$$

satisfasse à l'inégalité $c(R) > \|F(0)\|$ pour un certain R.

3. $0 < \varepsilon_n \leqslant 1$, $1/2 \leqslant \varepsilon_n M_n \leqslant 1$, $M_n = \max\left[1, M(R_n)\right]$, $R_n = \|x_n\| + \|F(x_n)\|$

Alors le procès itératif

$$x_{n+1} = x_n - \varepsilon_n \, U \, F(x_n) \qquad (n = 1, 2, 3, \ldots). \qquad (4.5)$$

est relaxante pour $f(x)$ et la suite $\{x_n\}$ est minimisante et con‐

verge vers le point de minimum absolu de $f(x)$.

Démonstration. De l'inégalité (4.14) grâce au corollaire 1.2 on

déduit la convexité stricte de la fonctionnelle $f(x)$. Montrons

que $f(x)$ est une fonctionnelle croissante. En effet

$$f(x) - f(0) = \langle F(0), x \rangle + \int_0^1 \langle F(tx) - F(0), x \rangle \, dt$$

et pour la formule de Lagrange généralisée

$$f(x) - f(0) = F(0), x + \int_0^1 F'(\tau tx) tx, x \, dt .$$

D'ici et de l'inégalité (4.14) il suit

$$f(x) - f(0) \geqslant R \, (c(R) - \|F(0)\|)$$

donc

$$\lim_{R \to \infty} f(x) = +\infty \qquad (R = \|x\|)$$

Comme $f(x)$ est une fonctionnelle strictement convexe et croissante il existe pour le théorème 3.1 un seul point $x_o \in E$ tel que

$$f(x_o) = \inf_{x \in E} f(x) = d$$

et $f(x)$ n'a pas d'autres points de minimum.

En outre du théorème 2.1 on a $F(x_o) = 0$ et du lemma 4.3 dont les conditions sont toutes vérifiées on a

$$\lim_{n \to \infty} F(x_n) = 0$$

la suite $\{x_n\}$ étant définie par le procès (4.5).

Finalement, compte tenu que $F(x_o) = 0$ de la condition (4.14) on a

$$\| F(x_n) \| \geq \left\langle F(x_n) - F(x_o) , \frac{x_n - x_o}{\|x_n - x_o\|} \right\rangle =$$

$$= \left\langle F'(x_o + \tau(x_n - x_o))(x_n - x_o), \frac{x_n - x_o}{\|x_n - x_o\|} \right\rangle \geq \gamma(\|x_n - x_o\|) \quad (4.13)$$

Rappelons que dans (4.5) on suppose $x_n \neq x_o$, donc $F(x_n) \neq 0$ pour chaque n. Mais des conditions $\gamma(z)$ est une fonction continue croissante et pourtant elle admet une fonction inverse γ^{-1} qui est continue, croissante et $\gamma^{-1}(0) = 0$.

Compte tenu de cela de l'inégalité (4.15) il suit que

$$\|x_n - x_o\| \leq \gamma^{-1}(\|F(x_n)\|)$$

et comme $F(x_n) \to 0$ pour $n \to \infty$, alors $\lim_{n \to \infty} x_n = x_o$.

Le théorème est démontré.

RÉFÉRENCES

M. Altman [1], A. Goldstein [1], T. Lezanskii [1], M.M. Vainberg [4, 5, 6].

V. LA MÉTHODE DE RITZ.

On supposera ici que E soit un espace réel normé séparable et

f(x) une fonctionnelle réelle finie donnée sur E . Seulement

aux nos 5.4 - 5.5 on supposera E complet. Pour minimiser la fon

ctionnelle f(x) . lorsq'elle est bornée inférieurement on emplo

yera la méthode de Ritz. V. Ritz appliqua sa méthode à la résolu

tion de problèmes concrets. Ensuite sa méthode a été développée

dans les travaux de plusieurs auteurs.

Pour démontrer les propositions fondamentales de la méthode de

Ritz on utilisera la

Définition 5.1. La fonctionnelle f(x) est dite semicontinue su

périeurement (inférieurement) au point $x_o \in E$ si pour chaque

$\varepsilon > 0$ il existe $\delta > 0$ tel que $\|x - x_o\| < \delta$ implique

$$f(x_o) - f(x) > - \varepsilon \qquad (f(x_o) - f(x) < \varepsilon)$$

La fonctionnelle f(x) est dite semicontinue supérieurement

(inférieurement) dans l'ensemble $M \subset E$ si elle est semicontinue

supérieurement (inférieurement) à chaque point de M .

5.1 Approximations et systèmes de Ritz.

Soit f(x) une fonctionnelle réelle bornée inférieurement donnée dans un

espace normé E . La méthode de Ritz pour minimiser f(x)   est la

suivante. On se donne à l'avance un système de coordonnées, c'est-

a-dire un système de vecteurs linéairement indépendantes

$$\varphi_1, \varphi_2, \varphi_3, \dots, \varphi_n, \dots \qquad (5.1)$$

M. M. Vainberg

avec les proprietes: 1. $\varphi_n \in E$ pour chaque n naturel ; 2. L'en

semble de toutes les combinaisons linéaires de ces vecteurs du sy

stème (5.1) est complet dans E , c'est-à-dire le système (5.1)

est complet dans E . Après on construit une suite de sous-espace

$\{E_n\}$ de dimension finie, où E_n est l'espace de dimension n en

gendré par les vecteurs $\varphi_1, \varphi_2, \ldots, \varphi_n$.

Comme $f(x)$ est bornée inférieurement sur E il suit que $f(x)$

est bornée inférieurement sur E_n . Posons

$$d_n = \inf_{x \,\in\, E_n} f(x) \qquad (n = 1,2,3,\ldots)$$

Par construction $d_1 \geqslant d_2 \geqslant \ldots \geqslant d_n \geqslant \ldots$ Supposons que pour

chaque n il existe $x_n \in E_n$ tel que $f(x_n) = d_n$. Comme $x_n \in E_n$

on aura

$$x_n = \sum_{k=1}^{n} a_k \varphi_k \qquad\qquad (5.2)$$

où $a_k = a_k(n)$. Les vecteurs x_n sont dits les approximations

de Ritz. Dans ce n° et dans le suivant on étudie la question de

déterminer ces approximations.

Soit $f(x)$ différentiable au sens de Gateaux sur E et $F(x) =$

$= \operatorname{grad} f(x)$. Alors $f(x)$ est différentiable au sens de Gateaux

sur E_n et pour des vecteurs $x, h \in E_n$ arbitraires, c'est-à-dire

pour

$$x = \sum_{k=1}^{n} \alpha_k \varphi_k \,, \quad h = \sum_{k=1}^{n} \beta_k \varphi_k$$

avec α_k et β_k arbitraires, on a

$$\frac{d}{dt} f(x + th)\Big|_{t=0} = \langle F(x), h \rangle = \sum_{i=1}^{n} \beta_i \langle F(\sum_{k=1}^{n} a_k \varphi_k), \varphi_i \rangle$$

Du théorème 2.1 il suit que si x_n est un point de minimum absolu de $f(x)$ sur E_n alors

$$\langle F(\sum_{k=1}^{n} a_k \varphi_k), \varphi_i \rangle = 0 \ , \quad (i = 1,2,\ldots,n) \tag{5.3}$$

et si $f(x)$ est une fonctionnelle convexe, le système (5.3) donne une condition nécéssaire et suffisante afin que les vecteurs
x_n définis par la formule (5.2) donnent des approximations de
Ritz. Le système (5.3) qui définit les coefficients a_k est dit
un système de Ritz.

Remarquons que l'affirmation que dans le cas où $f(x)$ est convexe la solution du système (5.3) donne au moyen de la formule
(5.2) des approximations de Ritz, c'est-à-dire les points de minimum absolu de $f(x)$ sur E_n , répose sur la proposition suivante

Lemma 5.1. Si la fonctionnelle convexe $f(x)$ donnée sur un espace linéaire (non nécéssairement normé) admet deux points de minimum différents, alors ses valeurs à ces deux points sont égales.

Démonstration. Soient x_1 et x_2 deux points de minimum de
$f(x)$, $d_1 = f(x_1)$ et $d_2 = f(x_2)$. Supposons que $d_1 < d_2$. On peut
supposer que $x_2 = 0$ et $d_2 = 0$ car si ce n'était pas ainsi on
pourrait (Cfr. la démonstration du lemma 2.1) considérer la fonctionnelle $\varphi(z) = f(x_2 + z) - f(x_2)$. Alors $d_1 < 0 = d_2 = f(x_2)$.

Soit U un voisinage de l'origine dans lequel soit $f(x) \geqslant 0$

pour chaque $x \in U$. Si λ positif est suffisamment petit on au-

ra $0 = f(0) \leqslant f(\lambda x_1)$ et de la convexité de $f(x)$

$$f(\lambda x_1 + (1 - \lambda) \, 0) \leqslant \lambda f(x_1) = \lambda d_1 \, .$$

Par conséquent $0 = f(0) < \lambda d_1$, donc $d_1 \geqslant 0$ qui contredit

l'hypothèse $d_1 < d_2 = 0$. La contrédiction démontre le lemma.

5.2   Sur la résolubilité du système de Ritz.

On a vu que si on se donne sur E une fonctionnelle convexe et

différentiable au sens de Gateaux, alors afin que les vecteurs

x_n donnés par la formule (5.2) répresentent des approximations

de Ritz il est nécessaire et suffisant que les coefficients a_k

satisfassent au système de Ritz (5.3). Rappelons que si $f(x)$ est

une fonctionnelle strictement convexe sur E elle est stricte-

ment convexe sur $E_n \subset E$ aussi et pourtant grâce au corollaire

2.1 le système (5.3) ne peut avoir plus qu'une solution.

Nous nous occupons ici de la résolubilité du système (5.3).

En effet une solution $(a_1, a_2, \ldots, a_n)$ du systeme (5.3) donne lieu

au moyen de la formule (5.2) soit à un point critique de $f(x)$,

c'est-à-dire à un point où le gradient de $f(x)$ est nul, soit à

un point de minimum local de $f(x)$ dans E_n, soit à un point de

minimum absolu de $f(x)$ sur E_n. Seulement dans ce dernier cas

la solution du système (5.3) donne une approximation de Ritz.

Lorsque $f(x)$ est une fonctionnelle convexe alors (Cfr. Théorème 2.1) chacun de ses points critique est aussi un point de mini

mum et de plus (Cfr. Lemma 5.1) un point de minimum absolu.

On va donner des propositions sur la résolubilité du système (5.3).

Lemma 5.2. Si la fonctionnelle réelle, croissante au sens de la

définition 3.2 et différentiable au sens de Gateaux, donnée sur

E est semicontinue inférieurement sur chaque sous-espace

$E_n \subset E$ alors le système de Ritz (5.3) est résoluble pour chaque

n .

Démonstration. Comme $f(x)$ est une fonctionnelle croissante il

existe $r > 0$ tel que $\|x\| \geqslant r$ implique $f(x) > f(0)$.

Considérons dans E_n la boule $D_r^{(n)} = \left\{ x \in E_n : \|x\| \leqslant r \right\}$.

En vertu de la semicontinuité inférieure de $f(x)$ sur $D_r^{(n)}$ il

existe grâce à un théorème connu de l'Analyse classique un point

$x_0 \in D_r^{(n)}$ auquel $f(x)$ atteint sa valeur minimum, c'est-à-dire

$$f(x_0) = \inf_{x \in D_r^{(n)}} f(x) \; , \; x_0 = \sum_{k=1}^{n} a_k^{(0)} \varphi_k$$

Ce point ne peut pas appartenir à la surface de la boule $D_r^{(n)}$

car on a $f(x) > f(0)$ sur la surface même. Mais au dehors de la

boule $D_r^{(n)}$ on a $f(x) > f(0) \geqslant f(x_0)$ donc $f(x_0) = \inf_{x \in E_n} f(x)$

et comme on a vu $a_1^{(0)}, a_2^{(0)}, \ldots, a_n^{(0)}$ satisfont au système (5.3).

Le lemma est démontré.

Lemma 5.3. Si la fonctionnelle réelle $f(x)$ différentiable au

sens de Gateaux, donnée sur E , est semicontinue inférieurement sur chaque sous-espace $E_n \subset E$ et pour un certain $r > 0$

$$\langle F(x), x \rangle > 0 \ , \ \| x \| = r \ , \ F(x) = \text{grad } f(x) \tag{5.4}$$

alors le système de Ritz (5.3) est résoluble pour chaque n .

Démonstration. Considérons dans E_n la boule $D_r = \{ x \in E_n :$ $\| x \| < r \}$. Dans elle il existe un point de minimum absolu de $f(x)$ car $f(x)$ est semicontinue inférieurement dans E_n . Indiquons par x_o ce point. Du lemma 2.2 x_o est un point intérieur à la boule D_r donc pour chaque $h \in E_n$

$$\frac{d}{dt} f(x_o + th) \Big|_{t=0} = 0 \ .$$

D'ici, comme $h = \sum_{k=1}^{n} \beta_k \varphi_k$ avec β_k arbitraires

$$\langle F(x_o), \varphi_i \rangle = 0 \ , \qquad (i = 1, 2, \ldots, n)$$

où x_o comme un vecteur de E_n est de la forme

$$x_o = \sum_{k=1}^{n} a_k^{(o)} \varphi_k \tag{5.5}$$

donc le système (5.3) est résoluble et $a_k = a_k^{(o)} (k=1, 2, \ldots, n)$. Le lemma est démontré.

Remarque 5.1. Il faut noter que pour la validité de l'inégalité (5.4) il est suffisant, par exemple, que

$$\overline{\lim_{R \to \infty}} \ \langle F(x), x \rangle = + \infty \qquad (R = \| x \|) \ .$$

Lemma 5.4. Si la fonctionnelle différentiable au sens de Gateaux et convexe $f(x)$, donnée sur E , est croissante alors le système de Ritz (5.3) est résoluble pour chaque n .

Démonstration. De la convexité de la fonctionnelle finie f(x) il

suit (Cfr. Théorème 1.10) sa semi-continuité inférieure faible

sur E, et pourtant sa semicontinuité inférieure faible sur E_n

pour chaque n . Mais dans un espace de dimension finie la semi-

continuité faible coincide avec la semicontinuité, donc f(x) est

semicontinue inférieurement sur chaque $E_n \subset E$. Pourtant les con-

ditions du lemma 5.2 sont toutes vérifiées et il s'ensuit la thè-

se.

Il faut remarquer que parmi les conditions de ce lemma la conve-

xité peut etre remplacée par l'hypothèse de la semicontinuité in-

férieure faible.

Remarque 5.2. Dans les conditions du lemma 5.4 le système de Ritz

peut avoir plusieurs solution, mais en vertu du lemma 5.1 chaque solu-

tion du système (5.3) donne au moyen de la formule (5.2) un ve-

cteur auquel f(x) atteint le minimum absolu sur E_n.

Si au lieu de la convexité on admet la convexité stricte de f(x)

alors chaque système de Ritz admet une solution unique.

Il faut remarquer que l'hypothèse de la convexité (convexité stri-

cte) peut être remplacée par une hypothèse équivalente, par exem-

ple la monotonicité (monotonicité stricte) de l'opérateur F(x)=

= grad f(x) .

5.3 Sur la minimisation des fonctionnelles par les approximations de

Ritz.

Lemma 5.5. Supposons que les approximations de Ritz (5.2) pour la

fonctionnelle $f(x)$ donnée sur E et bornée inférieurement exi-

stent pour chaque n . Alors si $f(x)$ est semicontinue supérieu-

rement, les approximations de Ritz donnent une suite minimisante.

Démonstration. Soit $d = \inf\limits_{x \in E} f(x)$ et soit $\left\{u^{(n)}\right\} \subset E$ une suite

minimisante quelconque satisfaisant aux inégalités

$$f(u^{(n)}) \leqslant d + 1/n , \quad (n = 1,2,3,\dots)$$

Comme (5.1) est un système coordonné complet à chaque vecteur

$u^{(n)}$ et à chaque δ_n positif il correspond un vecteur $v^{(m)} \in E_m$

$$v^{(m)} = \sum a_k^{(m)} \varphi_k \quad (m = m(n) \geqslant n)$$

tel que

$$\| u^{(n)} - v^{(m)} \| < \delta_n .$$

Comme $f(x)$ est une fonctionnelle semicontinue supérieurement on

peut choisir les nombres positifs δ_n assez petits de telle fa-

çon que pour un vecteur $u \in E$ quelconque satisfaisant à l'inéga-

lité $\| u^{(n)} - u \| < \delta_n$ on ait

$$f(u^{(n)}) - f(u) \geqslant -1/n .$$

Prenant $u = v^{(m)}$ on aura donc

$$f(v^{(m)}) \leqslant f(u^{(n)}) + 1/n \leqslant d + 2/n$$

De cette inégalité il suit que $\left\{v^{(m)}\right\}$ est une suite minimisante.

Mais comme pour les approximations de Ritz on a

$$f(x_m) = d_m = \inf_{x \in E_m} f(x)$$

alors

$$f(x_m) \leqslant f(v^{(m)}) \leqslant d + 2/n \ .$$

Comme $f(x_m) \geqslant d$ on déduit

$$\lim_{n \to \infty} f(x_m) = d$$

Le lemme est démontré.

Il faut remarquer que pour l'existence des approximations de

Ritz (Cfr. la démonstration du lemme 5.2) il est suffisant que

$f(x)$ soit croissante et semicontinue inférieurement sur chaque

sous-espace $E_n \subset E$.

Des lemmes 5.2 et 5.5 on obtient le suivant

Théorème 5.1. Si la fonctionnelle réelle $f(x)$ donnée sur E ,

croissante et différentiable au sens de Gateaux est continue et

bornée inférieurement, alors le système de Ritz (5.3) est résolu-

ble pour chaque n et les approximations de Ritz donnent une sui

te minimisante.

Remarquons que la thèse du théorème reste valable en remplaçant

l'hypothèse de la continuité de la fonctionnelle par celle de sa

semicontinuité supérieure sur E et de sa semicontinuité infé-

rieure sur chaque sous-espace E_n engendré par les vecteurs

$\varphi_1, \varphi_2, \dots, \varphi_n$.

En outre, des lemmes 5.3. et 5.5 on déduit le

Théorème 5.2. Supposons que la fonctionnelle réelle f(x) diffé-
réntiable au sens de Gateaux, donnée et bornée inférieurement sur
E satisfasse à la condition: f(x) est semicontinue supérieurement
sur E , semicontinue inférieurement sur chaque sous-espace E_n
et elle satisfait à l'inégalité (5.4) pour un certain r > 0.
Alors le système de Ritz (5.3) est résoluble pour chaque n et
les approximations de Ritz donnent une suite minimisante.
Remarquons que dans la Remarque 5.1 on a indiqué une condition
suffisante pour la validité de (5.4).

Théorème 5.3. La fonctionnelle réelle f(x) différentiable au sens
de Gateaux, donnée et bornée inférieurement sur E soit convexe
et continue. Alors si elle satisfait la condition (5.4) ou bien
si elle est croissante au sens de la définition 3.2 le système de
Ritz (5.3) est résoluble pour chaque n et les approximations de
Ritz donnent une suite minimisante.
La démonstration de ce théorème est peu différente de celles des
deux théorèmes précedents.
Le théorème reste valable en remplaçant l'hypothèse de la convexité de f(x) par celle de sa semi-continuité inférieure faible
et l'hypothèse de la continuité par celle de la semicontinuité
supérieure.

Remarque 5.3. Si dans les théorèmes 5.1=5.3 on ne suppose pas la
différentiabilité au sens de Gateaux de la fonctionnelle on ne
peut pas parler de la résolubilité du système de Ritz, mais les
autres affirmations de ces théorèmes restent valables.
Précisement les approximations de Ritz (5.2) existent pour chaque

ņ et elles donnent une suite minimisante.

Remarquons encore que les conditions des théorèmes 5.1-5.3 n'assurent pas l'existence du minimum de la fonctionnelle ni la conver
gence vers un point de minimum (si cela existe) des approximations
de Ritz. Dans ces théorèmes on parle de l'existence de

$$d = \inf_{x \in E} f(x) \quad , \quad d_n = \inf_{x \in E_n} f(x) \; ,$$

et des $x_n \in E_n$ tels que $f(x_n) = d_n$ et de la propriété $d_n \to d$
pour $n \to \infty$.

Les conditions pour l'existence d'un point x_0 tel que $f(x_0) = d$ et
la convergence des approximations de Ritz (5.2) au vecteur x_0 se
ront considérées au n° suivant.

4 Sur la convergence faible des approximations de Ritz.

Dans ce n° on suppose que E soit un espace de Banach réel réflexif. La combinaison des propositions données dans 5.1-5.3 avec
celles de 3.3 donnent lieu à des théorèmes différents parmi lesquels nous choisissons les suivants.

Théorème 5.4. Si la fonctionnelle réelle $f(x)$ strictement convexe
et continue sur E est croissante alors on peut affirmer que: 1. Il
existe un seul point de minimum x_0 auquel $f(x)$ atteint sa borne
inférieure. Les approximations de Ritz (5.2) existent pour chaque
n et elles convergent faiblement vers x_0 .

Démonstration. Soit E_n le sous-espace engendré par $\varphi_1, \ldots, \varphi_n$ (Cfr.
5.1). Pour la convexité stricte et la croissance de $f(x)$ il suit
(Cfr. Théorème 3.1) l'existence d'un seul point $x_n \in E_n$ tel que

$$f(x_n) = d_n = \inf_{x \in E_n} f(x)$$

M. M. Vainberg

Ces approximations de Ritz en vertu du Lemma 5.5 donnent une sui-

te minimisante si

$$\inf_{x \in E} f(x) = d > - \infty$$

Mais en vertu du théorème 3.1 il existe un seul point $x_o \in E$

tel que $f(x_o) = d$ et chaque suite minimisante converge faible-

ment vers x_o, donc $x_n \longrightarrow x_o$. Le théorème est démontré.

Remarquons que dans ce théorème l'hypothèse de la convexité stric

te de $f(x)$ peut être remplacée par les hypothèses: a) $f(x)$ est

faiblement semicontinue inférieurement, b) il existe un seul point

de minimum de la fonctionnelle $f(x)$, coincidant avec le point

de minimum absolu.

Remarquons aussi que la Remarque faite après la démonstration du

théorème 3.1 reste valable ici.

Théorème 5.5. La·fonctionnelle $f(x)$ donnée dans E soit finie

croissante semicontinue supérieurement et faiblement semicontinue

inférieurement. Alors il existe au moins un point de minimum abso

lu pour $f(x)$ et de la suite des approximations de Ritz on peut

extraire une sous-suite faiblement convergente vers un point de

minimum absolu.

Démonstration. Comme $f(x)$ est croissante il existe $r > 0$ tel

que $f(x) > f(0)$ si $\|x\| \geqslant r$. En outre de la réflexivité de E

il suit que la boule $D_r = \left\{ x \in E : \|x\| \leqslant r \right\}$ est faiblement compa

cte, faiblement complète et faiblement fermée, donc grâce à la se mi-continuité inférieure faible de la fonctionnelle $f(x)$ (Cfr.

Théorème 2.2) il existe un vecteur $x_o \in D_r$ tel que

$$f(x_o) = \inf_{x \in E} f(x)$$

Ce vecteur x_o est intérieur à D_r car on a $f(x) > f(0)$ sur

la surface de D_r. Par conséquent

$$f(x_o) = d = \inf_{x \in E} f(x)$$

En appliquant ce résultat à chaque sous-espace $E_n \subset E$ on déduit

qu'il existe un vecteur $x_n \in E_n$ pour lequel

$$f(x_n) = d_n = \inf_{x \in E_n} f(x)$$

Pourtant les approximations de Ritz existent pour chaque n et

en vertu du lemma 5.5 elle donnent une suite minimisante. D'ici

en vertu du lemma 3.1 on déduit que la suite $\{x_n\}$ est bornée

donc on peut en extraire une sous-suite $y_n \longrightarrow y_o \in E$. De la

semicontinuité inférieure faible de $f(x)$ on a

$$f(y_o) \leqslant \lim_{n \to \infty} f(y_n)$$

mais comme $f(x_n) \to d = f(x_o)$ pour $n \to \infty$ on a

$$f(y_o) \leqslant \lim_{n \to \infty} f(y_n) = \lim_{n \to \infty} f(x_n) = f(x_o) = d$$

Comme d est la borne inférieure on a $f(y_o) = d$, donc y_o

est un point où $f(x)$ atteint le minimum absolu et $y_n \to y_o$.

Le théorème est démontré.

Théorème 5.6. La fonctionnelle réelle $f(x)$ soit continue, dif-

férentiable au sens de Gateaux et son gradient satisfasse aux con ditions:

1. la fonction de t , $\langle F(tx),x \rangle$, soit continue sur $[0,1]$ pour

   chaque $x \in E$;

2. $\langle F(x + h) - F(x), h \rangle \geqslant 0$ pour chaque x, h $\in$ E .

3. $\lim_{R \to \infty} \dfrac{\langle F(x), x \rangle}{R} = + \infty$ $(R = \|x\|)$.

Alors la thèse du théorème 5.5 est vérifiée. Si dans la condition

2 l'égalité a lieu seulement pour h = 0 alors est valable la thè

se du théorème 5.4.

Démonstration. De la condition 2 (Cfr. Théorème 1.4) il suit la

semicontinuité inférieure faible de la fonctionnelle f(x) et des

conditions 1,2 et 3 on obtient (Cfr. la démonstration du théorème

2.6) l'égalité

$$\lim_{R \to \infty} f(x) = + \infty \quad (R = \|x\|)$$

qui signifie que la fonctionnelle est croissante. Pourtant toutes

les hypothèses du théorème 5.5 sont vérifiées donc la thèse du mê-

me théorème est valable. Si l'égalité dans la condition 2 est pos-

sible seulement pour h = 0 alors f(x) est une fonctionnelle

strictement convexe, donc toutes les hypothèses du théorème 5.4

sont vérifiées. Le théorème est démontré.

Remarquons que les conditions du théorème 5.6 assurent la résolu-

bilité du système de Ritz (5.3) pour chaque n .

5.5 Sur la convergence forte des approximations de Ritz.

Théorème 5.7. La fonctionnelle réelle $f(x)$ donnée sur E soit différentiable au sens de Gateaux, bornée dans une certaine boule et son gradient $F(x)$ satisfasse aux conditions du théorème 3.4. Alors le système de Ritz (5.3) est résoluble pour chaque n et les approximations de Ritz (5.2) convergent vers le seul point x_o de minimum absolu de $f(x)$.

Démonstration. De l'inégalité (3.5) il suit la monotonicité stricte de $F(x)$ et d'ici la convexité stricte de $f(x)$. De l'inégalité (3.5) il suit (Cfr. la démonstration du théorème 3.4) que $f(x)$ est une fonctionnelle croissante et comme celle-ci est une fonctionnelle convexe bornée dans une certaine boule il suit comme on sait qu'elle est continue sur E . Pourtant toutes les hypothèses des théorèmes 5.1 et 5.4 sont vérifiées et il suit l'existence d'un seul point de minimum auquel $f(x_o) = \inf_{x \in E} f(x)$, la solution du système de Ritz est unique pour chaque n et les approximations de Ritz (5.2) donnent une suite minimisante. Cette suite minimisante converge vers x_o en vertu du théorème 3.4.

Le théorème est démontré.

Théorème 5.8. La fonctionnelle $f(x)$ donnée sur E soit bornée dans une certaine boule et elle satisfasse aux conditions du théorème 3.5. Alors la thèse du théorème 5.7 est valable.

M. M. Vainberg

Demonstration. De la condition 2 du théorème 3.5 (Cfr. la démonstration du théorème 3.5) il suit que f(x) est une fonctionnelle croissante et convexe ayant un seul point de minimum x_0, donc f(x) étant bornée toutes les conditions des théorèmes 5.1 et 5.4 sont satisfaites et il s'ensuit que le système de Ritz est résoluble pour chaque n et les approximations de Ritz donnent une suite minimisante. La convergence de cette suite minimisante vers le vecteur x_0 découle du théorème 3.5. Le théorème est démontré.

Théorème 5.9. La fonctionnelle f(x) différentiable deux fois au sens de Gateaux, donnée dans l'espace de Banach réflexif E satisfasse aux conditions 1 et 2 du théorème 2.10 et de plus la fonction $\gamma(z)$ qui entre dans la condition 2 soit continue et la fonctionnelle f(x) soit bornée dans une certaine boule.

Alors la thèse du théorème 5.7 est valable.

Démonstration. Des hypothèses (Cfr. les démonstrations des théorèmes 2.10 et 2.6) il suit que f(x) est une fonctionnelle croissante strictement convexe, de plus elle est continue étant convexe et bornée dans une certaine boule. Pourtant toutes les hypothèses des théorèmes 5.1 et 5.4 sont satisfaites et il suit l'existence d'un seul point x_0 de minimum absolu pour f(x), la résolubilité du système (5.3) pour chaque n et les approximations (5.2) donnent une suite minimisante. La convergence de cette suite vers le vecteur x_0 s'obtient du théorème 3.6. Le théorème est démontré.

M. M. Vainberg

Théorème 5.10. Les conditions du théorème 3.7. soient vérifiées.

Alors il existe un seul point x_0 tel que $f(x_0) = \inf_{x \in E} f(x)$,

les approximations de Ritz existent pour chaque n et elles convergent vers x_0.

Démonstration. Soit $\{E_n\}$ une suite de sous-espaces de dimension

finie (**Cfr.** 5.1). Comme dans la démonstration du théorème 3.7. or

obtient qu'il existe un seul vecteur $x_0 \in E$ et, pour chaque n

un seul vecteur $x_n \in E_n$ tels que

$$f(x_0) = \inf_{x \in E} f(x), \quad f(x_n) = \inf_{x \in E_n} f(x)$$

En outre $f(x)$ étant convexe et bornée dans une certaine boule

est une fonctionnelle continue. Par conséquent en vertu du Lemma

5.5 les approximations de Ritz $\{x_n\}$ donnent une suite minimisante

et elles convergent vers x_0 en vertu du théorème 5.7.

R É F É R E N C E S

V. Ritz [1, 2], S. G. Michlin [1, 2, 3].

M. M. Vainberg

VI. METHODE DE NEWTON.

La méthode Newton pour la solution de l'équation $\varphi(t) = 0$ où $\varphi(t)$ est une fonction réelle d'une variable réelle a été transpor- tée au cas abstrait et développée pour la solution des équations non linéaires dans des espaces divers. Ici nous utilisons cette méthode pour la minimisation des fonctionnelles.

6.1 Constructions des approximations de Newton.

Soit $f(x)$ une fonctionnelle réelle donnée dans un espace normé E, bornée inférieurement et trois fois différentiable au sens de Gateuaux. Cette hypothèse sera utilisée pour appliquer la formule de Taylor avec l'erreur sous forme intégrale. Si l'on faisait usa- ge de la formule avec l'erreur sous la forme de Peano il serait suffisant supposer $f(x)$ deux fois différentiable ou bien l'opé- rateur $F(x) = \text{grad } f(x)$ une fois différentiable.

Supposons que x_0 soit un point de minimum absolu de $f(x)$.

Si x_1 indique la première approximation à x_0 on a de la formu- le de Taylor

$$f(x) = f(x_1) + f'(x_1)(x - x_1) + \frac{1}{2} f''(x_1)(x - x_1)^2 + r_3(x_1, x)$$

ou

$$f(x) = f(x_1) + \langle F(x_1), x - x_1 \rangle + \frac{1}{2} \langle F'(x_1)(x - x_1), x - x_1 \rangle +$$
$$+ r_3(x_1, x) .$$

Si l'on substrait $r_3(x_1, x)$ on obtient une fonctionnelle quadratique

$$\varphi_1(x) = f(x_1) + \langle F(x_1), x - x_1 \rangle + \frac{1}{2} \langle F'(x_1)(x - x_1), x - x_1 \rangle$$

La méthode de Newton consiste à prendre comme deuxième approximation x_2 le point de minimum absolu de la fonctionnelle quadratique $\varphi_1(x)$. Pour obtenir ce point il faut connaître préliminairement le grad $\varphi_1(x)$. On écrit

$$\frac{d}{dt} \varphi_1(x + th)\Big|_{t=0} = \langle F(x_1), h \rangle + \frac{1}{2} \langle F'(x_1) h, x - x_1 \rangle +$$

$$+ \frac{1}{2} \langle F'(x_1)(x - x_1), h \rangle$$

Comme $F(x)$ est un opérateur potentiel (Cfr. M.M. Vainberg [2])

$$\langle F'(x_1) h, x - x_1 \rangle = \langle F'(x_1)(x - x_1), h \rangle$$

donc

$$\frac{d}{dt} \varphi_1(x + th)\Big|_{t=0} = F(x_1) + \langle F'(x_1)(x - x_1), h \rangle$$

et par conséquent

$$\phi_1(x) \equiv \text{grad } \varphi_1(x) = F(x_1) + F'(x_1)(x - x_1)$$

Si x_2 est un point de minimum absolu de $\varphi_1(x)$ alors du théorème 2.1 on a $\phi_1(x_2) = 0$, ou

$$F(x_1) + F'(x_1)(x_2 - x_1) = 0$$

et d'ici s'il existe l'operateur inverse $\Gamma(x_1) = [F'(x_1)]^{-1}$ on a

$$x_2 = x_1 - \Gamma(x_1) F(x_1)$$

Pour avoir les approximations suivantes on procède de la même façon. Précisement on approche la fonctionnelle $f(x)$ par la fon-

M. M. Vainberg

ctionnelle quadratique

$$\varphi_2(x) = f(x_2) + \langle F(x_2), x - x_2 \rangle + \frac{1}{2} \langle F'(x_2)(x - x_2), x - x_2 \rangle$$

et par l'application des raisonnements précedents on obtient

$$x_3 = x_2 - \Gamma(x_2) \, F(x_2) \ .$$

Continuant ce procédé on obtient

$$x_{n+1} = x_n - \Gamma(x_n) \, F(x_n), \ (n = 1,2,3,\ldots) \tag{6.1}$$

où $\Gamma(x) = F'(x)$.

Le procès (6.1) s'appelle procéssus itératif (ou méthode) de

Newton et les vecteurs x_n calculés par cette formule sont dits

les approximations de Newton. Chaque approximation x_{n+1} est don-

née comme point de minimum absolu de la fonctionnelle quadratique

$$\varphi_n = f(x_n) + \langle F(x_n), x - x_n \rangle + \frac{1}{2} \langle F'(x_n)(x - x_n), x - x_n \rangle \tag{6.2}$$

Pourtant le vecteur x_{n+1} est donné par l'équation

$$\Phi_n(x) = \mathrm{grad} \, \varphi_n(x) = F'(x_n)(x - x_n) = 0 \tag{6.3}$$

Cette équation est obtenu comme condition nécéssaire pour le mini-

mum de $\varphi_n(x)$. Toutefois si $f(x)$ est une fonctionnelle convexe

alors l'équation (6.3) est aussi suffisante pour l'existence du

minimum absolu de la fonctionnelle quadratique $\varphi_n(x)$ grâce au:

Lemma 6.1. Si $f(x)$ est une fonctionnelle convexe alors la fon-

ctionnelle quadratique $\varphi_n(x)$ donnée par la formule (6.2) est aus-

si convexe.

M. M. Vainberg

<u>Démonstration</u>. De la convexité de $f(x)$ en vertu du Corollaire

1.2 il suit $< F'(x) h, h > \geqslant 0$ pour chaque $x, h \in E$, pourtant

$$< \phi_n(x) - \phi_n(y), x - y > = < F'(x_n)(x - y), x - y > \geqslant 0$$

donc $\phi_n(x)$ est un opérateur monotone, c'est-à-dire $\varphi_n(x)$ est

une fonctionnelle convexe. Le lemma est démontré.

Si φ_n est une fonctionnelle convexe, en vertu du théorème 2.1

l'équation (6.3) est nécéssaire et suffisante pour l'existence d'u

point de minimum absolu (Cfr. Lemma 5.1).

En outre si l'opérateur $\Gamma(x) = \left[F'(x) \right]^{-1}$ existe pour chaque

$x \in E$ alors l'équation (6.3) est résoluble pour chaque n .

On est arrivé a la méthode de Newton (6.1) comme une conséquence

du problème de minimum de la fonctionnelle $f(x)$. Comme dans un

point de minimum de $f(x)$ son gradient est nul on peut arriver

au procédé (6.1) partant du problème de la solution approchée de

l'équation

$$F(x) = 0 \tag{6.4}$$

Soit $F(x)$ deux fois différentiable au sens de Gateaux. Alors de

la formule de Taylor on a

$$F(x) = F(x_1) + F'(x_1)(x - x_1) + r_2(x_1, x)$$

où x_1 est la première approximation aux racines de l'équation

(6.4). Si l'on substrait $r_2(x_1, x)$ on remplace l'équation (6.4)

par l'équation

$$F(x_1) + F'(x_1)(x - x_1) = 0$$

dont la solution x_2 sera prise comme solution approchée de l'é-

quation (6.4). S'il existe l'opérateur $\Gamma'(x_1) = \left[F'(x_1)\right]^{-1}$ on

obtient $x_2 = x_1 - \Gamma(x_1) F(x_1)$. Continuant ce raisonnement on

aboutit au procès itératif (6.1). Il faut rappeler que ce procès

s'arrête si l'on a $F(x_n) = 0$ pour un certain n . Alors x_n est

une racine de l'équation (6.4) et un point de minimum de la fon-

ctionnelle $f(x)$ si celle-ci est convexe.

6.2 Sur la propriété de relaxation du procédé.

On donne ici une condition suffisante simple afin que les approxi

mations de Newton (6.1) donnent un procéssus relaxant, c'est-à-di-

re tel que

$$f(x_{n+1}) \leqslant f(x_n) \quad (n = 1,2,3,\ldots)$$

Lemma 6.2. La fonctionnelle réelle $f(x)$ donnée dans un espace

normé E soit trois fois différentiable au sens de Gateaux, soit

$F(x) = \operatorname{grad} f(x)$ et la fonction $\langle F''(x + th) h^2, h \rangle$ soit conti

nue pour $t \in [0,1]$ et pour chaque $x, h \in E$. Alors si le condi-

tions suivantes sont satisfaites

$$\langle F'(x) h, h \rangle \geqslant m \| h \|^2 , \quad m > 0 \tag{6.5}$$

$$\langle F''(x) h^2, h \rangle \leqslant N \| h \|^3 \tag{6.6}$$

avec N tel que

$$N \| x_{n+1} - x_n \| = N \| \Gamma(x_n) F(x_n) \| \leqslant 3 m \tag{6.7}$$

pour les approximations (6.1), on a $f(x_{n+1}) \leqslant f(x_n)$.

M. M. Vainberg

<u>Démonstration</u>. De la formule de Taylor on peut écrire

$$f(x_{n+1}) - f(x_n) = \langle F(x_n), x_{n+1} - x_n \rangle +$$

$$+ \frac{1}{2} \langle F'(x_n)(x_{n+1} - x_n), x_{n+1} - x_n \rangle + r_3(x_n, x_{n+1})$$

où

$$r_3(x_n, x_{n+1}) = \frac{1}{2} \int_0^1 \langle F''(x_n + t(x_{n+1} - x_n))(x_{n+1} - x_n), x_{n+1} - x_n \rangle (1-t)^2 \, dt$$

où

$$f(x_{n+1}) - f(x_n) = \langle F(x_n) + F'(x_n)(x_{n+1} - x_n), x_{n+1} - x_n \rangle -$$

$$- \frac{1}{2} \langle F'(x_n)(x_{n+1} - x_n), x_{n+1} - x_n \rangle + r_3(x_n, x_{n+1}) \quad .$$

D'ici pour (6.1) nous avons

$$f(x_{n+1}) - f(x_n) = - \frac{1}{2} \langle F'(x_n)(x_{n+1} - x_n), x_{n+1} - x_n \rangle +$$

$$+ r_3(x_n, x_{n+1}) \quad .$$

Mais si $x_{n+1} \neq x_n$ de l'inégalité (6.5) on a

$$A = \frac{1}{2} \langle F'(x_n)(x_{n+1} - x_n), x_{n+1} - x_n \rangle \geqslant \frac{m}{2} \|x_{n+1} - x_n\|^2 > 0$$

et pour la condition (6.6)

$$| r_3(x_n, x_{n+1}) | \leqslant \frac{1}{6} N \|x_{n+1} - x_n\|^3$$

d'où en vertu de l'inégalité (6.7)

$$A^{-1} | r_3(x_n, x_{n+1}) | \leqslant \frac{N}{3m} \|x_{n+1} - x_n\| \leqslant 1$$

c'est-à-dire $| r_3(x_n, x_{n+1}) | \leqslant A$, et pourtant

$$f(x_{n+1}) - f(x_n) = - A + r_3(x_n, x_{n+1}) \leqslant 0 \quad .$$

Le lemma est démontré.

6.3 La convergence des approximations de Newton.

Théorème 6.1. La fonctionnelle réelle $f(x)$ donnée sur un espace de Banach réflexif E soit trois fois différentiable au sens de Gateaux, soit $F(x) = \text{grad } f(x)$ et la fonction $\langle F''(x + th)h^2, h \rangle$ soit intégrable en $t \in (0,1)$ pour chaque $x, h \in E$. Si les conditions suivantes sont satisfaites

$$\langle F'(x) h, h \rangle \geqslant m \|h\|^2 \qquad (m > 0) \qquad (6.5)$$

$$\langle F''(x) h^2, h_1 \rangle \leqslant N \|h\|^2 \|h_1\| \qquad (6.6')$$

$$d = \frac{N}{m} \|x_2 - x_1\| < 1 \qquad (6.8)$$

et si $\Gamma(x) F(x) = [F'(x)]^{-1} F(x)$ est un opérateur continu, alors la fonctionnelle $f(x)$ admet un seul point de minimum x_0 auquel converge la suite des approximations de Newton (6.1) et la rapidité de convergence est donnée par la formule

$$\|x_n - x_0\| \leqslant \frac{m}{N} \sum d^{2^i} \ .$$

Démonstration. De l'inégalité (6.5) en vertu du Corollaire 1.2 il suit que $f(x)$ est une fonctionnelle strictement convexe. En outre

$$f(x) = f(0) = \langle F(0), x \rangle + \int_0^1 \langle F(tx) - F(0), x \rangle \, dt =$$

$$= \langle F(0), x \rangle + \int_0^1 \langle F'(\tau t x) tx, x \rangle \, dt$$

D'ici et de (6.5) il suit que

$$f(x) - f(0) \geqslant \langle F(0), x \rangle + \frac{1}{2} m \|x\|^2 \geqslant \|x\| \left(\frac{m}{2} \|x\| - \|F(0)\| \right)$$

M. M. Vainberg

c'est-à-dire $f(x)$ a la m-propriété et pourtant en vertu du Corollaire 2.1 il existe un seul point de minimum x_0 auquel $f(x)$
atteint le minimum absolu.

Pour démontrer la convergence vers x_0 des approximations de
Newton nous considérons la fonctionnelle quadratique (6.2) et nous
écrivons

$$\varphi_n(x_{n+1}) = f(x_n) + \langle F(x_n), x_{n+1} - x_n \rangle +$$
$$+ \frac{1}{2} \langle F'(x_n)(x_{n+1} - x_n), x_{n+1} - x_n \rangle$$

d'où en vertu de (6.1) on a

$$\varphi_n(x_{n+1}) - f(x_n) = - \frac{1}{2} \langle F'(x_n)(x_{n+1} - x_n), x_{n+1} - x_n \rangle$$

Faisant usage de l'inégalité (6.5) nous avons maintenant

$$f(x_n) - \varphi_n(x_{n+1}) \geq \frac{m}{2} \|x_{n+1} - x_n\|^2$$

ou

$$\|x_{n+1} - x_n\|^2 \leq \frac{2}{m} \left[f(x_n) - \varphi_n(x_{n+1}) \right] \tag{6.9}.$$

Mais en vertu de (6.5)

$$\varphi_n(x_{n+1}) = f(x_n) + \langle F(x_n), x_{n+1} - x_n \rangle +$$
$$+ \frac{1}{2} \langle F'(x_n)(x_{n+1} - x_n), x_{n+1} - x_n \rangle \geq$$
$$\geq f(x_n) + \langle F(x_n), x_{n+1} - x_n \rangle + \frac{m}{2} \|x_{n+1} - x_n\|^2 \geq$$
$$\geq f(x_n) + \langle F(x_n), x_{n+1} - x_n \rangle .$$

Remplaçons $F(x_n)$ par sa valeur donnée par la formule de Taylor

$$F(x_n) = F(x_{n-1}) + F'(x_{n-1})(x_n - x_{n-1}) + r_2(x_{n-1}, x_n)$$

ou

$$\langle r_2(x_{n-1}- x_n),x_{n+1}- x_n\rangle = \int_0^1 \langle F^*(x_{n-1}+ t(x_n - x_{n-1})(x_n - x_{n-1})^2,$$

$$, x_{n+1} - x_n\rangle (1 - t)\, dt$$

donc en vertu de (6.6')

$$|\langle r_2(x_{n-1},x_n),x_{n+1}- x_n\rangle|\leq \frac{1}{2} N \|x_n - x_{n-1}\|^2 \|x_{n+1} - x_n\|$$

Compte **tenu** que en vertu de (6.1) $F(x_{n-1}) + F'(x_{n-1})(x_n - x_{n-1})$

$= 0$ on obtient

$$\varphi_n(x_{n+1})\geq f(x_n)+ \langle F(x_n),x_{n+1}- x_n\rangle = f(x_n) + \langle r_2(x_{n-1},x_n),x_{n+1}-x_n\rangle$$

ou

$$f(x_n) - \varphi_n(x_{n+1}) \leq \langle r_2(x_{n-1},x_n),x_{n+1} - x_n\rangle \leq$$

$$\leq \frac{1}{2} N \|x_n - x_{n-1}\|^2 \|x_{n+1} - x_n\|$$

D'ici et de (6.9) il suit que

$$\|x_{n+1} - x_n\|\leq \frac{N}{m} \|x_n - x_{n-1}\|^2$$

et par induction

$$\|x_{n+1} - x_n\| \leq \frac{M}{N}\left[\frac{M}{m} \|x_2 - x_1\|\right]^{2^{n-1}} = \frac{m}{N} d^{2^{n-1}} \quad .$$

Par conséquent pour chaque n et p naturels

$$\|x_{n+p} - x_n\| \leq \frac{m}{N} \sum_{k=1}^{p} d^{2^{n+k-2}}$$

donc les approximations de Newton donnent une suite fondamentale.

Comme l'espace E est complet $\lim\limits_{n \to \infty} x_n = y_0 \in E$.

En outre de (6.1) il suit que

M. M. Vainberg

$$\lim_{n \to \infty} \Gamma(x_n) = \lim_{n \to \infty} (x_{n+1} - x_n) = 0$$

et d'ici pour les conditions du théorème

$$\Gamma(x_o) \, F(x_o) = 0 \qquad\qquad (6.11)$$

Remarquons que de (6.5 il suit pour chaque x fixe et h arbitraire

$$\| F'(x) \, h \| \geqslant m \, \| h \| \quad , \quad m > 0$$

donc pour un théorème connu (Cfr. par exemple, E. Hille - R. Phillips [1] , Theorem 2.11.6) il existe, borné, l'opérateur inverse $[F'(x)]^{-1} = \Gamma(x)$. Pourtant les approximations de Newton existent pour chaque n et de l'équation (6.11) il suit $F(y_o) = 0$.

(Il faut remarquer que l'hypothèse de la continuité de $\Gamma(x)F(x)$ peut être remplacée par l'hypothèse que $\| F'(x) \| \leqslant C < +\infty$, car alors de (6.1) il suit que $F(x_n) = - F'(x_n)(x_{n+1} - x_n)$, donc $\| F'(x_n)(x_{n+1} - x_n) \| \leqslant C \| x_{n+1} - x_n \| \to 0$ pour $n \to \infty$ car $x_n \to y_o$. D'ici $\| F(x_n) \| \to 0$. Mais de l'hypothèse $\| F'(x) \| \leqslant C$ il suit la continuité de $F(x)$, donc $F(y_o) = 0$.

Comme $f(x)$ est une fonctionnelle strictement convexe, son gradient peut s'annuler seulement dans un point (Cfr. Lemma 2.1 et Théorème 2.1) et pourtant $y_o = x_o$. Pourtant les approximations (6.1) convergent vers x_o. Passant à la limite maintenant dans (6.10) pour $p \to \infty$ on a

$$\| x_o - x_n \| \leqslant \frac{m}{N} \sum_{i=n-1}^{\infty} d^{2^i}$$

Le théorème est démontré.

M. M. Vainberg

Remarquons que si la fonctionnelle $f(x)$ a un minimum à un certain

point $x_o \in E$ alors on peut supprimer parmi les hypothèses du

théorème 6.1 celui de la réflexivité de E et dans ce cas il est

suffisant que les inégalités (6.5) et (6.6') soient satisfaites

pour $x \in D$ où $D = \left\{ x \in E : \|x\| \leqslant r , \quad r > \|x_o\| \right\}$.

Remarquons encore que l'hypothèse de la réflexivité de E dans

les hypothèses du théorème 6.1 peut être remplacée par l'hypothè-

se que E soit le dual d'un espace normé séparable (Cfr. Théorè-

me 2.3).

On connait beaucoup de propositions sur la convergence de la mé-

thode de Newton (6.1) vers la solution de l'équation $F(x) = 0$

(Cfr. L. V. Kantorovich - G.P. Akilov [1]). Toutefois, déjà au cas

où $F(x)$ est un opérateur potentiel les racines de l'équation

$F(x)=0$ ne sont pas néccéssairement des points de minimum du poten-

tiel $f(x)$ de l'opérateur $F(x)$, ni moins encore des points de

minimum absolu de $f(x)$. La chose est différente lorsque $F(x)$

est le gradient d'une fonctionnelle convexe. Dans ce cas les raci-

nes de l'équation $F(x) = 0$ et elles seulement sont les points

de minimum de $f(x)$ (Cfr. Théorème 2.1 et Lemma 5.1). Pourtant si

$f(x)$ est une fonctionnelle convexe et $F(x) = \text{grad } f(x)$ alors

chaque théorème sur la convergence de la méthode de Newton

vers une racine de équation $f(x)$

M. M. Vainberg

est aussi un théorème de convergence de cette méthode vers un point de minimum absolu de $f(x)$. On pourra pourtant utiliser les propositions connues.

RÉFÉRENCES

A. Bennett [1], L. V. Kantorovich [1] , L. V. Kantorovich - G. P. Akilov [1] .

BIBLIOGRAPHYE

Altman M.

1. Generalized gradient methods of minimizing a functional.
   Bull. Acad. Polon Sci., 14, N 6 (1966)

Bennett A.A.

1. Newtons method in general analysis. Proc. Nat. Ac. Sci.,
   USA, 2, N 10 (1916).

Dunford N. and Schwartz Ja.T.

1. Linear operators. General theory. New York, London, 1958

Goldstein A.A.

1. On steepest descent. J. SIAM Ser. A, 3 (1965)

Graves L.M.

1. Riemann integration and Taylor's theorem in general analysis.
   Trans. Amer. Math. Soc., 29 (1927)

Hille E. and Phillips R.S.

1. Functional analysis semi-groups. Amer. Math. Soc. Colloq. Pub.
   Providence, 1957

Kantorovich L.V.

1. On functional equations. Leningrad. Gos. Univ. Uc. Zap. Ser.
   Mat. Nauk 3(17) (1937)
2. Some further applications of Newton's method. Vestnik Leningrad,
   Univ. 2, N 7 (1957)

Kantorovich L.V. and Akilov G.P.

1. Functional analysis in normed spaces. Moscow, 1959

Lezanski T.

1. Uber das Minimumproblem für Funktionale in Banachschen Räumen.
   Math. Annalen, 152 (1966)

Lyubich Yu.I. and Maistrowsky G.D.

1. General theory the relaxation process for the convex
   functionals. Uspehi Mat. Nauk 25, 1(1970)

Mikhlin S.G.

1. The problem of the minimum of a quadratic functional. Mosoow,
   1952. English translation, Holden-Day, 1964

2. Variational methods in mathematical physics. Moscow, 1957

3. The numerical realization of the variational methods. Moscow,
   1966

Neumann J.

1. Zur algebra der Funktionaloperationen und Theorie der normalen
   Operatoren. Math. Ann. 102(1929)

Ritz W.

1. Uber eine neue Methode zur Lösungen Variationsprobleme der
   mathematischen physik. Journ. fur die reine und angewandte
   Mathematik, Bd. 135, H 1(1908)

2. Theorie der Transwersalschwingungen einer quadratischen Platte
   mit freie Ränderen. Gesammelte Werke, Paris 1911

Vainberg M.M.

1. On the solvability of certain operator equations. Dokl. Akad.
   Nauk SSSR, 92, N 3 (1953)

2. Variational methods for the study of nonlinear operators.
   GITTL, Moskow, 1956; English transl., Holden-Day, San Franci-
   sco, Calif., 1964

3. New theorems for nonlinear operators and equations. Moskov.
   Oblast. Ped. Inst. Ucen. Zap., 77, 5(1959)

4. On the convergence of the method of steepest descent for non
   linear equations. Dokl. Akad. Nauk, SSSR, 130, N 1(1960)

5. On certain new principles in the theory of nonlinear equations.
   Uspehi Mat. Nauk 15, 1 (1960)

M. M. Vainberg

6. On the convergence of the process of steepest descent for non linear equations. Sibir. Mat. J. 2, N 2 (1961)

7. On the minimum of the convex functionals.. Uspehi Mat. Nauk 20, 1 (1965)

8. Nonlinear equations with potential and monotonic operators. Dokl. Akad. Nauk SSSR 183, N 4 (1968). Soviet Math. Dokl. 9, N 6 (1968)

9. Metodo variazionale e metodo di Caccioppoli nella teoria delle equazioni funzionali non lineari. Istituto nazionale di alta matematica, Symposia mathematica, v. 2 (1968).
   Academic Press London - New York, 1969

10. On the minimum of certain nonlinear functionals. Moskov. Oblast. Ped. Inst. Ucen. Zap. 225, 12 (1969)

Vainberg M.M. and Engelson Ia.L.

1. On the conditional extremum of functionals in linear topologi cal spaces. Mat. Sb. 45, 4 (1958).

M. M. Vainberg

T A B L E   D E S   M A T I E R E S

M. M. Vainberg

Editoriale Grafica - Roma